SCIENTIFIC
REALISM

■

Selected Essays of
MARIO BUNGE

SCIENTIFIC
REALISM

■

MARTIN MAHNER

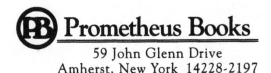
Prometheus Books
59 John Glenn Drive
Amherst, New York 14228-2197

Published 2001 by Prometheus Books

Inquiries should be addressed to
Prometheus Books
59 John Glenn Drive
Amherst, New York 14228–2197
VOICE: 716–691–0133, ext. 207
FAX: 716–564–2711
WWW.PROMETHEUSBOOKS.COM

05 04 03 02 01   5 4 3 2 1

Library of Congress Cataloging-in-Publication Data

Bunge, Mario Augusto.
    [Essays. Selections]
    Scientific realism : selected essays of Mario Bunge / edited by Martin Mahner.
        p. cm.
    Includes bibliographical references and index.
    ISBN 1–57392–892–5
    1. Realism. 2. Science—Philosophy. I. Mahner, Martin, 1958–

Q175.32 .R42 B87 2001
501—dc21   2001019544

CONTENTS

■

5

II. METHODOLOGY AND PHILOSOPHY OF SCIENCE

III. PHILOSOPHY OF MATHEMATICS

IV. PHILOSOPHY OF PHYSICS

V. PHILOSOPHY OF PSYCHOLOGY

VI. PHILOSOPHY OF SOCIAL SCIENCE

VII. PHILOSOPHY OF TECHNOLOGY

VIII. MORAL PHILOSOPHY

IX. SOCIAL AND POLITICAL PHILOSOPHY

PREFACE

■

On September 21, 1999, the physicist and philosopher Mario Bunge celebrated his eightieth birthday. I take this opportunity to edit—instead of a traditional *Festschrift* such as those previously edited by Agassi and Cohen (1982) as well as Weingartner and Dorn (1990)—a selection of his papers. These papers have been selected so as to give a panorama of the philosophical system that Bunge has developed over the past fifty years, and published in about thirty-five books, not to mention reprints and translations, as well as uncounted articles. The most important part of Bunge's œuvre is his monumental eight-volume *Treatise on Basic Philosophy* (1974–1989). In this work Bunge treats semantics, ontology, epistemology, philosophy of science, and ethics from a unified and unique perspective. In other words, he presents these topics as parts of a philosophical system. This system is anchored in *scientific realism* and *materialism*, and is thus an instance of what is also known as *philosophical naturalism* (see the introduction).

Now, not everyone has the time or the interest or the patience to work their way through an eight-volume treatise, let alone twenty-five further books—at least not unless they have a rough idea of what to expect and gain from so doing. Also, the junior undergraduate student of science or philosophy may not yet have the resources to do so, even if principally interested in this philosophy. Finally, it is practically impossible to cover eight volumes in class, unless one is prepared to spend several semesters on a single philosophical system. For all these reasons, this collection, containing thirty articles which illustrate most aspects of his philosophy, from metaphysics to philosophy of science to moral and social philosophy, provide a useful entry to Bunge's philosophical system. Such a collection can be managed in a reasonable time, not only by individuals but also in class. Equally, for courses not specifically directed to Bunge's philosophy, a teacher may use the book as a philosophically coherent source of interesting papers on a variety of topics, such as for class presentations by students, in the context of a larger discussion.

The papers in this selection were written between 1954 and 1999. Some of them are original contributions to this collection. That is, this selection has something to offer not only the beginner, but also those who are already familiar with Bunge's work. The papers are unequal in both length and complexity. Whereas some should be accessible to everyone, some of them make use of some elementary formal tools, such as mathematical logic and set theory. However, even the reader not familiar with such formal tools should be able to follow most of those articles. In any case, he or she need not be scared by a few formulas here and there.

Many of the papers have been edited or even revised to a greater or lesser extent, either by the editor or by Mario Bunge himself. In particular, I have eliminated some redundancies as well as some inessential formalizations, and prepared a master bibliography. However, it was impossible to remove all redundancies, because some papers would otherwise be badly distorted, and some concepts need to be explained and discussed whenever required by the context. After all, it should also be possible to read only a single or a couple of papers in isolation, rather than the entire anthology.

Finally, I wish Mario Bunge many more years of fruitful and productive work to come.

Martin Mahner
December 1999/April 2001

ACKNOWLEDGMENTS

■

Most papers in this collection are reprinted from the books and journals, respectively, listed in the following. I am very grateful to the publishers for having kindly granted reprint permissions. Articles not listed here are original contributions to this collection.

"How Do Realism, Materialism, and Dialectics Fare in Contemporary Science?" In M. Bunge (1973) *Method, Model, and Matter*, pp. 169–85. Dordrecht: D. Reidel (© Kluwer Academic Publishers, Dordrecht)

"New Dialogues Between Hylas and Philonous." *Philosophy and Phenomenological Research* 15 (1954): 192–99. (© Philosophy and Phenomenological Research, Brown University, Providence, R.I.)

"Energy: Between Physics and Metaphysics." *Science & Education* 9 (2000): 457–61. (© Kluwer Academic Publishers, Dordrecht)

"The Revival of Causality." In G. Fløistad (ed.) (1982) *Contemporary Philosophy. A New Survey*, vol. 2, pp. 133–55. The Hague: Martinus Nijhoff (© Kluwer Academic Publishers, Dordrecht)

"Emergence and the Mind." *Neuroscience* 2 (1977): 501–509. (© Elsevier Science, Oxford)

"The Status of Concepts." In M. Bunge (1981) *Scientific Materialism*, pp. 161–74. Dordrecht: D. Reidel (© Kluwer Academic Publishers)

"Popper's Unworldy World 3." In M. Bunge (1981) *Scientific Materialism*, pp. 137–57. Dordrecht: D. Reidel (© Kluwer Academic Publishers, Dordrecht)

"On Method in the Philosophy of Science." In M. Bunge (1973) *Method, Model, and Matter*, pp. 1–23. Dordrecht: D. Reidel (© Kluwer Academic Publishers, Dordrecht)

"Induction in Science." In M. Bunge (1963) *The Myth of Simplicity*, pp. 137–52. Englewood Cliffs, N.J.: Prentice-Hall (© M. Bunge)

"The GST Challenge to the Classical Philosophies of Science." *International Journal of General Sytems* 4 (1977): 29–37. (© Gordon & Breach Science Publishers, Lausanne)

"The Power and Limits of Reduction." In E. Agazzi (ed.) (1991) *The Problem of Reductionism in Science*, pp. 31–49. Dordrecht: Kluwer (© Kluwer Academic Publishers, Dordrecht)

"Moderate Mathematical Fictionism." In E. Agazzi and G. Darvas (eds.) (1997) *Philosophy of Mathematics Today*, pp. 51–72. Dordrecht: Kluwer (© Kluwer Academic Publishers)

"The Gap Between Mathematics and Reality." Translated by M. Bunge from: "L'écart entre les mathématiques et le réel." In M. Porte (ed.) (1994) *Passions des Formes, dynamique qualitative, sémiophysique et intelligibilité. À René Thom*, pp. 165–73. Fontenay-St. Cloud: ENS éditions (© ENS éditions)

"Two Faces and Three Masks of Probability." In E. Agazzi (ed.) (1988) *Probability in the Sciences*, pp. 27–49. Dordrecht: Kluwer (© Kluwer Academic Publishers)

"Relativity and Philosophy." In J. Bärmark (ed.) (1979) *Perspectives in Metascience*, pp. 75–88. Göteborg: Acta Regiae Societatis Scientiarum et Litterarum Gothoburgensis, Interdisciplinaria 2 (© University of Göteborg)

"Hidden Variables, Separability, and Realism." In M. Marion and R. S. Cohen (eds.) (1995) *Québec Studies in the Philosophy of Science I*, pp. 217–27. Dordrecht: Kluwer (© Kluwer Academic Publishers)

"From Mindless Neuroscience and Brainless Psychology to Neuropsychology." *Annals of Theoretical Psychology* 3 (1985): 115–33. (© Plenum Press, New York)

"Explaining Creativity." In J. Brzezinski et al. (eds.) (1993) *Creativity and Consciousness: Philosophical and Psychological Dimensions*, pp. 299–304. Amsterdam: Rodopi (© Rodopi)

"Analytic Philosophy of Society and Social Science: The Systemic Approach as an Alternative to Holism and Individualism." *Revue internationale de systématique* 2 (1988): 1–13. (© Masson Editeur, Paris)

"Rational Choice Theory: A Critical Look at Its Foundations." In J. Götschl (ed.) (1995) *Revolutionary Changes in Understanding Man and Society*, pp. 211–27. Dordrecht: Kluwer (© Kluwer Academic Publishers)

"Realism and Antirealism in Social Science." *Theory and Decision* 35 (1993): 207–35. (© Kluwer Academic Publishers, Dordrecht)

"The Nature of Applied Science and Technology." In V. Cauchy (ed.) (1988) *Philosophie et Culture*, pp. 599–605. Laval: Éditions Montmorency (© M. Bunge)

"The Technology-Science-Philosophy Triangle in Its Social Context." Translated and revised by M. Bunge from: "El sistema tecnica-ciencia-filosofía: un triangulo fuerte." *Telos* 24 (1991): 13–22. (© Fundación Telefónica, Madrid)

"A New Look at Moral Realism." In E. Garzón Valdés, W. Krawietz, G. H. von Wright, and R. Zimmerling (eds.) (1997) *Normative Systems in Legal and Moral Theory*, pp. 17–26. Berlin: Duncker & Humblot (© Duncker & Humblot)

"Morality Is the Basis of Legal and Political Legitimacy." In W. Krawietz and G.H. von Wright (eds.) (1992) *Öffentliche oder private Moral?* pp. 379–86. Berlin: Duncker & Humblot (© Duncker & Humblot)

"Technoholodemocracy: An Alternative to Capitalism and Socialism." *Concordia* 25 (1994): 93–99. (© Verlag Mainz, Aachen)

I am grateful to Alexander Paul (Hannover) for having redrawn the figures, and to Leonard Burtscher (Habach) for having scanned many of the articles not available as electronic files. As usual, I have benefitted from the regular exchanges with my friend Michael Kary (Montreal). Finally, it is a pleasure to thank my teacher and friend Mario Bunge for providing his files, for offering some of his unpublished material, for revising some of the papers, and for answering all those irksome, though important, little questions arising from my editing this book.

INTRODUCTION

■

Mario Bunge's philosophy is a paragon of both *realism* and *naturalism*—more precisely, *materialism*. That is, he assumes that there is a real world outside the mind of the thinking subject, and that that world is inhabited not just by natural (in contradistinction to supernatural) entities, but exclusively by concrete or material (in contradistinction to abstract or immaterial) entities. Moreover, Bunge's philosophy is intimately tied to science, which is why he prefers to call his position *scientific realism* and *scientific materialism*, respectively. This collection will give an impression of how Bunge conceives of these positions.

1. METAPHYSICS

We begin with a paper examining the relation between realism, materialism, and science ("How Do Realism, Materialism, and Dialectics Fare in Contemporary Science?" 1973). It examines whether science rests in fact on a realist and materialist outlook. It also asks whether the materialism inherent in science is dialectical materialism, which is, or rather used to be, one of the most prominent versions of materialism. Bunge answers this question in the negative. At the end of this article Bunge states the central goal of his philosophy, namely to build a realist and materialist philosophy.

The second essay is a defense of realism. The target is the epitome of subjective idealism, namely the philosophy of George Berkeley. Paraphrasing Berkeley's famous dialogues between Hylas and Philonous ("New Dialogues between Hylas and Philonous," 1954), Bunge shows why realism wins over subjective idealism: only realism can explain why scientific theories sometimes fail. Thus, Bunge turns the *failure* of scientific theories into an argument against subjective idealism, and not, as one might have expected, the success of scientific theories.

In the third paper, we turn to materialism again. Some people have argued against materialism that the world obviously not only consists of matter, but also of energy, calling the basic stuff the world is made of *matter-energy*. Some even have added information as a third "substance". In his paper "Energy: Between Physics and Metaphysics" (2000), Bunge shows why this view is mistaken: it confuses matter with mass, and conceives of energy as a substance, although it is a *property* of things. In so doing, this paper also clarifies the notion of a concrete object as opposed to an abstract one.

Essay 3 invites us to pursue two lines of thought. The first is to further explore the use of the notion of energy. Indeed, the latter figures prominently in Bunge's view of causality. Obviously, the realist and materialist cannot share the view of Hume and Kant (and their followers) that causation is an epistemological category, i.e., something that we impose on the world, rather than something that exists independently of the inquiring subject. To the materialist and realist, causation is an *ontological* category—one that can be elucidated by regarding causation as a mode of energy transfer from one thing to another, something that produces changes in other things. Essay 4, "The Revival of Causality" (1982), reviews various notions of causation and eventually introduces Bunge's own conception of it. (More on this topic in his 1959a.)

The second line of thought concerns the conceptions of abstract or immaterial objects. As is well known, whereas the Platonist believes that ideas exist in themselves, in an immaterial realm of their own, and that only such immaterial objects constitute the ultimate reality, the materialist cannot admit the autonomous or real existence of immaterial entities. But how does he or she conceive of such objects as numbers and theories, or poems and symphonies? The materialist may regard such entities as *objects of the mind*, as *constructs*, as *conceptual objects*; in other words, as objects whose existence depends on some mind or other, as objects that exist only in some mind or other. This view, however, raises the immediate question of whether the mind is not itself an immaterial object. If so, Bunge's materialist conception of abstract objects would fail.

For these reasons, we first need to take a look at what Bunge has to say about the mind before we proceed with the status of abstract objects. Essay 5 ("Emergence and the Mind," 1977) gives an impression of how Bunge views the mind-body problem. To him, the mind is of course no immaterial entity either, but instead the function of a material entity, namely the brain. In addition, "Emergence and the Mind" introduces many other important ontological notions, such as those of state, event, process, level, and emergence, which are all going to appear again in many other papers in this book. (More on the mind-body problem in his 1980a.)

Having learned that the mind is what the brain does—more precisely, what certain neural subsystems of the brain do—we are ready to tackle Bunge's conception of abstract objects. In "The Status of Concepts" (1981) we shall learn more about the difference between real (or material) and conceptual existence. In particular, we shall see that constructs can be construed as equivalence classes of brain processes. We shall also learn that, contrary to what some philosophers (in particular Quine) have maintained, the so-called existential quantifier does not take care of the notion of existence in general, for it does not allow us to distinguish between real and conceptual existence. These distinctions also bear on Bunge's philosophy of mathematics (see section 3).

In the light of this analysis, "Popper's Unworldly World 3" (1981) criticizes one of the most famous contemporary idealist views on the status of conceptual objects, namely Popper's Three-Worlds-Doctrine. This paper makes the contrast between Bunge's and Popper's ontological views particularly clear.

This paper concludes the section on metaphysics. However, ontological aspects will resurface in many of the following articles. The reader interested in the details of Bunge's ontology may consult volumes 3 and 4 of his *Treatise on Basic Philosophy* (1977a, 1979a).

2. METHODOLOGY AND PHILOSOPHY OF SCIENCE

This section begins with a paper entitled "On Method in the Philosophy of Science" (1973). In his characteristically outspoken manner—which has often incurred his colleagues' displeasure—Bunge examines and criticizes various approaches in the philosophy of science, most of which he finds wanting. One of the most famous victims of this criticism is Nelson

Goodman's grue paradox. Bunge regards it as a glaring example of a philosophy of science that is "out of touch with science", for it is based on a phenomenalist epistemology.

The next paper, entitled "Induction in Science" (1963), examines the role and status of induction and inductivism in the philosophy of science. Bunge argues that, although there is no such thing as inductive logic, and although advanced science proceeds mostly in a hypothetico-deductive way, there is room for induction in the method of science. Moreover, not all inductions are of equal value: whereas some are nothing but unsupported generalizations from observed cases, others are much better grounded because they are justified by our knowledge of laws and mechanisms. For example, the prediction that the sun will rise tomorrow is not merely an unjustified inference from age-long observed successions of day and night, but can be trusted because we know the laws and mechanisms that underlie the dynamic stability of the solar system.

In his paper "The GST Challenge to the Classical Philosophies of Science" (1977), Bunge examines both confirmationism (inductivism, empiricism) and falsificationism (rationalism) as to their ability to accommodate generic scientific theories. He shows that, according to both confirmationism and falsificationism, generic scientific theories, such as information theory and general systems theory (GST), are non-scientific. Regarding this as a clear failure of these classical philosophies of science, Bunge concludes that we need broader criteria of scientificity than those proposed by either positivism or critical rationalism, and proceeds to formulate his own suggestions.

The preceding papers make it clear that Bunge cannot be simply classified as either a neopositivist or a critical rationalist. Indeed, Bunge has always gone his own way, regardless of any particular philosophical school. At most, he has borrowed from other philosophies whatever he regarded as their sound and viable ideas.

With "The Power and Limits of Reduction" (1991) we proceed with a different topic, namely the problem of reductionism. This paper is the epistemological complement to essay 5, which has explicated the (ontological) notion of emergence. Can we explain the emergent properties of systems in terms of the properties of their parts? If not, does this commit us to antireductionism? Or is there a middle way between radical reductionism and antireductionism? These are some of the questions tackled in this article.

The final paper in this section examines the role of metaphors in science. Indeed, scientists do use metaphors in their writings. But does this prove that scientific theories are nothing but metaphors, or that the advancement of science merely consists in the replacement of one metaphor by another, as some philosophers and scientists believe?

"Thinking in Metaphors" (1999), an original contribution to this collection, shows why this view of science is seriously mistaken.

The reader interested in a deeper look at Bunge's epistemology and methodology may consult his two-volume *Scientific Research* (1967a, b), reissued in 1998 under the title *Philosophy of Science*, as well as volumes 5 and 6 of his *Treatise on Basic Philosophy* (1983a, b).

In the following sections, we turn to some of Bunge's contributions to the special philosophies of science, beginning with the philosophy of mathematics.

3. PHILOSOPHY OF MATHEMATICS

The first paper in this section, "Moderate Mathematical Fictionism" (1997), takes up again the problem of conceptual existence (see essay 6). Mathematical objects have always seemed to be quite unlike other abstract objects, such as centaurs and fairies. Unlike the latter, mathematical objects seem not to be mere constructs but objective ideas existing independently of human thought. Not surprisingly, Platonism is still going strong among mathematicians. In this contribution Bunge explicates the differences between formal and factual truth, shows why his constructivist or fictionist view of conceptual objects also applies to mathematics, and argues that mathematics is ontologically neutral.

As is well known, mathematics can be employed in models and theories referring to real systems of all kinds. In fact, the match between mathematics and reality has appeared so perfect that already Galileo had thought that the "book of nature" is written in mathematical terms. In his paper "The Gap Between Mathematics and Reality" (1994), Bunge argues that there is no such perfect match: the idealizations and approximations involved in every factual theory, the dimensionality of most magnitudes, the existence of redundant solutions, and the ontological neutrality of pure mathematics prevent an ideal fit between mathematics and reality.

A good example of a mathematical theory that has been used to exactify many different concepts is the probability calculus. Indeed, philosophers have used the concept of probability to elucidate notions as diverse as those of uncertainty, plausibility, credibility, confirmation, truth, information, and causality. In his article "Two Faces and Three Masks of Probability" (1988), Bunge examines the logical, subjectivist (Bayesian), frequency, and propensity interpretations of probability. He concludes that only the latter is a legitimate interpretation of the mathematical theory of probability.

More on Bunge's philosophy of mathematics can be found in volume 7, part 1 of his *Treatise on Basic Philosophy* (1985a).

4. PHILOSOPHY OF PHYSICS

In this section we take a look at Bunge's view of the two most famous theories of physics: the theory of relativity and the quantum theory. In his paper "Relativity and Philosophy" (1979) (renamed "Physical Relativity and Philosophy" in this collection) Bunge examines various philosophical problems raised by the theory of relativity, such as the relation between relativity and philosophical relativism as well as conventionalism. He also criticizes various misconceptions about the theory of relativity, such as the ideas that reference frames are observers or that the theory of relativity has relativized all the concepts of classical physics.

In "Hidden Variables, Separability, and Realism" (1995), Bunge turns to the question of whether the quantum theory implies a refutation of realism. He concludes that it is not realism that is the casualty of the quantum theory, but a position he calls *classicism*, i.e., the idea that quantum objects should have the same basic properties as the long-known objects of classical physics. Once this view is given up, quantum theory fits in well with realism and materialism.

Since some people, following von Neumann, believe that the quantum theory applies to all concrete objects, they are inclined to take Schrödinger's famous cat paradox seriously. In an original contribution to this collection, entitled "Schrödinger's Cat Is Dead," Bunge examines Schrödinger's cat paradox, von Neumann's general measurement theory, and other related notions. Indeed, while people like Schrödinger and Einstein took the cat paradox for an indication that there is something seriously wrong with quantum mechanics and, in particular, the superposition principle, the received opinion is that, far from being paradoxical, Schrödinger's "smeared cat" illustrates the standard theory of measurement. Bunge concludes that both these ideas are mistaken: Schrödinger's paradox rests on false assumptions, and von Neumann's all-purpose theory of measurement is invalid.

For Bunge's philosophy of physics see his books *Foundations of Physics* (1967c), *Philosophy of Physics* (1973b), and volume 7, part 1 of his *Treatise on Basic Philosophy* (1985a). In the latter we also find a section on some philosophical aspects of chemistry. However, we shall not pursue this matter in this collection. We also have to omit Bunge's work in the philosophy of biology, for which see volumes 4 (1979a) and 7 (part 2, 1985b) of his *Treatise on Basic Philosophy*, as well as the book *Foundations of Biophilosophy* (Mahner and Bunge 1997). Instead, we proceed with a subject matter rooted in biology, namely psychology.

5. PHILOSOPHY OF PSYCHOLOGY

Essay 5, "Emergence and the Mind," has already introduced some of the central, in particular ontological, ideas of Bunge's philosophy of psychology. We are thus ready to take a look at some of the methodological consequences of this view. In his paper "From Mindless Neuroscience and Brainless Psychology to Neuropsychology" (1985), Bunge examines two classical approaches to the mind: behaviorism and mentalism (including functionalism). He criticizes both for ignoring the very substrate of the mental—the brain—and concludes that only a psychobiological approach to the mental holds promise.

In his little paper "Explaining Creativity" (1993), Bunge applies the psychobiological approach to a property that has appeared to some as being outside the grasp of science: creativity. He examines which entities or processes can be rightfully considered to be creative, and suggests a way human creativity may be accounted for in psychobiological terms.

For more on Bunge's philosophy of psychology see his books *The Mind-Body Problem* (1980a), *Treatise on Basic Philosophy*, volumes 4 (1979a) and 7, part 2 (1985b), and *Philosophy of Psychology* (Bunge and Ardila 1987).

6. PHILOSOPHY OF SOCIAL SCIENCE

As his interests have always been genuinely universal, Mario Bunge has not only dealt with the philosophy of the natural sciences including psychology, but also with that of the social sciences (Bunge 1979a, 1996, 1998, 1999). Three papers shall exemplify his work in this field.

The first paper, entitled *Analytic Philosophy of Society and Social Science: The Systemic Approach as an Alternative to Holism and Individualism* (1988), applies Bunge's systemic-materialist ontology to social systems. He shows why the two traditional approaches, holism and individualism (or atomism), fail to account for social systems. Hence, they cannot serve as the basis for any fertile philosophy of the social sciences. It is only what Bunge calls the *systemic* approach that has all the virtues necessary for such a philosophy to be realistic and useful.

The second paper takes a closer look at a popular theory in the social sciences that is based on an individualist outlook: rational choice theory (*Rational Choice Theory: A Critical Look at its Foundations*, 1995). It examines the ontological and methodological presuppositions of rational choice theory and concludes that the latter is not a solid theory of society.

In the third paper of this section, "Realism and Antirealism in Social Science" (1993), Bunge examines the fashionable subjectivist, constructivist, and relativist—in short, postmodernist—approaches to social science. He concludes that these philosophies are not only false, but are causing serious damage to social studies. In social science, as everywhere, a realist philosophy is called for, not an antirealist one.

7. PHILOSOPHY OF TECHNOLOGY

Having dealt with basic factual science in sections 4–6, we proceed to take a look at technology. The first question that springs to mind is of course: 'What, if any, is the difference between science and technology?'. Bunge answers this question in his article "The Nature of Applied Science and Technology" (1988).

Another question remains to be answered: 'What, if anything, has technology to do with philosophy?'. Indeed, as an explicit philosophical discipline, the philosophy of science is much older and wider known than the philosophy of technology. One of the philosophers who have first drawn their colleagues' attention to the fact that there is such a thing as a philosophy of technology has been Mario Bunge. Thus, one of Bunge's classic papers is entitled "The Philosophical Richness of Technology" (1976c). (See also volume 7, part 2 of Bunge's *Treatise on Basic Philosophy*, 1985b.) However, we shall not reprint this paper here, but instead a more recent one, which explores the manifold connections between technology, science, and philosophy: "The Technology-Science-Philosophy Triangle in Its Social Context." The author has kindly translated it for this collection from its original Spanish version published in 1991.

In this paper Bunge mentions that he also considers ethics as kind of a technology, more precisely as a philosophical technology. After all, ethics has long been subsumed under the heading of 'practical philosophy'. To learn more about what a philosophical technology might be, we close this section with an article entitled *The Technologies in Philosophy*, an original contribution to this collection. According to Bunge, not only ethics, but also axiology (or value theory), praxiology (or action theory), methodology (or normative epistemology), and political philosophy are philosophical technologies. Thus, this paper functions as a natural bridge to the two final sections, which are devoted to moral and political philosophy, respectively.

8. MORAL PHILOSOPHY

Being a realist, it comes as no surprise that Bunge also defends moral realism. This is the thesis that there are moral facts and, thus, moral truths: "A New Look at Moral Realism" (1997). These facts and truths, however, are not quite like scientific facts and truths. Rather, they are social facts, and the truths involved are contextual or relative. This article, then, is an attempt to defend moral realism in the context of an analytical and naturalist philosophy and, of course, in the light of Bunge's own ethical system, which he calls *agathonism* (Bunge 1989a).

A central idea of this system is that every right ought to be balanced by a duty, and vice versa. As a consequence, Bunge rejects all rights-only and duties-only moral philosophies, such as libertarianism and Kantianism, respectively. In his paper "Rights Imply Duties," an original contribution to this collection, Bunge proposes a (mildly formalized) moral theory in which the claim that every right implies a duty is deduced as a theorem.

9. SOCIAL AND POLITICAL PHILOSOPHY

It is quite obvious from the preceding papers that, for Bunge, the most pressing moral questions are not so much questions of individual morality, but social and political problems. We therefore complete this selection with a brief look at his social and political philosophy. A paper which bridges or, if preferred, unites the moral and the sociopolitical domains is his "Morality Is the Basis of Legal and Political Legitimacy" (1992). In this article Bunge maintains that a notion of full legitimacy coincides with that of justice, which in turn is based on the notion of the good. He proceeds to examine which kind of social order complies with this analysis, and concludes that only a combination of what he calls 'integral democracy' together with technical expertise can promise a just and legitimate social order—a view which he calls *technoholodemocracy*.

The final paper of this collection, "Technoholodemocracy: An Alternative to Capitalism and Socialism" (1994), allows the reader to take a closer look at this concept—and thus also at Bunge's political credo.

■ ■ ■

The reader who manages to work his or her way through this entire collection should get a glimpse of a unique philosophical system, consisting of a metaphysics, semantics, epistemology, methodology, philosophy of science and technology, as well as a moral, social, and political philosophy. Those who disagree with it should nonetheless find plenty of food for thought, and those who essentially agree with it may find various ideas on which to improve and build on.

Finally, I need to point out a notational convention. The reader will soon notice that the book contains both words or phrases in single and in double quotation marks. This is as a result of a convention adopted by Bunge to distinguish signs or symbols (e.g., words, terms, or sentences) from the concepts or propositions they stand for. The former are enclosed within single quote marks, the latter within double. For example, the words 'book', 'Buch', 'livre', 'liber', and 'βίβλος' all signify the same concept, namely "book". Note that philosophical precision requires some deviation from the *Chicago Manual of Style* adopted by many American publishers, including Prometheus Books, which directs that punctuation be contained inside any quotation marks, such as "in this example." For, of course, neither periods nor commas belong to any given concept or proposition, nor to most symbols. Note also that double quote marks are also used in the traditional way, e.g., for ordinary quotations. Surely, the usage will be clear from the context.

I

METAPHYSICS

■

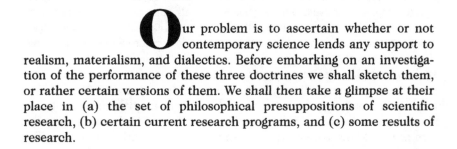

1

HOW DO REALISM, MATERIALISM, AND DIALECTICS FARE IN CONTEMPORARY SCIENCE?

(1973)

Our problem is to ascertain whether or not contemporary science lends any support to realism, materialism, and dialectics. Before embarking on an investigation of the performance of these three doctrines we shall sketch them, or rather certain versions of them. We shall then take a glimpse at their place in (a) the set of philosophical presuppositions of scientific research, (b) certain current research programs, and (c) some results of research.

1. CRITICAL REALISM

We shall begin by formulating the main theses of critical realism as distinct from naive and dogmatic realism. In the second subsection we shall see how critical realism fares in science.

1.1. The Main Theses of Critical Realism

The main theses, or rather hypotheses, of critical realism are as follows.

R1. There are things in themselves, i.e., objects the existence of which does not depend upon any mind. (Note that the quantifier is existential, not universal: artifacts, though nonmental, depend on minds for their design and operation.)

R2. Things in themselves are knowable, though only partially and by successive approximations rather than exhaustively and at one stroke.

R3. Any knowledge of a thing in itself is attained jointly by experience (in particular experiment) and by reason (in particular theorizing). However, none of these can pronounce final verdicts on anything.

R4. Factual knowledge is hypothetical rather than apodictic, hence it is corrigible and not final.

R5. The scientific knowledge of a thing in itself, far from being direct and pictorial, is roundabout and symbolic.

Note that R1 is an ontological (or metaphysical) thesis, while R2–R5 are epistemological theses. These theses are all that critical realism *lato sensu* commits itself to, and all that an analysis of actual scientific research supports. Beyond this, there is ample room to work out genuine theories (hypothetico-deductive systems) of knowledge, preserving and spelling out the preceding hypotheses. Critical realism thus keeps the seventeenth-century distinction, exploited by Kant and denied by empiricism, between the thing in itself (such as it exists) and the thing for us (as known to us). But critical realism drops Kant's theses that the former is unknowable and that the thing for us is identical with the phenomenal object, i.e., with appearance. Indeed, critical realism maintains: (a) that the thing in itself can be known (in a gradual fashion), and (b) that the thing for us is not the one presented to the senses but the one characterized by the scientific theories.

Moreover, critical realism does not assume that the thing in itself is knowable as such, i.e., without our introducing any distortion (removal and/or addition of traits). What distinguishes critical realism from other varieties of realism is precisely the recognition that such a distortion is unavoidable, for ideas are not found ready made: we think them out laboriously and correct them endlessly or even give them up altogether. There would be no point in going through this process if we were able to grasp external objects, such as electrons and societies, exactly as they are—even less if we were able to create such objects in the act of thinking about them.

1.2. How Does Critical Realism Fare in Contemporary Science?

The philosophical assumptions of the existence of external objects (R1) and of their knowability (R2) are not just hypotheses confirmed by a few cases but running the risk of overthrow by counterexamples: they are no less than *presuppositions of scientific research*. Indeed, there would be no point in investigating things if we could invent them or if they were unknowable. Nor would there be any point in searching for new theories and new methods bearing on old acquaintances, such as water and humans, if they could be known at one stroke. R1 and R2 are then as much a part of the background of scientific research as are logic and mathematics.

The philosophical hypotheses R3, R4, and R5 are not presuppositions of scientific research, but they are *inherent in the methodology of modern science*. If neither experience nor reason were necessary to conduct scientific research, we could resort to wild intuition or to mystic communion. If theory were sufficient, we would waste no time with empirical tests, and would give the triumph to idealism. If scientific experience were in no need of theory, empiricism would win. As it is, factual knowledge consists of a set of theories and a set of data, such that the former must be compatible with at least some of the latter, while the data must be sought and processed with the help of some theories. Moreover, data are in principle as corrigible as theories, in the light of both further data and other theories. The best theory of a concrete system we can build is not a copy but a theoretical model or conceptual reconstruction containing concepts without a concrete counterpart (such as the logical concepts) as well as hypotheses that, at best, are approximately true.

All five hypotheses of critical realism occur also, in however an implicit way, in any research program. Thus, any genuine research program will contain provisions for the checking and evaluation of the outcome of the research process as well as for the eventual correction and even replacement of some of the assumptions. This would be unnecessary if we did not want our ideas to correspond to reality, or if we believed them to be mere copies of reality. On the other hand, the hypotheses of critical realism do not occur in the output of scientific research. Consequently, the philosophies of science that focus on the net results of scientific research, with neglect of the process itself, are bound to miss those philosophical presuppositions of scientific research, claiming instead that science has no such presuppositions.

In sum, critical realism is inherent in contemporary scientific research. Some apparent exceptions to it will be dealt with in section 3.

2. Dynamical and Pluralist Materialism

Let us begin by taking a quick look at the central theses of a particular kind of materialism, then investigate how it is doing in contemporary science.

2.1. The Main Theses of Dynamical and Pluralist Materialism

Materialism is, of course, the ontological doctrine that whatever exists in the external world is material or, to put it negatively, nonmental. Note the differences between materialism and ontological realism. Ontological realism is compatible with immaterialism: one might posit that external objects are ghostly or immaterial. Materialism is compatible with irrealism: a solipsist could hold that he is a material system. It is only when we turn to scientific knowledge for a verdict that we are led to subscribe to both realism and materialism. Neither of these two philosophical views is self-validating.

Dynamical materialism asserts that every existent is in a process of change in some respect or other. (More on this in essay 3.) Pluralism is the ontological view that qualitative variety is irreducible: that things are not just aggregates of units of some single basic substance. It may also assert that there are different levels of complexity and organization. (More on this in essay 5.) The view we are interested in combines all these ingredients: it holds that whatever exists is both material and in flux, that things fall into different levels, and that levels are interrelated and the outcome of evolutionary processes.

More explicitly, the main theses of dynamical (or systemic or emergentist) materialism combined with integrated pluralism are as follows.

M1. Every existent is a material system. The world is a system of material systems. The world is one and all there is.

M2. Every material system is the outcome of a process of self-assembly (or self-organization) or a result of a process of decomposition of another material system.

M3. Everything interacts with something else and changes forever in some respect or other. Some changes involve the emergence of new qualities and new laws. Furthermore, there are evolutionary changes. (Note the existential quantifier.) Every structure and every change is lawful.

M4. Organisms are material systems satisfying not only physical and chemical laws, but also certain emergent laws (genetic, ecological, etc.). The mind is the activity of the central nervous system—hence not a prerogative of humans. Society is a system of organisms. There is no disembodied mind, and there is no nation or some other whole, above and beyond society.

M5. The world has a multilevel structure. Every level of complexity or organization has its peculiar properties and laws. No level is totally independent from its adjoining levels.

Just as with our treatment of critical realism, the above is nothing but a sketch of a view that ought to be formulated in detail. (For such an attempt see Bunge 1977a, 1979a.) This view emphasizes the materiality of the world, but recognizes its qualitative variety. It stresses the relatedness of levels but avoids reductionism. In particular, it does not regard organisms and societies as mere physicochemical systems. It emphasizes change without assuming any universal pattern of change, be it accretion, evolution, or the struggle of opposites: this kind of materialism leaves the investigation of the laws of change to science instead of wishing to compete with the latter. Rather, this kind of materialism seeks support in science and is prepared to have any of its hypotheses refuted by science: in other words, it is not naive or dogmatic materialism.

2.2. How Does Dynamical and Pluralist Materialism Fare in Contemporary Science?

Unlike realism, materialism is not an ingredient of the scientific method. Therefore, a physicist can make distinguished contributions to science even if he believes, as Heisenberg (1969) does, that the so-called elementary particles are the embodiments of symmetries conceived of platonistically. He can afford this extravagance so long as he grants the existence and the empirical investigability of such "particles". And a dualist (such as Sherrington or Penfield or Eccles) can make decisive contributions to the unveiling of the material "basis" of the mind. Materialism, then, is not indispensable for doing normal science.

On the other hand, materialism, if nondogmatic, has the following uses in scientific research. First, it helps avoid wasting one's time chasing ghosts like the vital force, the disembodied soul (or the id, the ego, and the superego), the archetypal memory, the spirit of the nations, or historical necessity. Second, materialism helps suggest certain research programs that may ensue in scientific breakthroughs such as the synthesis of a whole out of its lower-level components, and the tracing of a higher-level function or process back to processes on some lower level. In other words, although materialism is not a precondition of scientific research, it is an efficient vaccine against charlatanism as well as a good compass for scientific revolutions.

In addition to helping science, dynamical and pluralist materialism seems to be overwhelmingly confirmed by contemporary science, provided the results of the latter are formulated in methodologically correct terms or, more precisely, as long as both subject-centered and inscrutable concepts are avoided. This requirement does not amount to the tautology that science is materialist provided it is materialist, for it happens to be a methodological, not a metaphysical condition. Indeed, the requirement that no subject-centered objects be employed is necessary for attaining objectivity in reference and in test. (It is always easy to rephrase an objective statement, such as "That atom is in its ground state", in subject-centered terms, e.g., "I believe that atom is in its ground state". The converse translation is not always possible. (See Bunge 1967d, 1969.) The requirement that inscrutable predicates be avoided is necessary for testability. (It is always possible to assume that anything is dependent on some immaterial object that will forever escape detection: but such an assumption is of course untestable.) These two conditions are not quite obvious and they are often violated in contemporary microphysics, which therefore seems to contradict materialism. But it can be shown that this impression stems from semantical and methodological mistakes. More on this in section 3.

It would be interesting to review the achievements of the science of the day and check whether or not they confirm dynamical materialism. We have room only for commenting on two such breakthroughs, one in biology and the other in psychology. The former is, of course, the very constitution of molecular biology, in particular the molecular or chemical theory of heredity, and the theory of the spontaneous generation of life in its contemporary version. The synthesis of proteins and even more the recent synthesis of a gene out of their lower-level components score dramatic victories for dynamical and pluralistic materialism: they do not just confirm it, but they were possible because of that general outlook, which inspired Oparin, Urey, Calvin, Fox, Khorana, and others the very design of those research programs. Indeed, the latter were based on the hypothesis that such self-assembly and self-organizing processes could be induced if only the requisite ingredients and the favorable physical conditions were provided (Oparin 1938; Calvin 1969). The forecast that the syntheses would take place under such conditions was thus of the self-fulfilling kind.

Recent studies of the chemistry and physiology of the central nervous system have provided further confirmations of materialism, in the sense that they have shown the mental to be a function of that system rather than a separate substance. I mention at random three of the main achievements along this line: (1) Hebb's theory of the thought process as an activity of neuron assemblies (Hebb 1949) and Griffith's more detailed model of thinking and consciousness (Griffith 1967); (2) the chemical theory of memory; and (3) the control of feelings and behavior by means of drugs and electric pulses, and the chemical basis of psychoses. This does not mean that most workers in the field are convinced materialists: they are not, and they won't become materialists as long as (a) materialism be presented to them as a dead dogma and, moreover, as the philosophy of a political party, and (b) functions continue to be investigated separately from structures, as is done by behaviorists, and structures be investigated apart from their higher-level functions, as it often happens in neurology. But these obstacles are probably declining in importance: psychology, a traditional bulwark of idealism and dualism, is now embarking on research programs inspired in dynamical and pluralistic materialism and is reaping revolutionary results from such research programs.

3. APPARENT REFUTATIONS OF REALISM AND MATERIALISM

We must be prepared to see our philosophical hypotheses refuted by science and we should be willing to correct them or give them up in the face of such adversity. But we must also learn to mistrust the evidence when it points against well-corroborated hypotheses, be it philosophical or scientific: the evidence itself may be wrong. This is indeed the case with certain developments in physics and in psychology which constitute only apparent exceptions to realism and materialism. Let us take a quick look at some of them.

3.1. The Ghostly in Physics

The orthodox or Copenhagen interpretation of the quantum theory is neither realist nor materialist. It is not realist because it holds that the quantum event does not exist in the absence of some observer equipped with a measuring instrument. It is not materialist because it claims that the objects or referents of the quantum theory are not thoroughly material or physical but have a mental component: they would consist of indivisible blocks constituted by an object lacking independent existence (e.g., an atom), the observation set-up, and the human observer. Moreover, according to this view, it is impossible to estimate the relative weights of these three components: the frontier or "cut" between object and subject can be shifted arbitrarily, though always keeping the subject (von Neumann 1932).

If the Copenhagen interpretation were correct, then realism and materialism would be false. But it is not. First, because quantum theories are sometimes applied in circumstances where instruments and observers are absent—as in astrophysics. Second, because the very first thing an investigator does is to individualize the objects and the means of his investigation—making sure that she won't swallow or be swallowed by either. Third, and more basic, because no physical theory can be clearly formulated in the strict Copenhagen spirit. Indeed, in a theory built in the pure Copenhagen style there should be a single reference class, namely the set of sealed units constituted by the object, the apparatus, and the observer. But the refusal to analyze the referent *unum et trinum* renders the interpretation task hopeless, for the properties to be assigned to that referent are neither strictly physical nor strictly psychological. Take, for instance, the notion of state occurring in quantum mechanics. While in the realist and materialist interpretation a state function represents a state of a system of a definite kind, in the Copenhagen interpretation that same function should represent a psychophysical state of a system-apparatus-observer block. However, no existing science accounts for such complex (yet unitary) entities. In particular, quantum mechanics does not tell us anything about the subject: if it did it would encompass the whole of biology, psychology, and sociology. In conclusion, it is impossible to build a consistent theory in the Copenhagen style. Therefore, the Copenhagen interpretation of quantum theory does not count as a refuter of either realism or materialism. (For details see Bunge 1969, 1973b; see also essay 17.)

3.2. The Ghostly in Psychology

Realism is not challenged in contemporary psychology, but materialism is. I do not think parapsychology is such a challenge: a purely methodological investigation shows it to be a hoax. On the other hand, dualists and classical behaviorists are worthy opponents. They should be taken seriously and criticized in the light of their own work rather than by being confronted with quotations from materialist philosophers.

One possible line of attack on dualism is the semantic analysis of psychological hypotheses and theories. All these hypotheses and theories involve psychological concepts which, if clear enough, can be assigned a definite mathematical status. Many, perhaps most such concepts turn out to be functions, such as the probability of learning a task in a given number of trials. Now, for any such function to be well defined, its domain must be fully characterized. For the function to represent a mental property of some organism, the domain of the function must contain the name of that organism. In other words, a function in psychology is a map on the set of organisms of a kind (e.g., albino rats) into some other set (e.g., the set of real numbers). It is only when attention is focused on the value of the function, with neglect of the whole concept, that talk about nonorganismic states and processes becomes possible. A semantic analysis restores the reference to the organism or a part of it, and thus reestablishes the function-structure unity.

The result is that mental states and mental processes are states and processes of the central nervous system. So much for the psychology of the whole organism. Neurology allows us to take a further step, and assume that mental states and processes are states and processes of portions of the central nervous system—e.g., that pain consists in simultaneous firings of groups of neurons at a certain brain level. (See essay 5.)

In conclusion, just as physical properties are properties of physical systems, so mental properties are properties of organisms endowed with a central nervous system. To speak of the mind-body duality, or psychophysical parallelism, is as incorrect as speaking of a mass-body duality or of any other parallelism between functions and structures. Semantic analysis, rather than a recourse to alternative theories or to new data, is sufficient to refute dualism and confirm materialism in the field of psychology.

As for orthodox behaviorism, it does not deny that behavior is the outcome of neuronal processes: it merely abstains from affirming this. Hence, its findings cannot be construed as a refutation of materialism. But it does refute the contention that an explicit acceptance of materialism is necessary to do scientific research. In any case, orthodox behaviorism can be said to be (a) limited, because it does not recognize the legitimacy of typical psychological concepts such as those of attention and ideation; (b) superficial, because it takes no advantage of the advances of neurophysiology; and (c) mistaken insofar as it insists on one way (stimulus-response) connections, which incidentally have no neurological basis. For these reasons, behaviorism is fading out and being replaced by psychological theories that admit and work out the hypothesis of the mind-body unity. (See also essay 19.)

In sum, contemporary psychology is coming closer and closer to dynamical and pluralist materialism. But it will tolerate no dogma, not even if originally propounded by the great Pavlov: faithfulness to the scientific method comes before philosophy. Philosophy can be embraced only insofar as it is consistent with scientific methodology and is furthermore heuristically instrumental in the formulation and evaluation of scientific hypotheses and research programs.

4. DIALECTICS

Modern dialectics, a product of pre-Socratic cosmology and German mysticism and idealism, is not easy to discuss because it has not been formulated in clear terms: it is ambiguous and therefore invites endless debate. Since dialectics can be construed in different ways, let us begin by picking a definite version of it. In the second subsection we shall investigate how this kind of dialectics is doing in contemporary science.

4.1. The Main Theses of Dialectics

Dialectics, as formulated by Hegel and his materialist disciples, is a collection of ontological hypotheses of purported universal validity: they would be true of every constituent of the world and would be confirmed by every chapter of science and mathematics. Prominent among those hypotheses are the following.

D1. Everything has an opposite.

D2. Every object is inherently contradictory, i.e., constituted by mutually opposing components and/or aspects.

D3. Every change is an outcome of the tension or struggle of opposites, whether within a system or among different systems.

D4. Progress is a helix every level of which contains, and at the same time negates, the previous rung.

D5. Every quantitative change ends up in some qualitative change, and every new quality brings about new modes of quantitative change.

All of the above statements are more or less metaphorical and, so far as I know, they have not been formulated in a literal and exact fashion. There are two possibilities: either this vagueness is incurable, or dialectics can be formulated exactly and thus be rendered communicable without ambiguity. In the first case, dialectics belongs in traditional, inexact philosophy. In the second case, it must surrender its claim to supersede formal logic. If the former is the case, then we should not bother with dialectics any longer. But if dialectics can be formulated with the help of ordinary logic and mathematics, then it loses interest as well. That is, dialectics is doomed either way.

Take, for instance, the typical dialectical concept of dialectical negation, or sublation, or *Aufhebung*. This concept, which is usually characterized in an obscure way, can be given an exact elucidation, namely thus. Let '$x > y$' stand for "x contains y" and let '$K(x)$' represent "the kind (or quality) of x". (The former concept can be explicated within mereology, whereas K can be elucidated in a theory of natural kinds. Whereas $>$ is a binary relation on the set S of all things, K may be construed as a set-valued function on that same set S. In any case, such elucidations call for formal concepts not available in dialectics.) Finally, let 'Nxy' mean "x negates (dialectically) y". Then this third concept, namely N, can be defined explicitly in terms of the *nondialectical* concepts $>$ and K, namely thus

$$Nxy =_{df} x > y \ \& \ K(x) = \bar{K}(y).$$

That is, x negates (dialectically) y equals, by definition: x contains y, and x and y are of opposite (complementary) kinds. (It has been assumed that the opposite of a class is its complement in the given universe.) The relation thus defined is antisymmetric and nonreflexive. Furthermore, N is not transitive. On the other hand, N has the required periodicity property: if Nxy and Nyz and Nzw, then Nxw. Hence, the proposed formalization captures and clarifies the notion of dialectical negation—but by the same token it emasculates that notion by showing that it depends on nondialectical concepts.

We have formalized and, as it were, dedialectized just one concept of dialectics, not the whole of dialectics. The latter remains as a whole outside logic and mathematics insofar as the dialectical thesis of the strife and unity of opposites is incompatible with the logical principle of noncontradiction. If a piece of philosophy is inconsistent with logic, then it is automatically inconsistent with factual science as well, because the latter presupposes ordinary formal logic.

If, on the other hand, it be argued that dialectics is not a kind of logic and has no quarrel with logic, but is meant to be a general cosmology or ontology concerned with real things rather than with conceptual objects (such as propositions), then this will not exempt it from the obligation of getting rid of vagueness with the help of the formal tools wrought outside dialectics over the past century. Nonetheless, a philosophy might have some heuristic value even if it does not qualify as a piece of exact philosophy—just as there is any number of fragments of exact philosophy that are sterile for science because of their lack of concern with it. Whether that is the case with dialectics, i.e., whether dialectics is compatible with science and, moreover, useful to it despite its obscurities, will be investigated next.

4.2. How Does Dialectics Fare in Contemporary Science?

The usual argument for dialectics is citing evidence apparently favorable to it, without any concern for unfavorable evidence (counterexamples). This procedure of selective induction is of course inadmissible in science and in scientific philosophy. The more so with the case of hypotheses formulated in such a vague way that they can be twisted to accommodate any fact. Let us try a more objective procedure for evaluating the evidence for and against dialectics.

The first (usually unstated) principle of dialectics, namely that everything has an opposite, is a relic of the archaic "thinking in opposites". It is nearly untestable because the very notion of an opposite is never carefully characterized. If one does try to characterize it, one comes up with as many different concepts of opposite as there are kinds of object. Thus, in some algebraic theories, such as group theory, the "opposite" (inverse) of an individual x is the (unique) individual $\bar{x}$ such that combined with x yields the neutral (or identity) individual e, i.e., $x \circ \bar{x} = \bar{x} \circ x = e$. In alternative mathematical structures the concept of an opposite is defined differently—or it just does not occur. Thus, in a semigroup and in a noncomplemented lattice, the concept of an opposite makes no sense. In sum, in mathematics polarity (a) is not ubiquitous, and (b) it does not follow the same pattern wherever it does happen to exist. Since mathematics is the language of science and of exact philosophy, the very first hypothesis of dialectics (the law D1 of polarity) is neither universal nor unambiguous. Moreover, where it does hold, as in sets (where the opposite is the complement) and in propositions (where the opposite is the negate), it is *a feature of human thought* rather than a trait of the world. Thus, if a fact f is represented by a proposition p, the opposite of the latter, i.e., not-p, does not express the opposite (or "negative") fact not-f (or anti-f), but just the denial of the first proposition. Hence, propositional contradiction may not "reflect" any ontic opposition or contradiction *in re*: it is basically a conceptual affair.

The second hypothesis of dialectics, namely the conjecture that everything consists of a unity of opposites, can be exemplified, provided the opposite is adequately identified and characterized. But it is far from universal. The positron is the opposite of the electron (with respect to charge only), but the system consisting of a positron and an electron is a rarity. The electron and the positron are not known to be themselves unities of opposites. Nor is an electric field, or a gravitational field, or a molecule, or an organism, a unity of opposites. Systems consisting of opposites are the exception rather than the rule. There are many more things constituted by similar entities, such as hydrogen molecules, polymer molecules, and solid bodies. In conclusion, the empirical evidence points away from the second principle of dialectics.

The third principle of dialectics, namely the dialectical law of the mechanism of every change, is not universal either. No struggle of opposites is discernible in mechanical motion, in the propagation of light, in heating, in chemical reactions, or in biological growth. Surely *some* processes, such as the antigen-antibody reaction and the inhibition of neural excitations, are dialectical or, rather, have dialectical *aspects*. But this does not establish the universality of dialectics. In sum, the hypothesis of the strife of opposites (D3) is not universal: dialectical change is a proper subset of change.

The fourth principle of dialectics, the law D4 of spiral progress, can be exemplified, provided the word 'negation' is given an ad hoc interpretation every time—as when it is said that the seed "negates" its plant. Such shifts of meaning intended to save the hypothesis are a flagrant violation of the scientific method. Even so, it has yet to be established that every progressive sequence consists of dialectical negations fitting that hypothesis. For one thing, scientific progress, which is the most uncontroversial type of progress, does not fit that pattern. A hypothesis that does not fit the facts, or fits them provided it is conveniently re-adapted to each fact, cannot be regarded as a law. In short, D4 is not a law.

The only hypothesis of dialectics that would seem to hold universally is the quantity-quality law D5. The first part of it is a plausible generalization, though there may be exceptions to it. The second part is almost trivial: if a property A goes over into a property B, then the rate of change of B is bound to differ from the rate of change of A. We may then give our assent to D5. But notice that, since this hypothesis does not involve the typically dialectical concept of opposition, it is not a typically dialectical statement itself—unless of course one makes the mistake of regarding quality and quantity as mutually opposed.

In conclusion: (a) only one of the five chief hypotheses of dialectics (namely D5) seems to be generally true, and this one can easily be detached from the rest because it does not involve the key dialectical concept of opposite; (b) the four remaining principles of dialectics can be exemplified, but they are not universal and, moreover, they concern exceptional cases—whence any intelligible theory of dialectical change must be a chapter of a more inclusive theory of change; (c) all the examples of dialectics involve considerable variations in the meaning of 'opposite' and 'negation', so much so that one wonders whether there are any examples at all. The last objection disqualifies dialectics as a piece of exact philosophy in agreement with science. For this reason the marriage of dialectics with materialism, which is compatible with logic and contiguous to science, must be an unhappy one.

On the other hand, a watered-down version of dialectics, one that emphasizes variety and change but does not insist on the pervasiveness of polarity, contradiction, and the strife and unity of opposites, can be stated in intelligible terms and be shown to be supported by contemporary science. Such a dynamical but nondialectical worldview is of course dynamical materialism. It respects logic and science and is therefore at variance with Hegelianism. By the same token, it is nondogmatic and subject to correction.

5. UPSHOT

The preceding discussion, though extremely sketchy, supports the following conclusions.

C1. *Scientific research presupposes critical realism.* This epistemology holds, among other things, that scientific theories purport to represent things out there, although, being creations of the mind, those theories are symbolic and partial rather than iconic and complete. However, critical realism is not yet a systematic body of theory and it will not become one unless it unites with semantics. Moreover, critical realism is full of holes. Among the unfulfilled tasks we may mention (a) an elucidation of the model-thing correspondence (i.e., of the representation relation); (b) an elucidation of the concept of factual meaning; (c) a theory of factual and partial truth; and (d) an adequate philosophy of logic and mathematics, one accounting for the formal (nonmaterial) nature of logical and mathematical objects without falling into idealism. (More on this in Bunge 1974a, b, 1983a, and essay 13.)

C2. *Contemporary science supports both dynamical materialism and integrated pluralism.* This ontology conceives of every existent as an ever-changing system; it emphasizes the variety and lawfulness of the world; it recognizes the emergence and multiplicity of levels of complexity or organization, but it assumes the lawfulness of emergence as well as the articulation of the whole-level system. But dynamical and pluralist materialism is, like critical realism, a program rather than a full-fledged theory. We still do not have theories, both clear and compatible with contemporary science, elucidating the following key concepts of that ontology: (a) thing and complex system; (b) connection and interaction; (c) possibility and coming into being; (d) event and process; (e) the various determination categories; (f) life, mind, and society. We shall not get such theories if we do not believe we need them, or if we make no use of mathematics and science in building them. The time of the stray and wild metaphysical conjecture is past: we need ontological systems capable of being checked. (See Bunge 1977a, 1979a.)

C3. *Contemporary science calls for a critical philosophy.* Science can both lend its support to, and be assisted by, a philosophy that, in addition to containing a realist epistemology and a dynamical materialist ontology, satisfies the following methodological conditions:

(a) It regards philosophical theses as hypotheses to be assayed and discussed, expanded or incorporated into hypothetico-deductive systems, or dropped altogether if defeated by logic or by science—not as articles of faith.

(b) It adopts a critical and self-critical approach, looking for counterexamples as well as examples, for reasons instead of quotations, and for new problems instead of old solutions.

(c) It is creative, i.e., engaged in advancing philosophy instead of being intent on keeping it frozen; in other words, it is reformist and even revolutionary rather than conservative.

(d) It makes use of the formal tools evolved in logic, mathematics, metamathematics, semantics, etc., for the analysis and systematization of philosophical concepts and hypotheses.

(e) It keeps close to natural and social science: it stays near the method of science and its results, learning from it rather than censoring it—without, however, neglecting the legitimate critical analysis of the broad presuppositions and procedures of science.

Any philosophy that fails to satisfy either of the conditions C1, C2, and C3 is out of touch with contemporary science. Consequently, if adhered to it may hinder the advancement of science. Only a philosophy satisfying all three conditions, one that sheds outdated tenets, proposes new philosophical theories, and checks them in the light of formal and factual science, is alive and likely to inspire scientific research programs as well as an adequate worldview. If we insist on naming this philosophy we may call it *logical materialism*. This philosophy would presuppose mathematical logic, and it would include both dynamical materialism and critical realism. Let us build this philosophy, or rather unending sequence of philosophies.

2

NEW DIALOGUES
BETWEEN
HYLAS AND PHILONOUS

(1954)

■

FOREWORD

The *Three Dialogues between Hylas and Philonous* were written by George Berkeley in 1713 as a popular exposition of his no less famous *Treatise Concerning the Principles of Human Knowledge*, published three years earlier. Since then, the basic philosophical ideas of the Bishop of Cloyne do not seem to have been satisfactorily refuted, although they have been and are still being abundantly criticized. The most usual argument against Berkeley's idealistic empiricism is still the stick employed by Molière to convince Pyrrhonists. Of course, the argument of practice is historically important and psychologically effective, but, after all, it is along the line of Berkeley's ideas, its success being thus at the same time a paradoxical triumph of his own empiricism.

It is probable that one reason why no conclusive logical arguments against Berkeleyanism seem to have been given, is that most of the bishop's opponents, when thinking of refuting him, tacitly accept that solely experience—for instance, experience with sticks—is able to verify or falsify a proposition. For, as long as it is believed that only facts can compete with facts; that reason is solely able to reflect or at most to combine sense data, being incapable of creation; and as long as one clings to the belief that there cannot be rational proofs of empirical facts, it is only natural that Berkeley's chains of thoughts should remain substantially untouched. The bishop knew it too well, and that is why he based his system on the negation of abstract thought, on the thesis that abstract ideas are nothing but vices of language.

The present dialogues are an endeavor to refute Berkeley's philosophy from a new standpoint. They are offered as a homage to his astuteness, the occasion of the second centenary of his death, in 1953. (For a different and more recent criticism of Berkeley's idealism see Stove 1991.)

THE FIRST DIALOGUE

PHILONOUS. Good morrow, Hylas: I did not expect to find you alive. We have not seen each other for centuries.

HYLAS. To be exact, for two hundred and forty years. By the bye: How do you explain it that, notwithstanding, we have continued existing?

PHILONOUS. You know that the age of philosophies must be measured in centuries, not in years.

HYLAS. I did not mean that but the following: We have not perceived each other during more than two centuries; nevertheless, each of us is certain that the other has been alive during that lapse of time.

PHILONOUS. Now I see what you mean. But your irony is out of place. The distinctive axiom of my philosophy is *To be is to perceive or to be perceived*. I did not perceive you, nor did you perceive me during that time; but you have perceived other things, and so have I—therefore, we have both existed.

HYLAS. Allow me to dwell on this. I did not see you, nor did you perceive me; so that you can be certain that you have existed, but you could not assure that I was alive, say, one hundred years ago.

PHILONOUS. Why not? I am verifying it now. If I had not seen you alive, I could not maintain it with certainty. But I see you now, and since I absolutely rely on my senses, I assure it.

HYLAS. Yes, you *verify* that now. But it has not been enough for you to see me again in order to know that I have been alive one century ago: this knowledge was not contained within your perceptions, because senses do not make inferences. It is a product of reasoning.

PHILONOUS. I grant it. But, instead of speaking about rational inferences, I would prefer to speak about sequences of images. I do not find it difficult to imagine that you have existed one hundred years ago—or, for that matter, to imagine a hippogriph.

HYLAS. Of course you can imagine it, but you cannot demonstrate it unless you are able to ascend from images to concepts, because no sequence of images will ever make a proof. This is just why we form concepts and perform inferences: in order to know and to prove whenever the evidences of the senses are not enough to know and to prove—and they never are.

PHILONOUS. Could you prove, perhaps, the fact that you were alive one century ago, without resorting to evidences of a historical sort?

HYLAS. Certainly, though not by purely logical means. In order to turn it into the result of an inference, I need one more premise beside *Hylas existed in 1753 and exists now, in 1953*—the latter being true according to your own criterion of truth, since it has been warranted by the senses.

PHILONOUS. What is that new premise you need?

HYLAS. A law of nature known by induction from numberless singular cases, namely, that the life cycle of every individual is uninterrupted. In this way, the logical demonstration that I was alive in 1853 reduces itself to a simple syllogism.

PHILONOUS. I must avow that this assumption, which you wrongly call a law of nature, was beneath my inference. Of course, I will not call it a law of nature, but a rule ordained by the eternal Mind.

HYLAS. Be that as it may, the net result of our conversation is that reason is not only able to reflect sensible things, but is also capable of proving, or at least suggesting, the existence of empirical facts otherwise unknown by immediate perception. Such is the case with the fact expressed in the sentence *Hylas was alive one century ago*—which I would call a rational truth of fact.

PHILONOUS. Methinks I agree. But, in return, you must avow that it was not pure logic which gave us this result, since you had to use a so-called law of nature.

HYLAS. I acknowledge it with pleasure, because I am not for a sharp dichotomy between theory and experiment. It was not pure logic which gave us that result; but it was a non-empirical procedure—profiting, of course, from previous experience and being an experience itself—wherein the laws of thought bring together facts and laws of nature. But let us return to our argument.

PHILONOUS. We had come to agree that I must accept as a truth that *Hylas existed in 1853*, notwithstanding the fact that it did not pass through my senses.

HYLAS. That was it. Now, if you admit it, you will have to avow that the famous saying that the intellect contains nothing that has previously not been in the senses, is at least partially untrue. In other words, you must acknowledge that reason is a kind of practice capable of creating things—things of reason, or ideal objects which, of course, may refer to sensible things.

PHILONOUS. Slow, my friend. A single instance does not suffice to confirm a theory.

HYLAS. But it is enough to disprove a theory—yours, for instance. Besides, if you wish I can add a host of cases showing that we cannot dispense with ideal constructs unperceivable or as yet unperceived, existing in the outer reality or in thought alone.

PHILONOUS. I will content myself with a couple of samples.

HYLAS. First, neither you nor anybody else can perceive the facts that occurred millions of years ago, as reported by geology or by paleontology—which reports, incidentally, are believed by the petroleum companies. Second, physics, astronomy, and other sciences are engaged in making predictions of future events, of events known by us if only probably, but in any case anticipating perception and even, as in the case of scientific warfare, annihilating perception.

PHILONOUS. I own there is a great deal in what you say. But you must give me time to look about me, and recollect myself. If you do not mind, we will meet again tomorrow morning.

HYLAS. Agreed.

THE SECOND DIALOGUE

PHILONOUS. I beg your pardon, Hylas, for not meeting you sooner. All this morning I have been trying—alas, unsuccessfully—to refute your contention that the proofs of reason may be as acceptable as the evidences of the senses, and that even creations of thought may be as hard as facts.

HYLAS. I expected it.

PHILONOUS. Still, you did not convince me that we are able to form abstract ideas.

HYLAS. Nevertheless, it is an empirical fact that you cannot avoid employing abstract ideas, such as existence, being, idea, all, none, etc.—especially when trying to convey the abstraction that abstraction is a fiction.

PHILONOUS. Well, I might concede to you that we are able to form abstract ideas. But I hold that we derive them all from perceptions. To be more precise, mind can elaborate the raw material offered by the senses, but it cannot create new things, it cannot make objects that are unperceivable in principle.

HYLAS. You forget that we agreed yesterday that not every idea has a previous existence in immediate perception. Remember, this was your case with *Hylas existed in 1853*.

PHILONOUS. Yes, but that idea *might* have arisen from the senses, if only I had had a chance of seeing you one hundred years ago. Moreover, you must avow that the ideas of existence and being, which you deem to be abstract, are in any case derived by a sort of distillation of an enormous aggregate of concrete ideas of existent beings.

HYLAS. Naturally, I agree with you, and I am glad to detect a little germ of historical reasoning in you. This is just how most abstract ideas are formed: by a long distillation, though not a smooth one. But this is not the case with all abstract ideas: some are pure, though not free, creations of the human mind—even when they arise in the endeavor to grasp concrete things.

PHILONOUS. Do I hear well, Hylas? Are you defending spirit?

HYLAS. I never was his enemy. It was *you* who denied the existence of abstract ideas and, in general, the possibility for mind of creating new ideas without previously passing through experience—thus reducing your famous *nous* to a poor little thing.

PHILONOUS. It will not be long before you accuse me of atheism.

HYLAS. To be sure, I could do it. Think only of the imperfections of your God, who has neither the attributes of materiality nor of abstract thinking. But let us leave theology aside: I do not wish an easy victory over you. I accuse you of being an inconsistent empiricist, because you do not understand that abstract thinking is an activity, an experience. And I accuse you of being an inconsistent idealist, because you do not understand that mental activity is able to create new objects, things that are not to be found in perception.

PHILONOUS. Pray, give me a single instance of these ideal objects not developed from perceptions and moreover, as you would say, not having a material counterpart.

HYLAS. Mathematics is full of such entities not corresponding to any objective reality but which, nevertheless, are auxiliary instruments in the labor of explaining and mastering the world. Not to mention higher mathematics, let me only recall imaginary numbers. Or, better, an even simpler object, the square root of two, or any other irrational number, which you would never obtain by measurement, by perception.

PHILONOUS. As far as I am concerned, these inventions might as well not exist. You know, I struggled against such absurdities long ago.

HYLAS. I remember: You maintained that they were not only meaningless but also harmful—by the way, there you have two nice abstract concepts, which you have employed hundreds of times: meaningless and harmful. Still, as you ought to know, events have demonstrated that irrational numbers are one of the prides of reason, and that imaginary numbers are nearly as useful and practical as real ones.

PHILONOUS. Your arguments remain very poor if they cannot pass beyond the inventions of the analysts.

HYLAS. Do not worry. You may have plenty of abstract ideas outside of mathematics and logic. You always employ them: design, eternity, identity, whole. . . . The trouble with you, Philonous, is that you move so freely on the plane of abstractions that you do not notice it and take them for granted.

PHILONOUS. Perhaps a detailed historical investigation could show that they are really not *new* ideas, but mere refinements and combinations of percepts.

HYLAS. I would not advise it: history is precisely the most effective destroyer of errors. For instance, it is history that shows how concrete singulars become abstract universals, and how the latter enable us to discover and even to make new sensible things. But let us not go astray. I tried to convince you before that there are abstract ideas not made up of perceptions and lacking a material counterpart.

PHILONOUS. Right.

HYLAS. Now I wish to remind you that, conversely, there are real things that can solely be grasped by abstraction. I mean species, wholes, and structures, of which perception and even image provide us with only partial accounts. For instance, you cannot perceive the Irish people, or democracy, or the human species. Nor can you see or smell order, law, prosperity, and what not. Of course, you understand them all on the basis of percepts and with the aid of images; but they are, none the less, abstract ideas corresponding, this time, to objective wholes.

PHILONOUS. I suggest that we do not discuss my deep political convictions. Why do not we return to logic?

HYLAS. With delight. Since you feel so sure about your logic, allow me to put the following question: How do you know that perceptions are the sole ultimate and authentic source of knowledge, the sole factory of human knowledge, and the sole guarantee of reality?

PHILONOUS. Proceeding along empirical lines, I have found that this knowledge, and that one, and a third, they all derive from sense experience.

HYLAS. I very much doubt that you have actually proceeded that way. But, for the sake of discussion, I will assume that every knowledge stems from a corresponding percept; that every singular cognition derives from a corresponding experience—which is in itself singular, unless you should hold that experience is capable of yielding universals.

PHILONOUS. God forbid!

HYLAS. Well, then I further ask: Whence comes this new knowledge, this universal judgment which you take for true, namely *The source of all knowledge is experience?*

PHILONOUS. I am not sure that I have understood you.

HYLAS. I said that, for the sake of discussion, I might grant that every singular knowledge comes from experience. But whence comes your knowledge that all knowledge comes, came, and will come from sense experience? Does this new knowledge originate in experience too?

PHILONOUS. I avow that your argument embarrasses me. I should have dispensed with such general maxims, in the same way as I excluded abstract ideas. But, then, what would have remained of my teachings?

HYLAS. Nothing. And this is just Q.E.D. that your entire system is false, because it relies on a *contradictio in adjecto*.

PHILONOUS. Do not abuse me in Latin, please.

HYLAS. I will explain that to you, but tomorrow, if you please. I shall be glad to meet you again at about the same time.

PHILONOUS. I will not fail to attend you.

THE THIRD DIALOGUE

HYLAS. Tell me, Philonous, what are the fruits of yesterday's meditations?

PHILONOUS. I found that you were right in suggesting that the first axiom of empiricism—"The sole source of knowledge is experience"—is an abstract idea and, to make things worse, partially untrue.

HYLAS. And self-contradictory, as I told you yesterday with a Latin jargon. In effect, the starting-point of empiricism—as of every other philosophy—is not experience but a universal judgment, so that empiricism starts denying abstraction in abstract terms, with which it destroys itself.

PHILONOUS. Thus far I am forced to avow. But I challenge you to prove the falsity of my principle "To be is to perceive or to be perceived".

HYLAS. As far as I remember, I did it in our first dialogue. But, since you are now more used to abstractions, I will give you more refined arguments. In the first place, notice that you cannot maintain that saying on empirical grounds any longer, since we experience singulars only, never universals.

PHILONOUS. I agree.

HYLAS. My new proof runs as follows: If you admit that you are able to conceive at least one abstract idea, an idea not immediately derived from the senses—an idea by definition imperceptible and unimaginable—then your famous principle is done with.

PHILONOUS. I do believe now that I have always played with abstract ideas, but still I do not see your point.

HYLAS. That concession of yours implies two things. First, at least sometimes—while you are making abstractions, while you are working with concepts—you exist without being conscious of your sense impressions. And this destroys your *esse est percipere*. Second, granting the existence of abstract ideas you concede that not every thing consists in being perceived, since abstractions are unperceivable. And this does away with your *esse est percipi*.

PHILONOUS. I am forced to own it. But this new concession would demand only a slight change in my system: from now on I will say that existence is identical with any faculty of the mind.

HYLAS. You are wrong in supposing that you will manage to save your immaterialism after so many concessions as you have made.

PHILONOUS. Why not? So far, I have only admitted theses concerning mind.

HYLAS That is sufficient. As soon as you admit—as you have done it—that not every thing consists in being perceived; and as soon as you accept the validity of theoretical proof, you are forced to admit at least the possibility of theoretically demonstrating the existence of things out of the mind, that is to say, the reality of the extramental world. Whereas before your concessions, this possibility was ruled out.

PHILONOUS. I should agree with such a possibility. But you know how long a stretch there is between possibility and act.

HYLAS. Let us try. You have come to agree that reason is not passive and not confined to coordinating sense data, but is also able to create abstract ideas, and theories containing such abstract ideas.

PHILONOUS. I do.

HYLAS. Now, some of those theories, most of them, are designed to give account of experience. Thus, it is a fact that there are theories of matter, of life, of mind, and even theories of theories.

PHILONOUS. Yes. But why could reality not be a product of theoretical activity?

HYLAS. No, you cannot envisage the possibility of converting to objective idealism. It is true that every true theory enriches the previously existing reality. But not every theory is true.

PHILONOUS. It is a truism that the number of wrong theories is by far greater than that of true theories.

HYLAS. And *that* is just one of my theoretical proofs of the existence of an independent outer world. First, if to think were the same as to exist, most people would not exist. Second, error among the few chosen ones would be unknown and everybody would be a sage.

PHILONOUS. I must own that that is not the case.

HYLAS This lack of complete overlapping or harmony between thoughts and things; this fact that disagreement between thinking and its objects is more frequent than the corresponding agreement, suffices to prove that thought is not the same as matter. That there is a reality, existing out of the mind, and which we are pleased to call 'matter'.

PHILONOUS. I never expected to see unsuccessful theories of matter used to prove the reality of matter.

HYLAS Insofar as our theories of matter fail, they thereby demonstrate the reality of matter; and insofar as they succeed, they demonstrate that we are able to understand the world surrounding us.

PHILONOUS. I must confess that the mere applicability of the concepts of truth and error proves that reality and its theoretical reflections are not identical.

HYLAS. Then, it seems to me that I have succeeded in destroying your basic principles one by one.

PHILONOUS. I own that you have. I am now convinced that experience is not the sole source of knowledge, and that there are things beyond our perceptions, images, and concepts. Only one question now troubles my soul: What has become of the omnipresent, eternal Mind, which knows and comprehends all things?

HYLAS. If my memory is still faithful after so many years, your favorite argument ran as follows. I know by experience that there are other minds exterior to my own; and, since every thing exists in some mind, there must be a Mind wherein all minds exist.

PHILONOUS. Exactly.

HYLAS. But you have agreed today that some things exist *out* of mind, so the very basis of your argument is gone. As to your argument based on the supposed passivity of the mind, which passivity would require an external mover, it fell long ago, as soon as you conceded that minds can create new objects of their own.

PHILONOUS. You have satisfied me, Hylas. I acknowledge it, and I think I shall withdraw until some next centenary.

3

ENERGY: BETWEEN PHYSICS AND METAPHYSICS

(2000)

■

The word 'energy' has a long and chequered history. For example, it occurs in Aristotle's work, where it appears to stand some times for potency and at other times for act. As recently as in the eighteenth century, energy was sometimes identified with force and at other times with mechanical work. From the mid-nineteenth century onward, the word 'energy' designates of course a central physical concept. It is taught early on in school, measured in thousands of laboratories and workshops the world over, and the electric utility company charges us monthly for the energy we consume. And yet, the famed *Feynman Lectures on Physics* tell us that present-day physics does not know what energy is. True or false? Let us see.

It is well known that there are various kinds of energy: kinetic, elastic, thermal, gravitational, electric, magnetic, nuclear, chemical, etc. More precisely, there are as many kinds of energy as kinds of process. (I will ignore Freud's metaphorical talk of psychic energy, as well as New Age empty talk of crystal energy, positive and negative energy, and the like.) Unlike other species, the various kinds of energy are equivalent, in that they can be transmuted into one another. For example, when we tense a bow we store potential elastic energy in it; and when we release the bow, that energy is transformed into the arrow's kinetic energy. In this two-phase process, the type or quality of energy changes, but its quantity remains the same. Such quantitative conservation is the reason that we regard all the kinds of energy as mutually equivalent. In other words, the introduction of the general concept of energy is justified by the general principle of conservation of energy. But both the concept and the principle, though rooted in physics, overflow physics. Let me explain.

Every particular concept of energy is defined in a given chapter of physics. For example, "kinetic energy" is defined in dynamics; "thermal energy" in thermodynamics; "electromagnetic energy" in electrodynamics; and "nuclear energy" in nuclear physics. Every one of these fields of knowledge has its own concept of energy. Moreover, in all of these fields, save thermodynamics, a theorem of conservation of energy is provable from the corresponding equations of motion or field equations. To add energies of two or more types, we need to join the corresponding disciplines. For example, the total energy of a jet of an electrically charged fluid can only be calculated in the hybrid science of electro-magneto-thermo-hydrodynamics, which is needed to study sunspots.

All the sciences that study concrete or material things, from physics to biology to social science, use one or more concepts of energy. For example, cognitive neuroscientists want to measure the metabolic cost (in calories or, rather, joule) of a bit of information transmitted across a synapse; anthropologists, sociologists, and economists are interested in finding out the energy consumption per capita in a community, and whether people work so as to optimize their efficiency, or ratio of energy output to energy input.

Because it is ubiquitous, the concept of energy must be philosophical and, in particular, metaphysical (ontological). That is, it belongs in the same league as the concepts of thing and property, event and process, space and time, causation and chance, law and trend, and many others. Assuming that the general concept of energy is indeed philosophical, let us proceed to analyze it in philosophical terms. And, since the best analysis is synthesis, or the embedding of the construct of interest into a theory (or hypothetico-deductive system), let us craft a minitheory centered in the concept of energy.

We start by identifying energy with changeability. That is, we propose the following

DEFINITION. Energy $=_{df}$ changeability. (This convention may be rewritten as follows: For all x: x has energy $=_{df}$ x is changeable.)

Let us now put this definition to work. We begin by assuming

POSTULATE 1. All and only concrete (material) objects are changeable. (That is: For all x: x is concrete [material] if and only if x is changeable.)

REMARK 1. We have equated "concrete" with "material". This convention is more common in philosophy than in physics. According to it, fields are just as material as bodies. For example, photons are material in the philosophical sense of the word, although they lack mass, solidity, and a shape of their own—the attributes of matter according to pre-field physics.

The above Definition and Postulate 1 entail

THEOREM. For all x: if x is a material object, then x has energy, and vice versa.

This theorem has two immediate consequences:

COROLLARY 1. The abstract (nonconcrete) objects lack energy.

That is, the conceptual objects are not changeable. What do change are the brains that think them. For example, the concept of energy has no energy. To be sure, one may redefine to death the concept of energy. But every such concept is timeless. If preferred, one's successive conceptual creations do not change by themselves. The same holds for the concept of matter: matter, as a concept, is immaterial.

COROLLARY 2. Energy is a property, not a thing, state, or process.

REMARK 2. Because energy is a property, it can be represented by either a function or an operator. In classical physics one may say that $E(c, x, t, f, u)$ is an arbitrary value of the energy of a thing c situated at point x, and time t, relative to a frame f, and reckoned or measured in the energy unit u. The function in question has the general form

$$E: C \times E^3 \times T \times F \times U \longrightarrow \mathbb{R},$$

where C is a set of concrete things, E^3 the Euclidean space, T the set of (time) instants, F a set of reference frames, U a set of energy units, and

ℝ that of real numbers. In the case of an interaction energy, C will be replaced with the Cartesian product $C \times C$. In quantum physics, energy is represented by the Hamiltonian operator H. The corresponding property is the energy density $\psi*H\psi$, which has the same general form as the classical energy density.

REMARK 3. The kinetic energy of a particle, relative to a reference frame bound to the latter, is nil. Similarly, the total energy of a thing embedded in a field becomes zero when its kinetic energy equals its potential energy (as is the case with the outer electron of an atom at the time of ionization). However, null energy is not the same as lack of energy, just as zero temperature (on some scale) is not the same as lack of temperature. In these cases, unlike the case of the photon mass, zero is just a special numerical value.

REMARK 4. The expression 'free energy', which occurs in thermodynamics, may give the impression that it denotes a kind of stuff-free energy. This is not the case: the free energy of a macrophysical system is nothing but the portion of its energy available for mechanical work.

REMARK 5. Corollary 2 entails that the concept of a concrete or material thing cannot be replaced with that of energy. There is no such thing as energy in itself: every energy is the energy of some thing. In other words energetism, which a century ago was proposed as an alternative to materialism, was radically mistaken. However, the energetists, particularly the famous physical chemist Wilhelm Ostwald (1902), were right in holding that energy is universal—a sort of cross-disciplinary currency. They would have been even more right had they proposed

POSTULATE 2. Energy is the universal physical property: the only property common to all material things.

REMARK 6. One might think that position in spacetime is another universal physical property. It is, but, according to any relational (as opposed to absolute) theory of spacetime, the latter is not basic but derived: it is the basic structure of the collection of all things, every one of which possesses energy. Roughly, space is rooted to the spacing of things, and time to their change. No space without things, and no time without changing things.

REMARK 7. Postulate 2 does not state that every thing has a precise energy value at any given time and relative to any given reference frame. It does not, because a sharp energy value is the exception rather than the rule. Indeed, according to quantum theory, in general a quanton (quantum-mechanical thing) is in a superposition of infinitely many energy eigenfunctions, the corresponding eigenvalues of which scatter around some average value, such as an atomic energy level.

Our final assumption is

POSTULATE 3. The total energy of an isolated concrete object does not change in the course of time.

REMARK 8. This is of course the general principle of conservation of energy. It is so extremely general that it belongs in philosophy rather than in physics.

REMARK 9. In an expanding universe, energy is not conserved, but is slightly dissipated.

REMARK 10. According to quantum electrodynamics, the vacuum energy is not zero but fluctuates irregularly around zero. This result does not invalidate the characterization of energy as the universal property of all concrete things. All it does is to restrict the domain of validity of the classical definition of vacuum as the absence of matter endowed with mass. The electromagnetic field that remains in a region of space after all the electric charges have been neutralized, and all the electric currents have been switched off, is a concrete though tenuous thing. It is so concrete that it exerts a force upon the electrons of the atoms in the walls of the container: this is the Lamb shift, one of several measurable effects of the vacuum.

REMARK 11. New Age scribblers have no monopoly on nonsense about energy. Careless physicists have produced some such nonsense. In fact, energy is often confused with radiation, and matter with mass. Let us examine a small sample of mistakes of this kind found in physics textbooks and popular science publications.

Example 1. The expressions 'annihilation of matter' and 'materialization of energy' for electron pair destruction and formation, respectively, are incorrect. What gets annihilated when a pair of electrons of opposite charge gets transformed into a photon is not matter but mass, a property of particles and bodies but not of photons. Dually, matter endowed with mass emerges in "pair creation." (The two sides in the conservation equation, $2mc^2 = h\nu$ do not hold at the same time. The LHS holds before, and the RHS after the "annihilation.") Unlike energy, mass is not a universal property.

Example 2. When an antenna radiates electromagnetic waves for a total of E ergs, it loses E/c^2 grams of mass. This does not exemplify the conversion of matter into energy, but a transformation of part of the energy of the antenna (and the field around it) into radiation energy, with the concomitant decrease in mass. The total energy is conserved.

Example 3. The formula $E = mc^2$ does not say that mass and energy are the same modulo c^2. Indeed, E and m are very different properties. For one thing, E is the ability to change, whereas m is inertia, or the disposition to resist change in one respect (state of motion). For another, the formula holds only for things endowed with mass, that is, particles and bodies. So much so that it is a theorem in relativistic mechanics, not in field theories.

So much for our minitheory and a few elucidations and examples. We are finally in a position to answer the question that triggered this article, namely whether it is true that in physics one does not know what energy is.

This statement is partially true. Indeed, in each chapter of physics we can identify a special type of energy, and guess, prove, or check a particular energy conservation law. But the general concept of energy and the general law of conservation of energy belong in philosophy. Still, philosophy cannot elucidate either without the help of physics. It is only by joining the two fields that a precise answer can be given to the question in hand, namely this: Energy is the universal property of matter—the property of being capable of changing in some respect.

What complicates the problem, and on occasion misleads even the specialist, is that (a) there are as many species of energy as large classes (genera) of process; (b) there are as many concepts of energy as general physical theories; (c) the general concept of energy is so general that it belongs in metaphysics (ontology); and (d) consequently the general principle of conservation of energy is philosophical, though rooted in physics.

Moral: Physics cannot dispense with philosophy, just as the latter does not advance if it ignores physics and the other sciences. In other words, science and sound (i.e., scientific) philosophy overlap partially and, consequently, they can interact fruitfully. Without philosophy, science loses in depth; and without science, philosophy stagnates.

4

THE REVIVAL OF CAUSALITY

(1982)

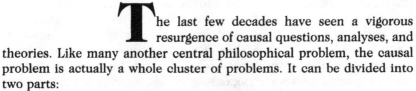

The last few decades have seen a vigorous resurgence of causal questions, analyses, and theories. Like many another central philosophical problem, the causal problem is actually a whole cluster of problems. It can be divided into two parts:

(i) *The ontological problem* of causation: What is causation? What are the relata of the causal relation (things, properties, states, events, mental objects)? What are the characteristics of the causal nexus? To what extent is the nexus real? Are there causal laws? How do causation and randomness intertwine?

(ii) *The methodological problem* of causality: What are the indicators and criteria of causation? How do we recognize a causal link, and how do we test for a causal hypothesis?

Every one of these subsystems is chockfull of problems, some old and others new. Just think of the difficult problem of unraveling causal relations beneath statistical correlations, and of the no less difficult problem of determining the causal range (or domain of validity) of a law statement. Moreover, the membership of every one of the above sub-clusters of problems keeps changing in the course of time. But somehow none of them seems to become depleted, even though once in a while a hasty philosopher will proclaim the "final" solution to "the" causal problem—perhaps without having bothered to propose a careful formulation of the causal principle. This is not surprising in philosophy: At some time or other, matter, property, space, time, law, determination, chance, life, mind, and many other metaphysical categories have been decreed dead until it was eventually discovered that they refused to lie down. Philosophical fashion seems to be as erratic as dress fashion, and in any case it is not a reliable indicator of the intrinsic worth of a problem or of a theory.

This review focuses on the ontological problem of causation, but it is not limited to the philosophical literature: it also attempts to uncover certain trends in recent science.

1. STATE OF THE CAUSAL PROBLEM IN MIDCENTURY

A perusal of the philosophical literature shows that the causal problem failed to attract the attention of philosophers during the '30s, '40s, and '50s. Surely, a few books and articles on the subject did see the light during that period. However, most of them were written by nonphilosophers, and they dealt either with the alleged death of causality in the hands of quantum mechanics, or with ways of inferring causal connections from statistical correlations. The general ontological problem of causation was usually neglected, as were in fact ontological questions in general.

Causality had been a casualty of two profound intellectual revolutions in the 1920s, namely the quantum theory and logical positivism. These were not mutually independent events. In fact, the quantum theory had been interpreted in the light of logical positivism, namely in terms of measurement operations, or even the experimenter's intentions. And this philosophy was the heir to Hume's empiricism, which involved a denial of causation as an objective connection beyond constant conjunction or regular succession. Besides, the quantum theory was basically probabilistic. No wonder, then, that causality seemed scientifically and philosophically dead.

Nor were such developments isolated. For one thing, the exact sciences are hardly interested in relations between whole events: instead, they are centrally interested in relations among properties represented by functions ("variables"). And functional relations among mathematical representations of properties are neither causal nor noncausal by themselves. For another, chance began to be recognized as an ontological category, not just as a cloak of ignorance. Think not only of random quantum jumps and of the random scattering of elementary particles by atomic nuclei. Think also of genomes, which are the product of random processes such as the random shuffling of genes occurring in recombination, and of the random mutations undergone by the nucleic acids in the cell. Nor were the quantum theory and genetics the only fields where probability laws occurred: probability has started to invade psychology (in particular learning theory) and sociology (in particular the theory of social mobility). The whole of contemporary science was shot through with the concept of randomness. Thus, by midcentury, scientists and philosophers had come to realize that chance is not just a name for our ignorance of causes: that chance is a mode of being and becoming. So, any theory of chance that ignores chance—and causalism does just this—had to be false, as Cournot and Peirce had known all along.

Causalism—the doctrine that all relations among events, except for the purely spatiotemporal ones, are causal—was then defunct at midcentury as a result of scientific developments as well as of the emergence of logical positivism. Any attempt to resume the discussion of causal questions seemed then quixotic and doomed to failure. But, as we know from the history of philosophy, certain problems are die-hard. This proved to be the case of causality.

2. THE RESURGENCE OF CAUSALITY

While philosophers rejoiced in the disappearance of one more troublesome problem, men and women of action and professionals continued to think unashamedly in causal terms. In particular, engineers cannot help doing so because it is their job to design systems whereby humans can cause changes in the environment. Likewise, physicians cannot help study the etiology of disease: they too are firm believers in causality. Nor do lawyers doubt it, for it behooves them to investigate the effects of certain human actions (or inactions). Unfortunately, the philosophy of technology and the philosophy of medicine were hardly in existence at midcentury, so philosophers took no notice of the discrepancy between the practice of engineers and physicians on the one hand, and the philosophy of causation on the other. But the philosophy of law was very much alive, and it did produce a major work on our subject, namely *Causation and the Law* by Hart and Honoré (1959).

The authors of this important and influential work show convincingly that certain causal questions occur often in every branch of jurisprudence; and moreover, that there is a large variety of causal concepts, some of them unsuspected by philosophers. Hart and Honoré characterize these concepts clearly (though not exactly) and illustrate them with a great many legal cases. A first result of their study is the great number of different causes that various legal experts may impute to one and the same human action. Take the case of murder with a gun. What or who caused the death of the victim? The coroner may declare that the cause of death was the shot wound; the public prosecutor, that it was the defendant's pulling the trigger of a gun; and the counsel for the defense, that it was the victim's harsh treatment of the defendant, or the climate of violence in which the latter had been brought up, or what have you. Even if all the experts agree on the facts, and act in good faith, every one of them is likely to select one member of the causal cluster, or one link of the causal chain, as the main or most relevant cause of the event—which is just as well as long as neither denies that every one of them may have been a contributory cause.

From an ontological point of view, perhaps the most intriguing causal questions arising in the law are not those concerning wrongdoing, but the cases of negligence or omission, i.e., the failure to do something when action was called for either morally or legally. Although omissions do not qualify as causes in ontology, they do so in ethics and in the law. Thus, the employee who fails to lock a door and thus makes a burglary possible or easier is a contributory cause of the crime and as such liable to retribution. That a nonevent, such as an omission, should count as a cause sounds absurd in ontology, not in ethics or in the law, and is therefore a good focus for investigating the difference between natural law on the one hand, and moral or legal norm on the other.

Another peculiarity of legal and ethical thought that distinguishes it from ontology is that it involves the notion of responsibility, absent from natural laws and typical of human sociality (and perhaps of mammalian sociality in general). Thus, it matters to both morals and the law who is morally responsible for a crime (e.g., as an instigator), not only who committed it, i.e., caused the wrong outcome. Ivan Karamazov is morally responsible for Smerdyakov's murder, and is therefore as guilty as he. Not surprisingly, the notion of responsibility and its relation to that of action plays a central role in the treatise of Hart and Honoré.

The same year saw the publication of the reviewer's own book *Causality: The Place of the Causal Principle in Modern Science* (1959a). This work seems to have appeared at the right moment, for it was reprinted and reviewed many times, is still in print, and has been translated into seven languages. It criticized a number of doctrines, among them the total denial of causation, its identification with uniform succession or with constant conjunction, and the view according to which causation is the sole mode of determination. Moreover, the book defended the following theses:

(i) *The causal relation is a relation among events*—not among properties, or states, let alone among ideas. Strictly speaking, causation is not even a relation among things (i.e., there are no material causes). When we say that thing A caused thing B to do C, we mean that a certain event (or set of events) in A generated a change C in the state of B.

(ii) Unlike other relations among events, the causal relation is not external to them, as are the relations of conjunction (or coincidence) and succession: *every effect is somehow produced (generated) by its cause(s)*. In other words, causation is a mode of event generation or, if preferred, of energy transfer.

(iii) *The causal generation of events is lawful* rather than capricious. That is, there are causal laws or, at least, laws with a causal range. (And such objective regularities must be distinguished from law statements, such as differential equations.)

(iv) *Causes can modify propensities* (in particular probabilities), *but they are not propensities.* In the expression 'Event X causes event Y with probability p' (or 'The probability that X causes Y equals p'), the terms 'causes' and 'probability' are not interdefinable. Moreover, strict causality is nonstochastic.

(v) *The world is not strictly causal although it is determinate*: not all interconnected events are causally related and not all regularities are causal. Causation is just one mode of determination among several. Hence, determinism should not be construed narrowly as causal determinism. Science is deterministic in a lax sense: it requires only lawfulness (of any kind), and nonmagic.

A spate of articles and books on causality followed the publication of Hart and Honoré (1959) as well as Bunge (1959a): the subject was no longer taboo. Thus, Lerner (1965) collected a number of lectures given during the 1960–61 academic year by such distinguished authors as Ernest Nagel, Ernst Mayr, Talcott Parsons, and Paul Samuelson. The *Encyclopaedia of Philosophy* devoted a long article by Taylor (1967) to the problem. Suppes published two works (1970, 1974) expounding a probabilistic theory of causality. Svechnikov (1971) approached the problem of physical causation from the point of view of dialectical materialism. Von Wright (1971) treated the matter of causation in human action from the point of view of hermeneutic philosophy. Wallace (1972, 1974) produced a scholarly historico-philosophical treatise on causality. Mackie (1975) reexamined Hume's doctrine and analyzed several notions of causation occurring in ordinary knowledge. Harré and Madden (1975) launched a vigorous attack against Hume and his successors, and attributed all things causal powers, i.e., powers to produce or generate something. Hence, they characterized the causal relation as one whereby "powerful particulars" produce or generate changes in other particulars. Finally, Puterman (1977), influenced by Polish and Soviet authors, studied the problem of physical causation regarded as a special form of determination.

All in all, the last few decades have seen a vigorous revival of philosophical studies of causation. However, we still lack a comprehensive work dealing with causal thinking in all fields, from the natural sciences to the social sciences to the various sectors of rational human action.

3. CAUSALITY IN PHYSICS

Half a century ago, causality seemed to have gone from physics, and this for two reasons. Firstly, because quantum mechanics and quantum electrodynamics, the two most fundamental theories of physics, are basically stochastic: their fundamental law statements are probabilistic. Secondly, because, according to the usual interpretation of these theories, called the *Copenhagen interpretation*, every quantum event is the outcome of some human operation or other, so that there are no subject-free causal chains. Moreover, quantum events are said to be "conjured up" at will by the experimenter.

Both arguments against causality have lost much of their force in recent years. For one thing, the Copenhagen interpretation of quantum mechanics is no longer accepted without qualifications: more and more physicists are becoming dissatisfied with its subjectivistic aspects (see, e.g., Bunge 1973b; Lévy-Leblond 1977). For another, it is being realized that randomness may alter the causal relation without eliminating it. For example, when a physicist calculates or measures the probability that an atom will scatter an incoming electron within a given solid angle, he treats this event as the atom's field causing a certain event (the electron scattering through the given angle) with the given probability. Thus, he determines the propensity of a given event (the two entities coming close) causing another event (the scattering).

Thus, causality, initially forced underground by the quantum revolution, is now back, though in a modified form. For one thing, all of the quantum theories retain the concept of force—which quantitates that of causation—although they define it differently from the way it is defined in classical physics. For another, they contain not only probabilities of spontaneous (uncaused) events, such as radioactive decays and radiative transitions of atoms in a void, but also probabilities of events induced by external agents such as fields. Only, the cause-effect relation is in such cases probabilistic rather than deterministic in the narrow sense.

Finally, the causal aspects of the quantum theories are increasingly regarded as representing causal aspects of the world.

What holds for physics holds a fortiori for astronomy and cosmology, as well as for chemistry and the earth sciences. In particular, although in principle theoretical chemistry is based on quantum physics, organic chemistry and biochemistry deal mostly with large molecules, and are therefore more receptive to causality than microphysics. Thus, the biochemist and the molecular biologist have few occasions to remember quantum randomness in their daily work: They reason not so much in terms of transition and scattering probabilities as in terms of chemical bonds and activation energies. Moreover, when dealing with large-scale chemical reactions, chemists use chemical kinetics, whose equations contain only concentrations and their rates of change (first-order derivatives). Because these equations are of the first order, they make no room for inertia. That is, chemical reactions lack the component of self-movement so obvious in other processes. In macrochemistry, then, the Aristotelian principle *causa cessante, cessat effectus* seems to hold (Bunge 1979a).

Another field where this principle holds is automata theory and its applications (e.g., to computer science). Indeed, a deterministic automaton stays in a given state unless it receives a stimulus; and it jumps to another state just in case it accepts an input: it lacks spontaneity. Hence it cannot possibly mimic the human brain. To put it formally, the next-state function M, characteristic of each automaton type, maps the set of ordered pairs (state, input) into the set of states, in such a way that $M (s, 0) = s$, where s is an arbitrary state and 0 the null input. (Incidentally, the M function constitutes a possible formalization of the causal relation.) Note that the concepts of space and time, essential to Hume's conception of causation, are absent from the above.

4. CAUSALITY IN BIOLOGY

Causal thinking is always at work in biology, but it has been particularly active during our period, especially in relation to evolution, morphogenesis, goal-directed behavior, and downward causation. Let us take up evolution first. Molecular biologists tend to believe that, since genic recombination and mutation are random processes, randomness is overriding on the biolevel and, moreover, that every biological novelty emerges by chance (see Monod 1970). However, whole-organism biologists know that this is only half of the story: that the environment selects phenotypes, not genotypes, and that such selection is not random. Indeed, the organisms that are not well adapted leave little offspring or none: a monster has no chance to start a new lineage. For this reason, it has been said that the environment is the antichance or causal aspect of evolution (Dobzhansky 1974).

As for final causation, laughed away by Rabelais, Bacon, and Spinoza, it is now back in biology under the name of *teleonomy* (cf. Mayr 1965; Ayala 1970; Monod 1970). Indeed, it is often claimed that the essential difference between the simplest of bacteria and a nonliving thing is that the former is goal-directed, or has a "teleonomic project" built into it. And, just like the teleologists of old, the teleonomists see in adaptation proof of goal or even design. Thus Ayala: "The features of organisms that can be said to be teleological are those that can be identified as adaptations" (in Dobzhansky et al. 1977, p. 498). But this identification of adaptedness with goal-directedness renders the theory of evolution by natural selection useless, for its gist is that adaptation, far from being the outcome of a goal-seeking process, is the (partially accidental) result of genic variation and environmental selection. If every organism behaved teleologically, then it would be perfectly well adapted, and so environmental catastrophes would be the only evolutionary factors.

The idea of downward causation is this. Whereas some geneticists claim that animal behavior is determined by the genetic blueprint ("upward causation"), some students of behavioral evolution point out that society influences the genotype by selecting the individuals that are best adapted to both the natural and the social environments ("downward causation"). Finally, others claim that both upward and downward causation are at work (e.g., Campbell 1974). Actually, the term 'causation' is misused in this context, for what is at stake is a multilevel system, such as a social animal, and two views of it. One is the "upward" view according to which the higher-level properties and laws of the system are determined by (hence reducible to) the properties and laws of its components (in particular its nucleic acid molecules). The rival view is the "downward" one according to which things proceed the other way around; that is, the behavior of the system on its higher levels (e.g., the social one) determines the behavior of its components. Thus, in selecting whole organisms, the environment (in particular the social environment) selects indirectly the genetic makeup, i.e., the RNA and DNA composition. The truth seems to lie in a synthesis of the upward and the downward views. However, neither of these should be formulated in terms of causation because levels, being sets, cannot act upon one another (Bunge 1977b, essay 5). What we do have here is not causal relations, but functional relations among properties and laws on different levels.

Finally, morphogenesis has made at least two contributions to causal thinking. One is the set of models of cell clumping and other morphogenetic processes in strictly physical or chemical terms, such as the diffusion of the products of chemical reactions and the mutual attraction of cells of the same type. A second contribution comes from catastrophe theory, which purports to deal with the emergence of novel forms in all realms, from hydrodynamics to biology to sociology to semantics. Its creator, the eminent topologist René Thom (1972), has unearthed Plato's Forms acting by themselves *ab extrinsico* upon inert matter. (He identifies forms with shapes, and endows them with a formative capacity, as well as with autonomous existence.) While this is nothing but Platonism in mathematical disguise, it has created an extraordinary stir among those mathematicians who, like Thom, could not care less for the mechanisms of processes—in particular quantum mechanical, molecular (e.g., hormonal), and selective. Predictably, the applications of catastrophe theory have been subjected to scathing criticism (cf. Zahler and Sussmann 1977).

In conclusion, causal questions have been intensely debated by biologists and biophilosophers during our period, in contrast to previous years, when the emphasis was on description and taxonomy. However, the resurrection of final and formal causes suggests that the level of sophistication of causal thinking with reference to organisms is rather modest.

5. CAUSALITY IN PSYCHOLOGY

Behaviorism was the golden age of causality in psychology and social science. Indeed, the environmentalist or stimulus-response schema and the unconditioned reflex do have the properties of the causal link, in particular the tacit pretense that the inner states of the organism do not matter for its response. Indeed, assuming that the animal is in any of a set of "normal" states, that it has no memory, and that a single stimulus operates at any given time, every S-R relationship should be constant. Thus, a prick on the hand should always produce the withdrawal of the hand—unless of course the animal has been anesthetized. The reason for the response constancy is that the excitation is transmitted by a direct pathway (or through-route) in the nervous system, without intervention of the higher centers. However, any departure from normality in the state of the nervous system will distort the S-R relationship. Thus, a curare or strychnine injection may put the pathway out of commission, so that the stimulus will fail to elicit the usual response. This shows that S-R relations have a limited causal range. More to the point, it shows that the state of the system is involved, only it may not matter precisely which state the input acts on as long as we do not care to discriminate among the possible responses. (In other words, the animal is not a black box, but is at the very least a gray box such as an automaton.)

As we move higher up on the ladder of nervous systems (or in the phylogenetic tree), we find that the S-R relationships acquire a narrower or even nil causal range. Indeed, eventually we meet the ideational processes that can delay or distort the responses, or even inhibit them altogether, or else produce novel (not-learned) responses. Thus, we can occasionally control sneezing, and armor ourselves against pain. Even more to the point: Some of these ideational processes are self-started rather than being evoked by sensory stimuli. They are not totally uncaused, though, but may be explained as the outcome of the activities of neural systems, or of the interactions among these and other body subsystems.

Because of its increasing concern with internal states, psychology has moved away from behaviorism, hence from causalism. Another factor in this trend has been the work in stochastic models of learning and in random neural networks. Even though most of the former have been behaviorist or at most neobehaviorist, they have centered on the probability of correct performance upon a given trial, and so have been stochastic rather than strictly causal. Yet they have retained the externality typical of causality.

The above-mentioned movement away from causality in psychology, particularly in learning theory and physiological psychology (or cognitive neuroscience), has had no profound influence on the philosophy of mind. In particular, the psychoneural dualists of the interactionist variety have continued to hold that brain processes can cause mental processes and conversely, although the brain and the mind are separate entities (see, e.g., Popper and Eccles 1977). But, of course, the concept of efficient cause occurring here is the vulgar one, not the one employed in science, where causation is a relation between changes occurring in concrete entities and moreover describable, at least in principle, in exact terms, not just in ordinary language. To a cognitive neuroscientist, there can be no action of the brain on an immaterial mind or conversely: he will admit only actions—some of them causal—of some subsystem of the neuroendocrine system upon another. For example, in some yoga experiences, the cerebral cortex can act on the autonomic system—much as it can "order" a limb to move. To the scientist, these are biological processes that can be explained, or ought to be explained, without resorting to an immaterial mind or self (essays 5 and 19, Bunge 1979a, 1980a).

Action theory too is rife with causal thinking. A landmark of our period is von Wright's *Explanation and Understanding* (1971), a study of the relations between human intentions, actions, and their results and consequences. Von Wright relies heavily on the non-Humean concept of causation as event generation or production, and he draws a neat distinction between the result and the consequence of a human action, which are generally lumped into the category of effect. However, the foil for his analysis of causation and causal explanation is the finite state probabilistic automaton (which he does not call by this name). This model might have served his purpose to some extent, for each state of the automaton (or node of its Moore graph) can be made to represent a "state of affairs" (or "possible world", or "scenario") that can be influenced by human action. However, all this gives is a fan of possible states: it does not tell us anything about the presumptive causal mechanism "responsible" for the unfolding of states—unless, of course, one adds the postulate that humans are the only causal agents in the universe. In any case, von Wright makes little use of the stochastic model of the world in discussing what interests him most, namely the causal relation between human actions and their consequences. Likewise, Tuomela (1977) adopts a causalist viewpoint and, more particularly, makes final or purposive causation the pivot of his theory, presumably on the dubious assumption that all human action is purposeful.

Finally, let us go back to psychology proper, to mention some recent studies on the psychology of causality in the wake of Piaget's classical investigations of the '20s. There are some interesting new results in this field, among them those of Shultz (1978), on whether children between two and thirteen years of age think of the causal relation as a mere conjunction (or as a succession) of contiguous events, or in terms of energy transmission (i.e., of production of events). Experiment points to the latter: children are not natural Humeans. The matter of understanding change in terms of causal mechanisms is of course quite different. (For example, Piaget 1974 finds that children under eleven years of age, i.e., before they can grasp formal operations, are unable to explain the sound emitted by a tuning fork excited by the vibrations of another tuning fork nearby.) Unfortunately, most psychologists have failed to analyze the notion of causation, and thus tend to lump the question of the "perception" (hypothesizing) of causation with the problems of conceiving of the world as lawful, of explanation, and of inference. The moral is clear, namely that one must read carefully their experiments rather than their conclusions.

6. CAUSALITY IN SOCIAL STUDIES

Accident and chance are ever-present in human affairs: were it not for them, we might be able to predict accurately the state of our economies, our polities, and even our cultures. However, causation is no less a factor to be reckoned with. Every person counts on certain causal links when intent on causing someone else to do something, or on preventing somebody from causing her to behave in a given manner. What holds for individual actions holds also for social action, or rather concerted individual actions. Surely, like in the cases of physics, biology, and psychology, often only the probabilities of certain causal links can be modified; for example social mobilities (which are probabilities) can be enhanced or weakened. But this may suffice for practical purposes and, in any case, it shows that the social sciences cannot ignore causation. Yet, this is exactly what they tried to do during the long period of positivist social science, which historians date back to Ranke.

Our period saw the beginning of the end of the long and boring night of narrative (or "objective") historiography in English-speaking countries. (The search for causal connections had been on the way long before, particularly among the French social and economic historians Fernand Braudel and other members of the Annales school. However, the work of these outstanding scholars was practically unknown in English-speaking countries until a few years ago.) The English manifesto of analytic (or interpretative) history was Carr's *What Is History?* (1967). This book, which devoted an entire chapter to the matter of causation, proclaimed the slogan "The study of history is a study of causes" (p. 113).

The historian, writes Carr, will commonly assign several causes to a given event, and will rank them either in order of importance or in temporal order. (Machiavelli nods.) Accidents, or chance events (in Aristotle's sense of the word), such as Antony's infatuation with Cleopatra, are not external to causal chains, but members of causal chains interfering with the one the historian happens to be interested in. Carr criticizes Karl Popper for claiming that almost anything is possible in human affairs, and for failing to realize the causal networks underlying historical processes. And he criticizes Isaiah Berlin for holding that historical determinism is morally objectionable because, in attempting to explain human actions in causal terms, it denies free will and encourages historians to evade their primary responsibility, which would be to condemn the Napoleons and Stalins of history. Rightly enough, Carr reminds us that the function of the historian is to uncover and explain facts, rather than to moralize.

The American historian Lawrence Stone concurs; and, just as Carr works on the Russian revolution, Stone investigates the *Great Rebellion* (1972), trying to uncover what he calls the preconditions, precipitants, and triggers of that complex process. Although he contrasts the approaches of the narrative and the analytic historians, Stone believes that the main difference between them is that "the former works within a framework of models and assumptions of which he is not always fully conscious, while the latter is aware of what he is doing, and says so explicitly" (p. xi).

A similar movement away from descriptivism, and towards the search for causal links, occurred at about the same time in anthropology and archaeology. The description of artifacts divorced from their social setting was replaced with the tentative explanation of their function or use. And the description of customs was supplemented by the conjecturing of social structures and economic determinants. Social anthropology became firmly established, and the need for some theoretical guidance began to be felt as field-workers started to use their brains in addition to spades and tape recorders. The landmark of the period is Harris's brilliant and eloquent *Rise of Anthropological Theory* (1968), a devastating critique of the antitheoretical bias as well as of cultural idealism. Harris defends instead cultural materialism, a variant of economic determinism that has recently attained prominence in anthropology. He also espouses operationalism and, like most workers in the social sciences, calls 'theories' what are for the most part hypotheses. Harris indicts the search for mere statistical correlations, which can be entrusted to computers and contribute little to our understanding of social reality, and asserts that "causality is alive and living everywhere" (p. 260).

Economists are divided as to the importance of causal hypotheses, as well as to their nature. Thus, Samuelson (1965) adopts the regular-succession-in-time view of causation. Consequently, he regards finite difference equations and differential equations with time as the independent variable as causal laws. However, there need be nothing causal about a regular sequence of events and, unless the equations describing it are enriched with semantic assumptions pointing to causal factors, the equations are neither causal nor noncausal (Bunge 1959a). Besides, no matter how unconcerned about causes theoretical economists may pretend to be, applied economists (for example, central bankers and ministers of finance) must know, in order to forecast and plan, which are the effects of printing more money in a given situation, which those of raising or lowering customs barriers, and so on.

The econometrician Wold (1969) emphasizes the importance of the search for causal links, and discusses some of the paradoxes that can result from a neglect of identifying correctly causes and effects. For example, it is possible to have, for a given pair of variables x and y (e.g., demand and price), two linear regression equations such as $y = ax + b$ and $x = cy + d$, such that the system they constitute is not compatible. But these are really different hypotheses, so they should not be treated jointly. In the first, y is causally influenced by x, whereas in the second the converse causal relation is assumed. The problem, then, cannot be solved unless one commits oneself to either hypothesis.

Finally, sociologists have continued to work on the methodological problem of inferring or, rather, conjecturing causal networks from statistical correlations. Two particularly distinguished workers in this area are Blalock and Blalock (1968) and Boudon (1967). The philosopher will be surprised to see that these and other investigators take properties ("variables"), such as age and sex, to be possible causes or effects. He may wish to interpret only (relative) differences in the values of such "variables" as causes or effects, because the latter are always changes in properties (hence states).

So much for a review of causal thinking in the sciences. Let us now return to the explicit philosophy of causation found in the recent literature.

7. ORDINARY LANGUAGE EXAMINATIONS OF CAUSATION

Most philosophers either do not analyze the notion of causal relation, or examine it within the context of common sense (or ordinary language), adding occasionally a pinch of logic. The latter serves often as a carpet under which the conceptual muddles of common sense are swept. An example of this kind of work is the much-quoted paper by Davidson (1967). This article purports to investigate the logical form of singular causal statements such as "The flood caused the famine", as well as of causal laws—no example of which is given. However, the very heart of the matter, namely the expression 'c causes e', is not analyzed other than linguistically. Nor is any attempt made to check whether there are any genuine law statements that fit the analysis that is being offered. Which illustrates the dogmatism to which the cult of common sense can lead.

In another popular paper, Scriven (1971) asserts that "A cause or an effect may be (at least) a state, an event, a relation, a configuration, a process, a thing, a possibility, a thought, or the absence of any of these" (p. 50). The reason for such permissiveness is that Scriven, like Davidson, Vendler (1967), and others, is interested mainly in ordinary language expressions, among which we find the likes of 'Susan was taken sick because she did not eat'—suggesting that a nonevent had causal efficacy.

But of course, expressions like this one are just handy abbreviations of long conjunctions of propositions concerning human metabolism under a variety of food inputs. It is these propositions, not their ordinary language summary, that should be investigated if one wishes to dig up causal connections, and thus find out what kind of objects can be related by such connections. Ordinary language is something to be analyzed, refined, or even discarded rather than being kept unkempt. In any case, it won't help us perform a causal analysis of scientific theories; a fortiori, it is incapable of identifying causal law statements, let alone their causal range.

Mackie (1975) points out that there are a number of causal concepts, and pleads for tolerance with regard to the invention of new ones. We can speak, if we like–he asserts—not only of necessary and sufficient causes, but also of causes that are only necessary, or only sufficient. And he claims that, in all cases of causal statements, we presuppose contrary to fact (counterfactual) conditionals. One might rejoin that the latter may be useful heuristic clues for inferring or suspecting causal links, as in "Surely, if *A* did not act, then *B* would have happened"; but they do not help formulate causal hypotheses. (Besides, we know what it means for one proposition to presuppose another, but not what it is for one proposition to presuppose a counterfactual sentence, which is not a proposition.)

Finally, some philosophers believe that the notion of causal relation can be formalized within the propositional calculus, namely thus: $C \Rightarrow E$ (sufficiency), $E \Rightarrow C$ (necessity), or $C \Leftrightarrow E$ (necessity and sufficiency), where C and E may in turn be analyzed into conjunctions or disjunctions. But this is plainly insufficient, since the causal connection is a relation, and so it calls for at least the predicate calculus. Thus, the common maxim "Every event has a cause" can be formalized as: "$(\forall x)(Ex \Rightarrow (\exists y)(Cyx))$". However, even this finer analysis is insufficient because it does not analyze the predicates C and E: it does not even tell us what sorts of objects they can apply to. So, the elementary analyses of the causal relation—which is the most we can get in ordinary language philosophy—are insufficient. Deeper approaches must be tried. Let us see whether probability can help.

8. THE PROBABILISTIC APPROACH

It has occurred to a number of thinkers that causation is just a particular case of propensity. Thus, consider a thing *b* at a place *p*, with two possibilities of motion or some other change: to the left or to the right. Moreover, suppose that *b*, when at *p*, can move to the left with probability *P*, or to the right with probability $1 - P$. If *P* is neither zero nor unity, either of the two changes may occur. But if $P = 0$, then *b* will move to the right, whereas if $P = 1$, *b* will move to the left. The generalization seems straightforward: determinacy is a limiting case of probability.

The above hint has been exploited systematically by Suppes (1970, 1974). A cause, in his view, is anything that probabilifies. More precisely, if A_t is an event at time *t* and $B_{t'}$ another event at a later time $t' > t$, then A_t is said to be a *prima facie cause* of $B_{t'}$, just in case the conditional probability of $B_{t'}$ given A_t is greater than the absolute probability of $B_{t'}$, that is, $P(B_{t'} \mid A_t) > P(B_{t'})$. A *sufficient* or *determinate cause* is defined as a cause that produces its effect with probability 1. And a spurious cause is a prima facie cause that has no influence on the outcome.

The whole theory consists in a careful exploitation of these definitions with the help of the probability calculus. The only (tacit) hypothesis is what physicists misleadingly call the 'causality condition', i.e., the assumption that influences precede their effects. There is no other assumption— not even, as Suppes himself emphasizes, the hypothesis that every event has some cause. The book expounds then a noncausal theory of causation: more precisely, it is an exact modern version of Hume's obsolete view of causation as constant conjunction or succession.

This analysis of causation is thoroughly inadequate for the following reasons. (For further criticisms see Bunge 1973d.) Firstly, it consecrates the *post hoc-ergo propter hoc* fallacy. Indeed, consider the events A_t = The barometer is falling (here) at time t, $B_{t'}$ = It rains (here) at time $t' > t$. Since the probability of rainfall at a certain place, given that the barometer has been falling, is greater than the absolute probability of rainfall, we are asked to regard the falling of the barometer as a cause of rain. Second, by the same token, the Hume-Suppes analysis fails to distinguish between sustained (but perhaps accidental) positive correlation, and causation. Third, Suppes adopts uncritically the vague notion of an event used in probability theory, where an event is just an element of the field of sets constituting the domain of definition of the probability measure. Consequently, he cannot distinguish events proper, i.e., changes in the states of things, from properties and states, or even from purely conceptual objects. Thus, he is led to speaking of "negative events", such as not catching a cold, and therefore also of "logical events", such as catching a cold or not catching a cold. This is all right as long as the mathematical construal of "event" is kept, but is of no help to the scientist and the philosopher, to whom there are only "positive" events and (affirmative or negative) propositions about events.

In conclusion, the probabilistic approach to the causal problem fails to come to grips with the very concept of causation as a mode of event generation. An alternative approach, equally exact but more profound, must therefore be tried. The one to be discussed presently seems to be the most promising.

9. THE STATE SPACE APPROACH

In the sciences both causes and effects are events, i.e., changes of state of concrete things, be they fields, organisms, or communities. Things, their properties and states do not qualify as either causes or effects, but rather as conditions, outcomes, or what have you. Only changes can cause further changes. But to say that the causal relation is a relation among events is insufficient. Nor does it suffice to add that the causal relation is irreflexive, asymmetric, and transitive, if only because there are infinitely many relations with these formal properties. We must specify, though not so much that the desired generality gets lost.

The way to deal with concrete (material) things and their changes in general, without presupposing any particular laws, is to adopt the state space approach. Consider any concrete thing—whether particle or field, atom or chemical system, cell or organism, ecosystem or society. We may assume that every thing is, at each instant, in some state or other relative to a given reference frame; and further, that no thing remains forever in the same state (Bunge 1977a).

One way of describing states and changes of state is as follows. (See also essay 5.) Start by drawing the list of all (known) properties of things of the kind concerned, and represent each property by some (mathematical) function. Such a list (or n-tuple) is the state function of the thing: call it F. The changeability of the thing is reflected in the time-dependence of F. As time goes by, the value of F at the instant t, i.e., $F(t)$, moves in an n-dimensional abstract space determined by the n properties that are being considered. This space is called the *state space* of the things of kind K, or S_K for short. Any event or change occurring in the thing can be represented by an ordered pair of points in S_K and visualized as an arrow joining those points. The collection of all such pairs of states is the *event space* of the things of kind K, or E_K. The set of all really possible (i.e., lawful not just conceivable) events in (or changes of) the things in question, i.e., E_K, is a subset of the Cartesian product of S_K by itself. The statement that event e happens in or to a thing of kind K is abbreviated: $e \in E_K$. A process in a thing of the given kind is representable as a sequence (or list) of states of the thing, or else as a list of events in it. A convenient representation of the set of all changes occurring in a thing over a given period of time is obtained by forming all the ordered pairs $\langle t, F(t) \rangle$ of instants of time and the corresponding states of the thing concerned. Such a set, or $h(x) = \{\langle t, F(t) \rangle \mid t \in T\}$, can be called the *history* of the thing x during the period T.

Consider now two different things or parts of a thing. Call them x and y, and call $h(x)$ and $h(y)$ their respective histories over a certain time interval. Moreover, call $h(y \mid x)$ the history of y when x acts on y. Then we can say that x acts on y if and only if $h(y) \neq h(y \mid x)$, i.e., if x induces changes in the states of y. The total action (or effect) of x on y is defined as the difference between the forced trajectory of y, i.e., $h(y \mid x)$, and its free trajectory $h(y)$ in the total state space. That is, $A(x, y) = h(y \mid x) \backslash h(y)$, or the intersection of the forced trajectory with the complement of the free trajectory in the given state space. Likewise for the reaction $A(y, x)$ of y upon x. And the interaction between things x and y is of course the set-theoretic union of $A(x, y)$ and $A(y, x)$.

Finally, consider an event e in a thing x at time t, and another event e' in a thing $y \neq x$ at time t'. (The events and the times are taken relative to one and the same reference frame.) Then we can say that e is a cause of e' just in case (a) t precedes t', i.e., $t < t'$, and (b) e' is in (belongs to) the total action $A(x, y)$ of x upon y. In this case e' is called an effect of e.

Having defined the notions of cause and effect, we may now state the strict *principle of causality*. A possible formulation of it is this: *Every event is caused by some other event.* More precisely: Let x represent a thing of kind K with event space E_K (relative to some reference frame). Then, for every $e \in E_K$, there is another thing $y \neq x$ of kind K' (equal to or different from K), with event space $E_{K'}$ relative to the same reference frame, such that $e' \in E_{K'}$ causes e'''. It is a thesis of Bunge (1959a) that the strict principle of causality is seldom, if ever, exactly true, because it neglects two ever-present features of becoming: spontaneity (or self-movement), as in inertia, and chance, as in particle scattering.

The matters of spacetime contiguity, continuity, etc., can be handled with the help of the above notions together with some theory of space and time (Bunge 1977a). It would be interesting to formulate all the conceivable causal doctrines within the state space framework. This would facilitate their mutual comparison as well as their confrontation with the most general principles of science. This is just one among the many open problems in the ontology of science.

10. CONCLUSION

Twenty years ago the causal problem was generally believed to have been buried once and for all. Now it is as lively as any philosophical problem can be. Philosophers have come back to it, some of them armed with new, more powerful analytic tools, and scientists speak unabashedly of causal factors, effects, and causal networks.

To be sure, few believe in the universal validity of the causal principle: we have curtailed it in view of the ubiquity of both chance and spontaneity (or self-determination). However, we continue to think of causes and of effects, as well as of both causal and noncausal relations among them. In particular, we can often compute or measure the probability that a given event will have a certain effect, or a certain cause— but we do not usually mistake probabilities for causal links.

We have abandoned radical causalism but, far from becoming indeterminists, we have enriched determinism with new, noncausal determination categories. And we are still busy characterizing our basic concepts and hypotheses concerning causation.

5

EMERGENCE
AND THE MIND

(1977)

■

1. THE PROBLEM AND ITS SETTING

This paper deals with the so-called mind-body problem. This is the set of questions about the nature of the mental and its relations to the bodily. For example, are mind and body two separate substances? If so, how are they held together in the living organism? These questions are rather difficult to answer. However, I submit that the difficulty is not wholly intrinsic, but has been compounded by hurdles such as the following. Firstly, several doctrines concerning the mind-body problem have some ideological bias or other—and ideologies are not particularly interested in fostering conceptual clarity and empirical investigation. Secondly, the very formulation of the mind-body problem employs certain concepts, such as those of substance, emergent property, state and event, which are far from clear. (These concepts occur in all sciences and are therefore elucidated by none: they belong in the branch of philosophy known as ontology or metaphysics.) In fact it is pointless to engage in an argument about whether or not there are mental states that are not brain states, or whether mental events have causal efficacy, unless one can make some sense of the very expressions 'mental state' and 'mental event', which in turn contain the philosophical concepts of state, event, and mind. Let us therefore start by trying to clarify these and a few other ontological concepts that occur in the discussions on the mind-body problem. (For a detailed mathematical treatment of these concepts see Bunge 1977a.)

A thing, or concrete object, may be characterized as whatever can join (or associate with) another thing to form a third thing. On the other hand, two concepts cannot always join to form a third concept, e.g., 'purple number' is not a concept although "purple" and "number" are. But things have of course many other properties in addition to that of joining to form other things. For instance, they can interact and get together, forming tightly knit complex things, i.e., systems; they can move about, change in kind, and so forth. We may then assume that every thing, no matter how simple it may look, has a large number n of properties. (We are here referring to general properties such as that of moving, not to particular properties such as that of moving from here to there with such and such an instantaneous velocity relative to a given frame.)

Now, every thing-property can be conceptualized as, or represented by, a function in the mathematical sense of the term and, in principle, by a real-valued function. And the n functions representing the properties of a concrete thing can be collected together into a single function in conformity with:

DEFINITION 1. Let each of the n properties of a concrete thing be represented by a real-valued function F_i of time, with $1 \leq i \leq n$. Then

(i) $\mathbb{F} = \langle F_1, F_2, ..., F_n \rangle: T \rightarrow \mathbb{R}^n$ is called the *state* function of the given thing;

(ii) the value $s = \mathbb{F}(t)$ of $\mathbb{F}$ at time t is called the *state* of the given thing at t;

(iii) the ordered pair $\langle s, s' \rangle$ of values of $\mathbb{F}$ at times t and t', respectively, is called an *event* occurring in the thing concerned between t and t';

(iv) the sequence of states joining two states s and s' of a given thing is called the *process* leading from s to s', or the history of the thing between t and t'.

Note that we have not been talking about properties, states, or changes in themselves: every property is a property of (possessed by) some thing or other; likewise, every state is a state of some thing, and every change of state is a change of or in some thing or other. Thus, physical states are states of physical things; chemical states, states of chemical systems; biological states, states of cells or organisms; social states, states of social systems; and so on. This manner of speaking, which is entrenched in modern science and which ignores the Platonic forms hovering above things, will prove of decisive importance in our discussion of the mind-body problem.

Now, the state function $\mathbb{F}$ describing the states and changes of state of a thing is not a priori: it is determined by the laws possessed by the thing. In other words, there are laws that restrict the possible forms of $\mathbb{F}$. These laws may take the form of mere restrictions on the range of $\mathbb{F}$, or of algebraic relations among the components of $\mathbb{F}$, or of differential equations satisfied by them, or what have you. By virtue of such restrictions, the tip of $\mathbb{F}$ spans not the totality of its codomain $\mathbb{R}^n$ but only a subset of it. This subset of the set of all logically possible states of the thing will be called the *state space* of the latter or, more precisely, its *lawful state space*. We designate it $S_L(x)$, where $\mathbb{L}$ is the set of laws possessed ("obeyed") by thing x (see fig. 1).

So much for the concepts of thing, property, state, and process. Let us now take a closer look at properties of a particular kind, namely emergent properties. They are of special interest to the neuroscientist who, while acknowledging that feeling, recalling, imagining, and reasoning are emergent properties of the brain, would like to explain them in terms of events occurring in certain subsystems of it.

2. RESULTANTS, EMERGENTS, AND LEVELS

Temperature and entropy are properties of an atomic aggregate, not possessed by any of its atomic components. These are examples of emergent properties, or properties characterizing a system as a whole and which the system components do not have. Emergence is conspicuous at all levels and, a fortiori, between levels. This much seems clear.

What are not clear at all are the various notions of emergence and level. There are several reasons for this. One is that most rationalist philosophers are radical reductionists and so have claimed that emergence is a myth. Another is that most emergentist philosophers are irrationalists and so have held that there is nothing to be explained: that emergence is as mysterious as it is real. A third reason for the obscurity of the notion is that scientists forever try to explain emergence and, when they succeed, give the impression that they have explained it away. But of course, things and their properties, even if radically new, do not go away just because scientists succeed in understanding them or philosophers pretend that they do not exist to begin with. We had therefore better face the task of elucidating first the elusive notion of emergence.

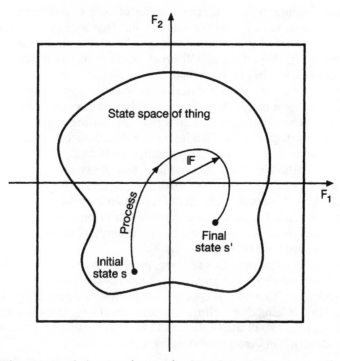

Fig. 1. The states and changes of state of a thing (e.g., neuron, neuronal circuit, subsystem of the CNS, or entire CNS) are representable in the state space of the latter, which is the space spanned by its state function, the ordered n-tuple of functions representing its various properties. (By contrast, according to eliminative and reductive materialism of the radical sort, i.e., mechanism, the states of any thing should be representable as points in spacetime.) In the diagram only two properties, represented by functions F_1 and F_2 (or rather their ranges), are shown. Actually, any realistic model of a complex thing will involve a state function with many more components. So, try to imagine the state of a thing as the tip of a vector in an n-dimensional state space.

We shall be concerned with complex things, in particular with systems, i.e., things the components of which are linked or coupled to one another. The properties of a complex thing are called *bulk* or *global* properties because they are possessed by the thing as a whole. Now, bulk properties are of two kinds: resultant and emergent. Energy is a resultant property, for it is possessed by every part of a thing. On the other hand, having a certain structure, being stable, being alive, and thinking are emergent or nonhereditary properties, for they are possessed by no component of the whole concerned. More precisely, we make:

DEFINITION 2. Let P be a property of a complex thing x other than the composition of x. Then

(i) P is *resultant* or *hereditary* if, and only if, P is a property of some components of x;

(ii) otherwise, i.e., if and only if no component of x possesses P, P is *emergent*, *collective*, *systemic*, or *gestalt*.

What holds for properties holds also, of course, for their carriers. Thus, a resultant thing (or just *resultant*) is one the properties of which are possessed also by some of its components. And an emergent thing (or just *emergent*) is one possessing properties that none of its components possesses. Note that emergence is relative. Thus, the ability to think is an emergent property of the primate brain relative to its component neurons, but it is a resultant property of the primate because it is possessed by one of the latter's components, namely its brain.

Radical monism, in particular mechanism, assumes all properties to be resultant or hereditary, hence explainable by straight reduction, as happens to be the case with the total energy and the total electric charge of a body. Radical pluralism, on the other hand, holds that there are emergent properties (an ontological hypothesis) and, moreover, that none of these is explainable in terms of the components and their links (an epistemological hypothesis). We take neither of these stands.

We recognize the fact of emergence but assume that every emergent can be accounted for in terms of a system's components and the couplings among them. For example, refraction is not a bulk property of transparent bodies: it is an emergent property relative to the atomic (or molecular) components of such bodies, for none of those components possesses the property of refrangibility. Yet this emergent property of the whole is explained by electrodynamics in terms of the electrical properties of atoms (or molecules) and light. However, this explanation is not reductive in a simple sense, as it does not consist in attributing refrangibility to individual atoms. It is reductive in consisting in the deduction of the formula for refractive power from premises concerning the interaction between electromagnetic waves and atomic lattices.

What holds for physical systems holds a fortiori for chemical, biochemical, biological, and social systems. For example, enzymatic catalysis is an emergent property of biochemical systems, sexuality an emergent property of some biosystems, and social cohesion an emergent property of sociosystems. However, these are not unintelligible properties: they can be and are being explained. (That no scientific explanation is likely to be definitive is beside the point.)

The foregoing assumptions can be compressed into two postulates, one ontological or concerning reality, the other epistemological or concerning our knowledge of reality. Here is the emergence postulate:

POSTULATE 1. Some of the properties of every system are emergent.

And here is the rationality postulate:

POSTULATE 2. Every emergent property of a system can be explained in terms of properties of its components and of the couplings amongst these.

These two postulates constitute the kernel of what may be called *rational emergentism*, a doctrine differing from both the irrationalist emergentism of the holists and the rationalist flattening (or leveling) by the mechanists, energetists, and idealists.

The last ontological concept we must handle before turning to the mind-body problem is that of level, particularly in view of the popular assumption that the mental constitutes a higher level than the biological one. First the intuitive idea.

Most biologists seem to agree that things, and in particular things of concern to biology, are found not pell-mell but rather in levels, and that these in turn constitute a sort of pyramid. Thus, one speaks of the atomic level and the molecular one, of the cellular level and the organ level, etc. And one assumes that the systems at any given higher level are composed of things belonging to some preceding level (see fig. 2). This suggests making:

DEFINITION 3. Let L be a family of nonempty sets L_i of things, with $1 \leq i \leq n$. Then, if L_i and L_j are members of L, L_i precedes L_j if and only if each member of L_j is composed exclusively of things in L_i.

Note two points. The first is that a level is not a thing but a set and therefore a concept, though certainly not an idle one. (Hence levels cannot act upon one another. In particular the higher levels cannot command or even obey the lower ones.) Second, the relation between levels is neither the part-whole relation nor the set inclusion relation, but a *sui generis* relation definable in terms of the composition function, which in turn is definable in terms of the part-whole relation.

The concept of a level occurs in the *levels hypothesis*, or

POSTULATE 3. Every thing belongs to some level or other.

Now, many components of a system are also its precursors in an evolutionary process. Thus, amino acids are at the same time the components and the precursors of proteins, and cells both compose a multicellular organism and give rise to it. (By contrast, according to holism, the whole precedes its parts and controls them.) One may wish to generalize, stating

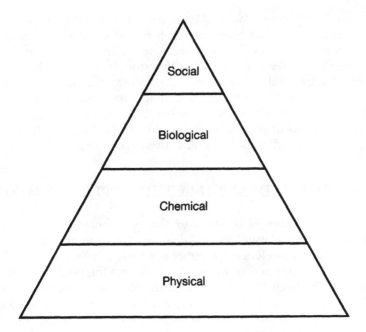

Fig. 2. The pyramid of levels of organization of the world. Each higher level consists of systems built with components belonging to some prior level. And each level splits into sublevels. For example, the biological level can be subdivided into the cell, organ, organism, population, and ecosystem sublevels. The pyramid suggests not subordination or excellence, but only that the higher a level the more dependent and the less populated it is.

POSTULATE 4. Every complex thing belonging to a given level has self-assembled or self-organized from things at some preceding level.

To put it metaphorically: higher levels emerge out of lower ones in a natural process of self-assembly or self-organization. Consequently, radical novelties emerge out of previously existing things. Therefore, emergence and levels, far from forming a static order, are features of an evolutionary process. And, according to our assumption, this process is natural or spontaneous: notice the term 'self' in Postulate 4.

Postulates 1, 2, and 3 form the nucleus of rationalist pluralism. Postulate 4 renders this ontology dynamicist and naturalist (instead of supernaturalist). In fact, the whole thing is a sort of generalization of the theory of evolution.

What has all this to do with the mind-body problem? Much, for the mental may be conceived of as an emergent relative to the physiological.

But the mind may be so conceived in either of two different ways: as an emergent *entity* or as an emergent *property* of entities of a certain kind—say vertebrates. In the first case, one might wish to claim that *minds* constitute a level of their own: this would be the thesis of psychoneural dualism embedded in an overall pluralistic ontology. In the second case, one would certainly hold that *organisms* endowed with mental abilities form a new level relative to mindless organisms: this would be the thesis of psychoneural unity embedded in the pluralistic ontology sketched earlier. Let us look at this problem more closely.

3. MONISM AND DUALISM IN THE MATTER OF MIND

There are two classes of solution to the mind-body problem: psychoneural monism and psychoneural dualism. And each of these classes contains at least five different doctrines: see table 1, where φ stands for *body* (or *the organic*) and ψ for *mind* (or *the mental*). (Cf. Armstrong 1968; Borst 1970; Feigl 1967; Feyerabend and Maxwell 1966; Hebb 1949; Hook 1960; Koestler and Smythies 1969; O'Connor 1969; Place 1956; Rosenblueth 1970; Smart 1959.) Let us examine them briefly, starting with the main varieties of psychoneural dualism.

We need not consider the independence thesis D1, as both introspection and neuroscience tell us that the bodily and the mental—whatever the latter may be—are interdependent. As for the parallelism or synchronization thesis D2, upheld by the Gestalt school, it begs the question instead of answering it, for what we want to know is precisely the mechanism responsible for the "parallel sequences" of mental and physiological states. To say that mental events have neural "correlates" is fine but not very informative, unless one states what a mental event is and the nature of its "correlation" with its neural "correlate". For these reasons, D2 is vague to the point of being confirmable by all possible empirical data. Hence, D2 is not a scientific hypothesis.

On the dualist side we are then left with either of the thesis acknowledging one substance's acting upon the other. However, in this case too only the physical is supposed to be knowable, whereas the mental is left in the dark or, at best, in the care of philosophy or even theology. We do indeed understand what it is for a given neuron, or neuron assembly, to be in such and such a state: a state of a thing is always an ordered n-tuple of the n properties we care to assign to it. And we understand what is a neural event or process, namely a change in the state of a neural unit (i.e., neuron or neuron assembly). Consequently, we know what it is for one neural unit to act upon another: A acts on B if and only if the states of B when it is connected with A are not the same as those of B when it

Table 1. Ten Views on the Mind-Body Problem

Psychoneural Monism		Psychoneural Dualism	
M1	Everything is ψ (*phenomenalism idealism*).	D1	φ and ψ are independent.
M2	φ and ψ are and so many aspects or manifestations of a single entity (*neutral monism*).	D2	φ and ψ are parallel (*psychophysical parallelism*).
		D3	φ affects (or even secretes) ψ (*epiphenomenalism*).
M3	Nothing is ψ (*eliminative materialism*).	D4	ψ affects (e.g., controls) φ (*animism*).
M4	The mind is corporeal (*reductive materialism*).	D5	φ and ψ interact (*interactionism*).
M5	ψ is a collection of emergent functions of φ (*emergentist materialism*).		

is not so connected. In short, we have some idea of neural functions (states, events, processes). Recall fig. 1.

But these ideas—common to all sciences—are not transferable to the mental "substance". If they are, nobody has shown how. In particular, attention, memory, and ideation have not been shown to be properties or changes of properties of a mental substance (mind, soul, or spirit). In sum, the concepts of mental state, event, and process do not fit within the general framework of contemporary science unless they are construed in neural terms, i.e., as, respectively, a state of the brain or an event or process in a brain. This is one of the reasons for dualism's inability to go beyond the stage of verbal and metaphorical formulations. This is why there is not a single dualistic model, in particular a mathematical model, in physiological psychology.

In short, interactionism is just as imprecise as parallelism—which is to be expected of a popular, i.e., nonscientific, view. And, not being a precise hypothesis, it can hardly be put to empirical tests. Moreover, even if parallelism and interactionism were to be formulated in a precise manner, it might not be possible to decide between them on the strength of empirical data. Indeed, it would seem that every psychological experience and every psychophysiological experiment could be interpreted (or misinterpreted) either in parallelist or in interactionist terms, since neural events are simultaneous with their mental "correlates".

We are led to the conclusion that the two main variants of psychoneural dualism, namely parallelism and interactionism, though conceptually different, are equally fuzzy and are empirically equivalent insofar as they accord (much too easily!) with the same empirical data. For these reasons, dualism is not scientifically viable. It is barren double-talk and, as Spinoza characterized it, a disguise for our ignorance. We are then left with psychoneural monism as the only scientifically and philosophically viable alternative.

But, as shown in table 1, psychoneural monism is a whole class of doctrines. Let us start with M1 or subjectivism. We can write it off without further ado because it is incompatible with physics, chemistry, molecular biology, and social science, all of which are busy hypothesizing and manipulating unobservables such as atoms, ecosystems, and societies. Moreover, all of these disciplines are supposed to abide by the scientific approach, which includes objectivity. As for neutral monism, it has yet to be formulated clearly and in agreement with the natural sciences. Even the least obscure and mystical of its versions, namely Ostwald's energetism, is vague. (Moreover, it rests on the mistaken reification of energy, which is actually a property of physical objects not a thing. Recall essay 3.) We may therefore dismiss M1 and M2, and examine materialism.

We distinguish three kinds of materialism, to wit, eliminative, reductive, and emergentist. Eliminative materialism holds that there is no such thing as the mental: that everything is material. There are two different versions of this thesis: the ancient thesis that all subjective phenomena are composed of particles, and the modern thesis that there are only neural facts (states, events, processes). Neither of these theses is capable of distinguishing between appearance and reality, i.e., between facts as perceived by a sentient being (i.e., phenomena) and facts as they are independently of the organism. Nor does eliminative materialism distinguish between *Homo sapiens* and its nearest cousin, the amazing chimpanzee, so similar at the cellular level and yet so different at the organismic level. In short, eliminative materialism can be eliminated.

The thesis of reductive materialism may be formulated thus: "Every mental state (or event or process) is a state (or event or process) of the central nervous system. Therefore, the mental is no different from the physical". While I have no quarrel with the premise, I submit that the conclusion is a non sequitur. But before arguing for emergence, let us examine this argument of the reductive materialists. It is an argument from analogy with other macrofacts rather than an independent examination of the models and the empirical evidence in physiological psychology. Let us dwell on it for a while.

Reductive materialists claim that the body-mind relation is just a particular case of the macroscopic-atomic relation, and that in both cases it is one of epistemological and ontological reduction. While I agree with the first contention, I disagree with half of the second, namely concerning the ontological reduction of the mental to the neural. The first thesis seems rather plausible: however localized some or even all mental facts may be, they always involve a large number of neurons, not to speak of blood cells and other nonneuronal components of the nervous system. As for the reducibility thesis, let us discuss it in the light of the stock-in-trade example of the reductionists, namely the alleged reduction of water to water molecules.

The microreductionists claim that water is just H_2O, which in turn they claim to be nothing but an aggregate of two hydrogen atoms and one oxygen atom. This they take to be a paradigm of microreduction. Epistemologically maybe so. (And I hope so, although as a matter of fact there exists no adequate theory of liquids, and a fortiori none has been deduced from quantum mechanics.) But the thesis of ontological reduction is obviously false. Indeed, to state that the *composition* of a body of water is a set of H_2O molecules is not to state that the former is nothing but the latter, any more than to say that the composition of a human society is a bunch of persons is to say that a society is nothing more than the set of its members. And this for the following reasons. First, a thing is not a set (which the composition of a thing is). Second, a body of water is a system, hence something with a structure, not only a composition. And that structure includes the hydrogen bonds among H_2O molecules. The result is a system with emergent properties such as fluidity, viscosity, transparency, and others, which its molecular components lack. Surely, one can (hope to) understand all of these emergent properties in terms of those of the water molecules and their interactions. That is, one can (hope to) "reduce" the macroscopic properties of water to the properties of its microcomponents. But such an explanation—which has yet to be provided—does not accompany ontological reduction: explained fluidity is still fluidity. Likewise, explained vision is still vision, explained imagination is still imagination, and explained consciousness is still consciousness. Therefore, ontological reductionism is just as untenable in the matter of mind as it was found to be in the matter of matter. This leaves us with psychoneural monism of the emergentist kind. Let us take a closer look at it.

4. EMERGENTIST PSYCHONEURAL MONISM

In this section we shall examine the strengths and the weaknesses of emergentist psychoneural monism, or M5 in table 1. This view boils down to

POSTULATE 5.

(i) All mental states, events, and processes are states of, or events and processes in, the central nervous systems of some "higher" vertebrates;

(ii) these states, events, and processes are emergent relative to those of the cellular components of the CNS;

(iii) the so-called psychophysical relations are interactions between different subsystems of the CNS, or between them and other components of the organism.

The first clause is the thesis of psychoneural monism of the materialist kind. The second clause is the emergence thesis. It states that mental facts are both organismic or biological, i.e., involve entire assemblies of interconnected cells. The third clause is a monistic version of the parallelist and interactionist myths.

If one accepts the above postulate, then one can talk about *mental phenomena* without jumping out of the biological level: the mentalistic vocabulary originally coined by religion and dualistic philosophies begins to make, or is hoped to make, neurological sense. (Equivalently: psychology becomes a neuroscience.) In particular, it now becomes possible to speak of *parallel sequences* of events, e.g., of processes in the visual system and in the motor system, or in the language system and in the cardiovascular system.

It also makes good scientific sense to speak of *psychosomatic interactions*, because these are now construed as reciprocal actions between different subsystems of one and the same organism, such as the neocortex and the sympathetic nervous system. For example, rather than say that love can color our reasonings, we may say that the right brain hemisphere affects the left one, and that sex hormones can act upon the cell assemblies that do the thinking. In short, ironic as it may sound, the dualistic modes of speech, which encapsulate our undigested introspective experience and which are but metaphorical and vague in the context of psychoneural dualism, become literal and precise in the context of emergentist materialism. The latter salvages whatever can be salvaged from the dualist myth.

Emergentist monism has many attractive features, the most important of which are that (1) it squares with the natural sciences by postulating that mental facts, far from being affections of an immaterial substance, are states of, or events and processes in, concrete organisms, whence (2) they can be investigated through the normal procedures of science—a feature which turns psychology into a natural science instead of a supernatural one.

Emergentist materialism holds then splendid promise and, moreover, has already rendered distinguished service by being the driving force behind physiological psychology. However, it has one important shortcoming, namely that it is still immature. In fact, emergentist materialism is not a *theory* proper, i.e., a hypothetical-deductive system containing precisely formulated and detailed hypotheses accounting for a wide range of psychoneural facts. It is instead a *programmatic hypothesis*—one both scientific and philosophical—in search of scientific theories embodying it. So much so that emergentist materialism can be summed up in a single sentence, to wit: *Mental states form a subset* (albeit a very distinguished one) *of brain states* (which in turn are a subset of the state space of the whole organism). This, however suggestive, is so little as to be representable in simple diagram: see fig. 3. (Dualism, on the other hand, cannot be diagrammed at all, except metaphorically, so it is even poorer.)

What is needed for implementing the program of emergentist materialism, i.e., for developing it into a mature scientific enterprise? Obviously, not more undigested data. What we do need are two different though complementary batches of theories: (1) *extremely general theories* (not just stray hypotheses or programs) of the mental conceived of as a collection of functions of the CNS; and (2) *specific* theories accounting for the functioning of the various subsystems of the CNS.

The general theories of psychoneural activity would belong to the intersection of ontology and psychology, while the specific theories of the psychoneural would be the exclusive property of physiological psychology. And all of them should be stated in precise terms, i.e., should be mathematical in form.

It may be argued that the preceding plea for intensifying theoretical work in the fields of psychophilosophy and physiopsychology are impertinent because there is no dearth of theories in both fields. Let us see about that.

Certainly, much has been written about the so-called identity theory over the past two and a half millennia. But none of the "theories" of the psychoneural that agree with the materialist hypothesis are *theories* proper, i.e., hypothetico-deductive systems, let alone mathematical ones. They are instead single and stray hypotheses. And they are verbal and often verbose. (This may be one of the reasons that most mathematical psychologists have not been attracted to materialism.) In other words, we still do not have a general materialist theory of the mind.

As for specific theories in physiological psychology, there is no doubt that many have been proposed, particularly over the past quarter of a century and largely thanks to Hebb's influence (Hebb 1949; Milner 1970; Bindra 1976). However, (a) there are too few of them, (b) those which are close to experiment are for the most part verbal, and (c) those which are mathematical are usually far removed from experiment. (Moreover, most theories in mathematical psychology are either (a) neobehaviorist learning theories disregarding the CNS, or (b) information-theoretic theories regarding the CNS as a computer rather than a biosystem. Both skip chemistry and biology.)

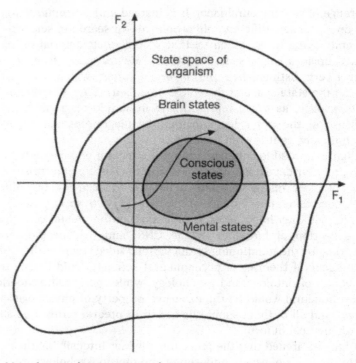

Fig. 3. Mental states form a subset of the collection of all brain states, which in turn are included in the set of possible states of the whole organism. Awareness (or the self) is conjectured to be a distinguished subset of the mental states and, therefore, a subset of the totality of organic states. The arc of curve represents a mental process, such as recollecting an experience, which is partly conscious. The diagram is programmatic: we still have to identify the properties represented by the blanks F_1 and F_2 in the state function of the CNS.

So much for the shortcomings of emergentist materialism in its infancy. However many and grave these may be, the emergentist materialist philosophy of mind seems to be the best we have, and this for the following reasons:

1. Because it eschews the mysterious mental substance without thereby denying the mental, emergentist materialism is *compatible with the scientific approach* far more than either dualism or eliminative and reductive materialism.

2. Emergentist materialism is *free from the fuzziness* that characterizes dualism with its talk of "correlations" between the mental and the physical relations that dualists do not care to clarify, perhaps because they cannot.

3. Unlike dualism, emergentist materialism is *consistent with the general concepts of state and event* that can be gleaned from all the sciences. (By contrast, according to dualism, mental states would be the only states that fail to be states of some thing, and mental events would be the *only* events that fail to be changes of state of some thing—this being why dualism agrees more closely with theology than with science.)

4. Unlike dualism, emergentist materialism *fosters interaction between psychology and the other sciences*, in particular neuroscience, and this precisely because it regards mental events as special biological events.

5. Unlike dualism, which digs an unbridgeable chasm between humans and (nonhuman) animals, emergentist materialism *jibes with evolutionary biology*, which, by exhibiting the gradual development of the mental faculties along certain lineages, refutes the superstition that only humans have been endowed with a mind.

6. Unlike dualism, which postulates an unchanging mind, emergentist materialism *accords with developmental psychology and neurophysiology*, which exhibit the gradual maturation of the brain.

None of the rivals of emergentist materialism can boast of so many and important supports, direct and indirect, scientific and philosophical. Therefore, it is worthwhile to try and implement the program of emergentist materialism, i.e., to attempt to build theories of various degrees of generality, mathematical in form and agreeing with the known facts, that construe the mind as a distinguished subset of the set of neural states and events.

5. CONCLUSION: PSYCHONEURAL MONISM *CUM* OVERALL PLURALISM

The first half of this paper argues for the reality of emergence and even for the plurality of levels: in fact it sketches a pluralist ontology. The second half defends a version of psychoneural monism. Contradiction? Not at all, for we take the mental to occur only at the organismic level: we assume that neurons are mindless as are populations, in particular societies. So we assert that organisms endowed with mental abilities constitute a level of their own, which can be called that of psychosystems. But we do not affirm that minds constitute a level of their own— and this simply because there are no disembodied minds. In short, in our ontology minds do not constitute a supra-organic level because they form no level at all. To repeat the same idea in different words: One can hold that the mind is emergent relative to the physical without reifying the former. That is, one can hold that the mind is not a thing composed of lower-level things—let alone a thing composed of no things what-ever—but a collection of functions of neuron assemblies, which individual neurons do not possess. (The brain and some of its subsystems can mind, i.e., be in mental states—but the mind cannot mind even its own business because it has no more an independent existence than does mass alongside bodies or history separately from people. Only the functioning—minding—brain can mind its business.) And so emergentist materialism is seen to be compatible with overall pluralism.

Our espousing emergentist materialism does not require affirming that it has in fact solved the mind-body problem. It has not. But it is working on it: witness the progress of physiological psychology. Moreover, we submit that emergentist materialism is the only philosophy of mind that enables a breakthrough in the scientific investigation of the mind-body (or rather brain-rest-of-the-body) problem. In fact, it is the only one that enjoys the support of all the life sciences, that does not promote a quixotic reductionism, and that defends neuroscience against obstruction by obsolete philosophies and ideologies.

Finally, three caveats are in order. The first is that to espouse emergentist materialism is not to deny subjective experience or even to disallow employing introspection as a tool in the scientific investigation of the mental. To espouse emergentist materialism is to favor the understanding of subjectivity in neural terms, and to encourage the control of subjectivity instead of allowing the latter to control the course of research. (In particular, hunches got by introspection must be regarded not as self-evident but as hypotheses to be subjected to objective tests.)

Second caveat: Emergentist materialism does not require one to investigate the mental in exactly the same way as one would investigate earthquakes or infections. Indeed, the psychologist is the luckiest of scientists in that she can tap a number of sources: she can learn from neurophysiology as it deals with the levels of the neuronal circuit, the brain subsystem (e.g., the brain stem), and the entire CNS; she can learn from introspection and the study of behavior, from neurosurgery and psychiatry, from comparative zoology and the study of cultures. She can command, then, many sources of hypotheses and data and just as many ways of checking her hypotheses. In this regard, then, the study of the psyche is unique. (It is not unique in its being subject to the standard canons of scientific research.)

Third and last caveat: To explain the mental in terms of the neural is not to rule out that the mental is a set of emergent functions of the brain, any more than explaining the formation of a liquid vortex rules out that it possesses properties beyond the properties of the individual atoms that take part in it. In other words, the ideal of rationality is consistent with pluralism: to explain is not necessarily to explain away.

Besides, the explanation of emergence is anything but straightforward: it is a matter not of deducing consequences from a theory concerning some lower level, but of suitably enriching the latter with new assumptions and data. Thus, the theory of neurons does not entail the theory of neuronal circuits, nor does the latter entail the theory of the reticular formation, and so on. (For the logic of reduction see Bunge 1977b as well as essay 11.) A psychophysiological theory, though concerned with some of the physical or chemical processes in the CNS, deals not just with them, but also with a distinguished subset of biological processes going on in neural assemblies, namely those processes that are commonly called 'mental'. To explain the mind in depth is to know it, not to ignore it. And to know something is to have adequate theories about it.

6

THE STATUS OF CONCEPTS
(1981)

■

It is well known that ideal (or conceptual or abstract) objects have been the undoing of traditional empiricism as well as of vulgar materialism, for those objects are neither distillates of ordinary experience nor material objects or properties thereof. Surely, the empiricist may claim that there are no conceptual objects aside from mental events. But he cannot explain how different minds can grasp the same conceptual objects, and why psychology is incapable of accounting for the logical, mathematical, and semantical properties of constructs. And the vulgar materialist (nominalist) will likewise discard conceptual objects, and speak instead of linguistic objects, e.g., of terms instead of concepts, and of sentences instead of propositions. But he is unable to explain the conceptual invariants of linguistic transformations (e.g., translations), as well as the fact that linguistics presupposes logic and semantics rather than the other way round. Therefore, we can accept neither the empiricist nor the nominalist reduction (elimination) of conceptual objects, any more than we can admit the idealist claim that they are ideal beings with an autonomous existence. We must look for an alternative consistent with both ontological naturalism and semantic realism.

The goals of this paper are (a) to stress the difference between conceptual objects and concrete things; (b) to characterize conceptual objects; (c) to define an existence predicate independent of the existential quantifier; and (d) to sketch a philosophy of mathematics that unites materialism with conceptualism and fictionism. All this will be done schematically and with the help of modest formal tools.

1. MATERIAL AND CONCEPTUAL OBJECTS

We shall take it for granted that there are things (or material objects), and constructs (or conceptual objects). We shall also postulate that no thing is a construct, and no construct is a thing. In other words, we shall divide every nonempty set of objects into two disjoint subsets: a set of material objects and another of conceptual objects (either of which may be empty). Moreover, we shall not assign to conceptual objects the same kind of existence that material objects possess. In fact, we shall hold that, whereas the existence of material objects is independent of the cognitive subject, that of conceptual objects consists in the possibility of being thought up by some living rational being. But this will come later, after having clarified the difference between material and conceptual objects.

The thing/construct dichotomy implies that constructs have properties essentially different from those possessed by things. In other words, material objects, whether natural or artificial, living or inanimate, share certain properties that no construct possesses. Among them we recall those of being able to change, of possessing energy, of associating with others to form concrete systems possessing emergent properties, and of being localizable (though not necessarily at a point in space).

The sciences that study material objects (or things) are the factual sciences and ontology. These disciplines try to find the laws of such objects, in particular their laws of change, such as equations of motion, field equations, chemical reaction equations, and social mobility matrices. These laws are formulated as nomological statements. And the latter may be conceived of as restrictions upon state variables, or predicates representing (potential or actual) properties possessed by the things concerned— whether atoms or fields, cells or societies (Bunge 1977a).

In other words, the law statements that factual scientists seek to express tell us what the really possible states of things are, as well as what the really possible changes of state (i.e., events or processes) of concrete objects are. On the other hand, the factual sciences do not attempt to attribute logical, mathematical, or semantical properties or laws to concrete objects. Only concepts and propositions have a sense; only propositions and (interpreted) theories can be true or false to some extent; and only theories can be logically consistent. Strictly speaking, being is meaningless. It makes no literal sense to speak of 'the meaning of life' or 'the meaning of history', unless a nonsemantic acceptance of 'meaning' is involved in these expressions. (Thus, Max Weber and the other "interpretivist" students of society spoke of the 'meaning of social actions,' to suggest that they are purposive. But this usage of the word 'meaning' is dangerous, because it suggests the wild hermeneutic thesis that social facts are texts.)

The notions of state and change of state are central to science and ontology, but alien to formal science (logic, mathematics, and semantics). For example, it makes no sense to ask about the state (mechanical, chemical, physiological, economic, political, and so forth) of the number 5, and even less about its changes of state. Nor does it make any sense to speak of the equation of motion of a Boolean algebra, or of the transmutation schema of a topological space. Not that these conceptual objects are eternal objects: what is true is that the categories of change do not apply to them, hence neither do the categories of nonchange. (Likewise, it is not that culture has a zero temperature: it has no temperature at all.)

In other words, constructs are in no state whatever, and therefore cannot change their state, i.e., the state space of any construct is empty (Bunge 1981a). Therefore, the laws of conceptual objects do not involve the notions of state or of change of state. Sets are neither at rest nor in motion; mathematical functions are neither fertile nor sterile; algebraic structures are neither hungry nor sated; theories are neither exploiters nor exploited. In Spanish one could say that, whereas material objects *son* (exist materially) as well as *estan* (in some state), constructs just *son* (exist conceptually). But this cannot be said in English.

Constructs have peculiar properties that no material object has, namely predicates. (Material objects have or possess properties, and some properties can be variously represented by predicates. In other words, the concept of property-representation can be elucidated as a partial function from the set of thing-properties to the power set of the set of predicates.) Predicates and the propositions formed with them have meanings, but meaning has no (physical) being. Consequently, the laws satisfied by constructs, such as those of associativity and commutativity, are very different from the physical, chemical, biological, or social laws. For one thing, the laws of conceptual objects make no reference to material objects and they involve no state variables. Examples: "If p, then p or q"; "If set A is included in set B, then the intersection of A and B is not empty, provided B is not empty"; "If m and n are any real numbers, then $(m + n) \cdot (m - n) = m^2 - n^2$; "If a function $f: \mathbb{R} \to \mathbb{R}$ is such that $f(x) = x^2$ for any $x \in \mathbb{R}$, then $df/dx = 2x$". There is nothing material about these laws; none of the objects occurring in them can be said to be in any state, let alone to undergo a change of state.

A class of constructs of particular interest to the philosopher is that of existence postulates. In mathematics an existence postulate is not an empirically refutable conjecture such as the physical hypothesis of the existence of quarks, gluons, or elementary particles of some other type. In mathematics an existence postulate *stipulates* that, in the context of some theory, there are objects with such and such characteristics. For example, in plane Euclidean geometry one postulates that, through a point not on a straight line, there is exactly one parallel to the given line. This postulate creates by fiat the object or objects concerned, subject only to the condition that the principle of noncontradiction be respected. (Axiomatic definitions are therefore sometimes said to be *creative*. For example, the axiomatic definition of an abstract group creates an arbitrary set equipped with a binary operation and a unary operation, and including a unit element.) It is only once the (conceptual) existence of certain mathematical objects has been posited that one can proceed to postulate or prove further properties. For example, before attempting to solve a difficult differential equation it may be convenient to make sure that it has solutions, i.e., that the latter exist (even though we may not know them yet).

But what is meant by the expression 'Such and such object exists conceptually', or by the phrase '*There is* an object with such and such properties'? This is what we shall try to find out presently.

2. WHAT IS CONCEPTUAL EXISTENCE?

The traditional philosophies of logic and mathematics are Platonism (or objective idealism), nominalism (or vulgar materialism), and empiricism (or psychologism). According to the former, conceptual objects exist by themselves and may be embodied and thought of. According to nominalism, conceptual objects are nothing but signs or marks such as written symbols. And according to psychologism, conceptual objects are thoughts.

Each of these philosophies has both virtues and fatal flaws. Platonism allows freedom, but it populates the universe with ghosts, and is therefore inconsistent with a naturalistic ontology. Nominalism avoids idealism and emphasizes rightly the importance of language, but it is incapable of distinguishing the meaningful signs from the meaningless ones. And psychologism reminds us that constructs are not found ready-made in some quarry, but it does not allow us to speak of such objects as actual infinities, since nobody can effectively think of each and every member of an infinite set. Therefore, none of the traditional philosophies of mathematics is viable. We must therefore explore alternatives.

The alternative I propose to explore here may be called *conceptualist and fictionist materialism*. The distinctive theses of this new philosophy of the conceptual are as follows:

1. Conceptual objects are neither material nor ideal in the Platonic fashion; nor are they psychical (neurophysiological) events or processes. Constructs have properties of their own, such as logical and semantical properties, which are neither material nor mental. This is a first *conceptualist thesis*.

2. Conceptual objects exist in a peculiar manner, namely conceptually. More precisely, a conceptual object exists just in case it belongs to some context, e.g., a theory. Moreover, it exists only as such. (For example, integers exist in number theory, but not in the abstract theory of groups.) This is a *second conceptualist thesis*.

3. Conceptual existence, far from being ideal (Platonism), material (nominalism), or mental (psychologism), is *fictitious*. We pretend that there are sets, relations, functions, algebraic structures, spaces, and so forth. That is, when inventing (or learning or making use of) conceptual objects, we assign to them their peculiar mode of being: we demand, stipulate, feign that they exist. This is the *fictionist thesis*.

4. Conceiving of a conceptual object and assigning conceptual existence to it are two sides of a single process occurring in some brain. Conceptual objects are thinkable, and their ontological status is the same as that of mythical characters: They exist the same way Minerva, Quetzalcoatl, or Donald Duck exist. They will cease to exist the day we stop thinking about them or stop imagining that they are thinkable—just as the gods of lost religions have ceased to exist. This does not entail that conceptual objects, be they mathematical, mythological, or of some other type, must be actually thought up by somebody: to exist conceptually, it is necessary and sufficient to be thinkable. For example, there are (conceptually) infinitely many integers that will never be thought of, although every one of them is thinkable. It is unlikely that the number 7,753,912,650,836,471, 580,077,231,724,333,019,010,832 has been thought before, but it existed by virtue of being thinkable. The same holds for all other conceptual objects. In short, the conceptual existence we assign to logical, mathematical, mythical, and other such objects consists in the possibility of *being thought of* by living beings. This is the *materialist thesis*.

The preceding four theses constitute only a sketch of our philosophy of the conceptual. This sketch should be expanded into a full-fledged theory, in response to questions such as the following: What are the peculiarities of the languages in which theories of the conceptual are couched? What are the formal (mathematical) peculiarities of the basic types of constructs, such as predicates, propositions, and theories? What are the semantic peculiarities of pure conceptual objects in contrast with those employed in the factual sciences (e.g., what are formal meaning and formal truth)? What is the difference between mathematical and mythological objects? What are the relations between the mathematical concepts of a factual theory and the things it refers to?

Mathematical logic and the theory of models (or the semantics of mathematics) answer some of the preceding questions, though perhaps not fully. The others are still open problems. They will have to be solved, at least temporarily, by any respectable philosophy of mathematics.

3. CONSTRUCTS AS CLASSES OF THOUGHTS

If we want to do logic or mathematics, we must pretend that there are constructs such as predicates. There is no harm in pretending that they exist, provided we do not mistake conceptual existence for material or real existence: what is real is the brain process of thinking of constructs or of any other immaterial objects, such as properties (rather than of propertied things) and mythological characters. In particular, the number 3 does not exist (by itself), though thinking of 3 is of course a process in a real thing, namely a brain.

We can do better than to espouse mathematical fictionism (for which see essay 13), namely try and give a psychobiological account of constructs on the basis of our theory of mind (see essay 5 as well as Bunge 1980a). Consider two or more different thoughts of one and the same object, say different instances of thinking of the number 3, or of a sphere. We can assume that all such instances, whether in the same brain or in different brains, differ in some respects only, i.e., they are equivalent. If they were not equivalent, then they would not be instances of thinking the same construct. That is, we say that *every construct is an equivalence class of thoughts* (brains processes of certain types). (We can make this idea more precise with the help of the concept of state function: see Bunge 1981a as well as essay 5).

This construal of constructs is materialist because it is rooted in the notions of thoughts as brain processes, but it is not psychologistic because constructs are not equated with thoughts, but rather with classes of (possible) brain processes. This secures the autonomy of logic, semantics, and mathematics vis à vis psychology. Indeed, constructs are atemporal objects: every thought process advances in time, but in defining a construct we abstract from time as well as from physiological details. So we attain a position similar to Plato's theory of ideas—except that now there are no ideas without brains.

It might be objected, though, that constructs do change. Thus, the von Neumann construction of natural numbers differs from that of Peano, which is in turn different from that of Pythagoras. which in turn differs from the notions held by his predecessors. Granted. But the point is that, whereas material objects change *by themselves* even if we attempt to keep them unchanging, constructs do not: it is people who change, thinking now of a construct in one fashion, now in a different one. In other words, conceptual change is ultimately a change in someone's brain. If conceptual objects could change by themselves, we should be able to write down (and check) their equations of evolution, e.g., the equation of motion of an equation of motion. But it makes no sense to ask what the rate of change of a construct is, let alone what forces elicit such change.

4. QUANTIFICATION AND EXISTENCE

Let us now approach the technical problem of formalizing the concepts of conceptual existence and material existence distinguished in the previous sections.

The ordinary language expressions 'there is', 'there are', and their tensed forms are ambiguous, for they designate two different concepts: the logical concept *some* and the ontological concept of *existence*. Logic takes care of the former and formalizes it as the existential quantifier $\exists$, which I prefer to call *particularizer* or *indeterminate quantifier*, to distinguish it from the *universalizer* (or universal quantifier) as well as from the *individualizer* (or *descriptor*). For example, the standard reading of the expression '$\exists x F x$' is: "There exists at least one individual with the property F". I prefer to read it "Some individuals are F's".

To be sure, all mathematical logicians, from Russell to Quine, have claimed that $\exists$ formalizes both the logical concept "some" and the ontological concept "exists". Unfortunately, (a) they offer no arguments for the thesis of the identity of both concepts; and (b) they are mistaken. One example will suffice to show the need for unfusing the two notions.

Consider the proposition "$(\exists x)(Sx \ \& \ Bx)$", where '$S$' is interpreted as "is a siren" and 'B' as "is beautiful". That formula is generally read in either of the following ways:

(1) "There are beautiful sirens".

(2) "The schema '$Sx \ \& \ Bx$' is satisfied (is true), under the given interpretation, for some values of x".

(3) "Some sirens are beautiful".

The statements (1) and (2) are different since, while the former is a statement, the latter is a metastatement. However, one might argue that they are equivalent. That is, if in actual fact there are beautiful sirens, then "*Sx & Bx*" is true (under the given interpretation) for some *x*, and conversely. For this reason, we shall say that either of them constitutes an *ontological interpretation* of the given formula.

But in this case the ontological interpretation has an obvious flaw: it suggests that the subject (writer, speaker) believes that there are sirens in reality. Most likely, she only wants to say "Some of the sirens *existing in Greek mythology* are beautiful". The particularizer ∃ formalizes the prefix 'some', but not the expression 'existing in Greek mythology'. (The juxtaposition of two particularizers bearing on the same variable generates an ill-formed formula.) We need then an exact concept of existence other than ∃ if we wish to formalize phrases such as "There are some beautiful sirens". Let us introduce it.

We shall define a concept of *relative* or *contextual existence*, which is the one occurring in the propositions "Birds exist in nature, but not in mathematics" and "Disjunction exists in logic, but not in nature". We can do this by making

DEFINITION 1. Let *A* be a well-formed set included in some nonempty set *X*, and χ_A the characteristic function of *A*, i.e., the function from *X* into $\{0, 1\}$ such that $\chi_A(x) = 1$ if and only if *x* is in *A*, and $\chi_A(x) = 0$ otherwise. Then

(i) *x exists in A* $=_{df} \chi_A(x) = 1$;
(ii) *x does not exist in A* $=_{df} \chi_A(x) = 0$.

To be sure, we could have stipulated simply that *x* exists in *A* if *x* belongs to *A*. But the membership relation is not a function, and therefore does not allow us to take the next step.

We now introduce an existence predicate:

DEFINITION 2. The *relative* (or *contextual*) *existence predicate* is the proposition-valued function E_A from an arbitrary set *A* to the set of all statements containing E_A, such that "$E_A(x)$" is true if, and only if, $\chi_A(x) = 1$.

Therefore, we can now see that the old and vexing question whether existence is a predicate is ambiguous: the answer depends on whether we refer to the prefix ∃ or to the function E_A. While the particularizer is not a predicate (or propositional function or statement-valued function), the concept of relative existence defined above is a genuine predicate.

5. OF HORSES AND CENTAURS

We are now in a position to distinguish two specific concepts of existence, namely those of conceptual existence (or existence in a conceptual context) and real existence (or existence in the world). We define them thus:

DEFINITION 3. If x is an object, then
(i) x *exists conceptually* $=_{df}$ for some set C of constructs, $E_C x$;
(ii) x *exists really* $=_{df}$ for some set F of things, $E_F x$.

For example, the Schrödinger equation exists in the sense that it belongs to quantum mechanics. (It would make no sense whatever if it were a stray formula. In general, stray formulas designate no conceptual objects: only systemicity confers meaning.) To be sure, it did not come into existence until Erwin Schrödinger invented it in 1926; but it has existed ever since, though of course only conceptually. Likewise, the concept of an electron has existed only since 1898, although its referent, the real electron, has presumably existed forever.

The following examples show how to handle the concepts of conceptual and material existence and how to combine them with the logical concept of some. In them M stands for the set of characters in Greek mythology, G for that of Greek history, c for Chiron (the wisest of centaurs), b for Bucephalus (Alexander's warhorse), C for "is a centaur", W for "is wise", and H for "is a horse".

The horse Bucephalus exists in Greek history.
Hb & $[\chi_G(b) = 1]$, or Hb & $E_G b$.

The centaur Chiron exists in Greek mythology.
Cc & $[\chi_M(c) = 1]$, or Cc & $E_M c$.

Some of the individuals (existing) in Greek history are horses.
$\exists x(Hx$ & $[\chi_G(x) = 1])$, or $\exists x(Hx$ & $E_G x)$.

Some of the centaurs (existing) in Greek mythology are wise.
$\exists x(Cx$ & Wx & $[\chi_M(x) = 1])$, or $\exists x(Cx$ & Wx & $E_M x)$.

All of the characters of Greek history are real and none is mythical.
$\forall x([\chi_G(x) = 1] \Rightarrow [\chi_M(x) = 0])$, or $\forall x(E_G x \Rightarrow \neg E_M x)$.

All of the centaurs exist in Greek mythology and none of them is real.
$\forall x(Cx \Rightarrow ([\chi_M(x) = 1]$ & $[\chi_F(x) = 0]))$, or $\forall x(Cx \Rightarrow (E_M x$ & $\neg E_F x))$.

So much for our analysis of the existence concepts designated by the ambiguous expressions 'there is' and 'there are'. The distinctions we have proposed allow one to remove such ambiguities. They are unnecessary when the context is fixed, as is the case with logic and mathematics, which deal only with conceptual objects. But they come in handy when the context is formed by both conceptual objects and material ones, as is the case with the factual sciences and philosophy, in particular ontology and epistemology. Thus, four of the most influential philosophies can be summarized as follows:

Vulgar materialism $\forall x E_F x$

Immaterialism $\forall x \neg E_F x$

Hylomorphism $\exists x(E_F x \,\&\, E_C x)$

Conceptualist materialism $\exists x(E_F x) \,\&\, \exists y(E_C y)$.

6. CONCLUDING REMARKS

We have sketched a philosophy of the conceptual, or conceptology, that is conceptualist, fictionist, and materialist. It is conceptualist because it admits the existence of conceptual objects distinct from material objects (e.g., signs of a language), mental objects (which to a materialist are brain events), and ideal or Platonic objects (which to a materialist are fictitious). Our conceptology is fictionist because, far from postulating the autonomous or independent existence of conceptual objects, it only postulates that such objects are fictitious, though not all of them idle or introduced only for the purpose of entertaining, moving, edifying, or scaring. And it is materialist because it assumes that such fictions are created and maintained by living beings, namely by merely being able to be thought about.

In order to be able to speak with some precision of the existence of conceptual objects, we have had to give up the received belief that the so-called existential quantifier exactifies the notion of existence, which in turn would be one. We have introduced an existence predicate that may be specified to indicate either conceptual existence (or belonging to some set of constructs) or material existence (or belonging to some set of concrete objects). While conceptual existence (e.g., of a function) is either postulated or proved, material existence (e.g., of a new particle or a new social system) is conjectured, and it is understood that such a hypothesis must eventually be put to empirical tests. In the former case, we make believe that something exists (belongs to some body of constructs), and in the latter case we assume and then verify (or refute) that some thing is part of the material world.

These differences notwithstanding, existence claims, whether in formal science or in factual science, are supposed to be responsible. That is, one does not waste one's time inventing idle constructs, i.e., concepts or propositions that discharge no function.

On the other hand, the differences between the conceptual and the material are profound with regard to the type of existence as well as the conditions and criteria of existence. It is not the same to claim that a given field equation has solutions of the wave type as to claim that such solutions represent real waves. While the former claim can be checked with paper and pencil (and some brains), the latter requires, in addition, inventing and constructing special detectors and performing observations with their help. The philosopher as such is not equipped to check on either claim, but she is equipped to discuss the existence concepts involved as well as the general nature of existence hypotheses and proofs in formal science and in factual science. And, unless she is intent on increasing the current confusion between conceptual existence and material existence, she will abide by the slogan that being is meaningless and meaning has no being.

7

POPPER'S
UNWORLDLY WORLD 3

(1981)

■

At the Third International Congress of Logic, Methodology, and Philosophy of Science, held in Amsterdam in August 1967, Karl Popper astonished the philosophical community by formulating his doctrine on "the third world", or world of ideas, or "objective mind", which he later called 'World 3'. In his paper, titled "Epistemology without a Knowing Subject", he wrote as follows: "Without taking the words 'world' or 'universe' too seriously, we may distinguish the following three worlds or universes: first, that world of physical objects or of physical states; secondly, the world of states of consciousness or of mental states, or perhaps of behavioral dispositions to act; and thirdly, the world of *objective contents of thought*, especially of scientific and poetic thoughts and of works of art" (Popper 1968, p. 333).

This thesis came as a surprise, not because it was new—which it was not—but because until then Popper had been a merciless critic of idealism. In particular he had criticized the objective idealism of Plato and Hegel (Popper 1945) as well as the subjective idealism of Berkeley

(Popper 1953), particularly for having influenced contemporary positivism. And now, without warning, Popper makes a right-about turn, or
so it seems, and adopts an idealistic stand. He owns the latter explicitly:
"What I call 'the third world' has admittedly much in common with
Plato's theory of forms or ideas, and therefore also with Hegel's objective
spirit, though my theory differs radically, in some decisive aspects, from
Plato's and Hegel's. It has more in common still with Bolzano's theory of
a universe of propositions in themselves and of truths in themselves,
though it differs from Bolzano's also. My third world resembles most
closely the universe of Frege's objective contents of thought" (Popper
1968, p. 333). In his *Autobiography* (1974) Popper has confirmed this
philosophical self-analysis.

1. A CONVERSION?

How is this sudden conversion of Popper's to objective idealism to be
explained? He himself does not explain it in his *Autobiography*. He tells
us only that, just like Bolzano, he had wondered for long years about the
ontological status of "propositions in themselves". He writes also that he
did not publish anything on the third world until he "arrived at the conclusion that its inmates were real; indeed, more or less as real as tables
and chairs" (Popper 1974, p. 146).

It would seem that Popper suffered a late conversion from the all-
round anti-Platonism of *The Open Society* (1945), but he himself holds
that this was not the case. In a letter of October 4, 1977, Sir Karl
informed me that "The Amsterdam paper of 1967 (published in 1968)
was a rewritten paper first read to my LSE [London School of Eco-
nomics] Seminar in 1959 or 1960. The ideas of this paper go back right
to the *Logik der Forschung*, and to the *Wahrheitsbegriff* of Tarski's
[1935/36]. Also, the difference between my views and those of Plato and
Hegel, etc., is very great; even that between my views and those of
Bolzano and Frege. In the Amsterdam paper, I stress similarities more
than differences. Main difference: world 3 is the *product* of human
minds. (But there is a strong feedback.) World 3 can act upon the physical world 1 (although only in an indirect way, through world 2)."

However, the doctrine of knowledge without a knowing subject,
which is part of the World 3 doctrine, is a direct generalization of the
thesis of the objectivity of scientific knowledge, maintained by Popper
himself with regard to the quantum theory, the same year he wrote his
Amsterdam paper (Popper 1967). Besides, Popper's letter poses the following problem to the historian of philosophy. If the doctrine was found
in nuce both in the *Logik der Forschung* and in Tarski's equally famous

paper, why did nobody seem to notice it in the course of three decades? Why has Tarski always regarded himself as a materialist and, more particularly, a nominalist? (Tarski, personal communication, Jerusalem, 1964.) And why did Popper start his Amsterdam lecture by stating: "I shall make an attempt to challenge you, and, if possible, to provoke you" (Popper 1968, p. 333)?

I suspect that what Tarski's paper on (formal) truth and the *Logik der Forschung* did contain is something very different from Plato's ideas and Hegel's objective spirit. (These two expressions occur in Popper's 1968 article as synonyms of 'the third world'. The synonymy is repeated next year in the lecture at the International Congress of Philosophy, entitled "On the theory of the objective mind".) What we do find here is a tacit rejection of psychologism, i.e., the doctrine according to which propositions are thoughts, and rules of inference laws of thought. We also find in these works a sort of formal objectivism consisting in the thesis that, once the rules of the game have been accepted, there is no arbitrariness, for everything proceeds according to rule. To be sure, both the logician and the mathematician treat propositions and inference rules *as if* they enjoyed an autonomous existence. But this may make them fictionists, not necessarily metaphysical (Platonic) realists. (See also essays 6 and 13.)

Be that as it may, the fact is that Popper has proposed the partition of the world into three, and that this doctrine revolves around psychophysical dualism. Let us then take a look at the latter.

2. MIND-BODY DUALISM IN POPPER'S PHILOSOPHY

Popper has become a staunch defender of psychophysical dualism. Jointly with his friend, the eminent neuroscientist Sir John Eccles, Popper has been looking for arguments in favor of the ancient thesis that the mind is an entity or substance separate from the body, though interacting with the latter. In particular, the Popper and Eccles volume *The Self and Its Brain* (1977) expounds the thesis that every one of us is an embodied mind. (Actually, Popper defends two different theses in that book, although he does not seem to realize that they are different: one is interactionism, the other is Plato's doctrine that the mind steers the body as the helmsman steers the ship.)

Psychophysical dualism is, of course, the most popular doctrine about mind, at least in the West. We adopt it tacitly in daily life when we speak of the influence of ideas on bodily states and behavior, and conversely. We espouse it when saying that a given state or process is psychological, not physiological, and when we distinguish a thought process from its "products". Dualism is inherent in psychoanalysis and spiritism, in Platonism and Cartesianism, in Christianity, Islam, and Buddhism. The doctrine is so entrenched, and it has been defended so zealously by conservative ideologies and the corresponding institutions, that we hardly realize it. In particular, the neuroscientist who writes about "the neurophysiological correlates" of mental states, without explaining what the latter are nor how they are "correlated" with their material counterparts, does not realize that he is the prisoner of a vulgar ideology.

This is not the place to perform a critical analysis of psychophysical dualism, a task done elsewhere (Bunge 1980a as well as essay 5). We must restrict ourselves here to stating dogmatically that the dualist describes the mental in vulgar terms, not in scientific ones, and that he resists any identification of mental states with brain states by resorting to equally vulgar arguments, such as the difference between the language of introspection and that of physiological psychology, and the difference between concepts and brain processes.

What we must point out here is the vagueness and irrefutability of psychophysical dualism, because both characteristics ought to have made it unacceptable to Popper. That psychophysical dualism is as imprecise as it is untestable seems clear from the following.

(i) Whereas the scientist takes it for granted that *every state is a state of some concrete entity* (physical, chemical, biological, social, or what have you), the dualist talks about mental states *in themselves*. (He may no longer dare to speak of states of a *res cogitans*, or mental substance, since anybody could ask him for his reasons for calling 'substance' that which lacks substantiality.) That this is a case both of reification and of confusion between distinction and detachment seems clear. It is also clear that dualism consecrates the traditional detachment of psychology and psychiatry from neuroscience.

(ii) The conceptual imprecision of psychophysical dualism is such that it lacks a theory proper. It confines itself to giving some examples of mental activity and to saying what is not mental, and in particular to insisting that the mental is not physical—which is obviously insufficient, since life too is not merely physical yet it is nonmental. Psychophysical dualism is so imprecise that it has not been mathematized.

(iii) Being imprecise, psychophysical dualism is untestable and, in particular, irrefutable, for it cannot issue precise predictions that can be put to the test. The fact that, when in an emotional state, one does not see well and does not coordinate one's movements well, so that one cannot drive correctly, can be explained in two ways. The dualist explanation is simple and therefore popular: your mind is acting upon your body—in a mysterious way, perhaps incomprehensibly. (Eccles 1951 suggested psychokinesis.) The psychophysiological explanation is more complex: it resorts, in particular, to the action of certain hormones on the synapses of the neurons of the motor system. By experimentally varying the concentration of such hormones one may control, at least in principle, the "somatic effect" of the "emotional state of mind" (as the dualist would put it). The interactionist dualist will tell us that this is nothing but an example of the action of body on mind—which he has never denied. And if the hormone injection were not to change perceptibly the subject's behavior, the dualist may argue that in some cases the mental states or processes are so intense that no neurophysiological process can alter them. He may cite the experiences of the martyrs who died singing, and of the yogis. There is no way of refuting those who hold an irrefutable doctrine—unless of course they realize that such a doctrine does not explain anything nor, consequently, contributes to posing or solving any scientific problems.

In conclusion, psychophysical dualism is inimical to Popper's own methodology. However, this contradiction is not as important as the possibility that it be one of the sources of his World 3 doctrine. And, no matter what the genesis of this doctrine may have been, we must examine it to find out whether it is true.

3. THE PLURALITY OF WORLDS

At the very beginning of his first paper on the tripartition of the world, Popper warns us that we must not take "too seriously" the words 'world' and 'universe', which occur in the formulation of his theory. (Incidentally, it is a thesis not a theory.) However, we cannot take that warning seriously: if we are asked to take seriously a given thesis, then we must take just as seriously the key terms occurring in its formulation. It won't do to tell us—as Popper has done repeatedly—that matters of words and their meanings are insignificant. They may be so in inexact philosophy, but they are just as important in exact philosophy as they are in mathematics and science.

Strictly speaking, *the world* (or universe) is the supreme concrete thing, i.e., that thing which contains (as parts) all other concrete things. This is the way physical scientists use the word. In a figurative sense, *a world* (or universe) is either a subsystem of the universe, such as our planet, or a structured set of objects, whether concrete or conceptual. Taken in its strict sense, the denotatum of 'the world' is concrete and, of course, unique. On the other hand, in a figurative sense there may be as many "worlds" as concrete systems and conceptual systems. The solar system is one of them, and the set of natural numbers is another. Likewise, we may talk of the "world" of the fish of a given species, or a given region, as well as the "world" of philosophical ideas.

There are many such partial "worlds", and not all of them are concrete or material. Undoubtedly, it may be convenient to use a single expression of the form the *world X* (or *the X world*) to denote some concrete system, such as an ecosystem, or a structured set, such as a topological space. However, verbal economy must often be paid for with conceptual confusion. This is exactly what happens with the term 'world' in Popper's thesis of the trinity: in it, the word 'world' is not used uniformly, i.e., with a single definite signification. Indeed, let us recall how Popper defines his "worlds".

World 1, or the physical world, is "the world of physical objects or of physical states". This is an ambiguous phrase. In fact, such a "world" may be a concrete individual, namely the system composed of all material things, or a set and therefore a concept. (If the components of "world 1" are physical states, then this "world" is a set and therefore an *être de raison*.) Clearly, any statement about "world 1" will depend critically on whether it is taken as a thing or as a set. In the first case, it may be attributed physical properties, not so in the second.

World 2, or the psychical world, is "the world of states of consciousness or of mental states, or perhaps of behavioral dispositions to act". It would seem that apes and, a fortiori, other higher vertebrates have no access to this "world", which is rather odd given the available evidence on their mental life and given the theory of evolution. It would also seem that this time the world is a set—unless the states of consciousness of one person are allowed to influence directly those of others. (Needless to say, to a materialist the elements of such a set are ghostly, for one can speak properly of brain states but not of states of consciousness as distinct entities.) Moreover, it would seem that, while world 1 is eternal (at least in the forward direction of time), world 2 is not so necessarily. Indeed, if no thinking beings were to remain, world 1 would continue to tick but world 2 would become extinct—though presumably leaving world 3 behind.

World 3, or the cultural world, is "the objective contents of thought, especially of scientific and poetic thoughts and of works of art". Among the "inmates" of this "world" Popper lists problems, critical arguments, and theories, as well as the "contents" of books, journals, and libraries. Sometimes the books, journals, and other material embodiments of intellectual and artistic work are included as well. Thus: "I regard books and journals and letters as typically third-world objects, especially if they develop or discuss a theory" (Popper 1974, p. 145). There are thus both *embodied* World 3 objects, such as gramophone records, and unembodied World 3 objects, such as numbers (e.g., Popper and Eccles 1977, p. 41). In short, World 3 is composed of all the "products" of mental activity, or World 2 inmates. And that "world", like World 2, is a set. But unlike the members of World 2, which are perishable, some of the inmates of World 3 are, or at least are very close to being, eternal objects in the manner of Plato, Hegel, Bolzano, Frege, Husserl, and Whitehead. Let us take a closer look at this idea.

To clarify his idea about World 3, and perhaps persuade us that it exists by itself once created by humans, Popper imagines two situations which many have envisaged since the beginning of the nuclear bomb age. In the first a worldwide blast destroys all of our cultural artifacts, except for the libraries and museums: these remain, and our ability to learn from them is left intact. It is certain, says Popper, that after long sufferings "our world" (in this case industrial civilization) may again be set in motion. (That this is doubtful rather than certain makes no difference to the philosophical argument.) In the second mental experiment there remain some human beings but all the libraries and museums are destroyed by nuclear bombs. In this case our ability to learn from books and journals would be useless, and it would take millennia to reconstruct civilization. (Why anybody would wish to reconstruct a civilization capable of self-destruction once in a while is not explained, but is not relevant to our problem either.)

The above "experiments" exhibit "the reality, significance, and the degree of autonomy of the third world (as well as its effects on the second and first worlds)" (Popper 1968, p. 335). Popper is satisfied with such *Gedankenexperimente* even though in his *Logic of Scientific Discovery* he had criticized, and rightly so, the physicists who claimed to prove theorems by imagining experiments. Others won't remain satisfied with a couple of science fiction stories, any more than they are persuaded by Walt Disney's movies that Mickey Mouse is real.

4. CRITICISM OF THE FANTASY

Only a crass materialist would dare deny the importance of ideas—or, rather, that of thinking brains. But this does not entail that ideation may constitute a world ("world 2") nor that the "products" of ideation (its "contents") constitute a "third world" enjoying autonomous existence from the moment of its coming into being. The least that can be objected to the thesis of the real existence of World 3 is that it is imprecise; the most, that it is groundless.

(i) Firstly, Popper does not tell us clearly what he means by 'real' and 'reality'. For one thing, he does not seem to regard the predicate "is real" as dichotomous: in fact he tells us that the inmates of World 3 are "more or less as real as tables and chairs" (Popper 1974, p. 146). To most other philosophers—the exceptions being some Thomists such as Jacques

Maritain—reality does not come in degrees: every object is either real or unreal. Also, many philosophers are careful to distinguish two concepts of existence or reality: material and conceptual, and to state that, whereas some objects exist materially, others exist conceptually. Finally, all philosophers interested in the problem of reality have attempted to clarify this concept. In particular, materialists equate "real" with "material", and in turn identify "material" with "changeable". Popper leaves us in the lurch on this point.

(ii) In the second place, it should be necessary that all the inmates of World 3 be ideal objects, and preferably also of the same kind, e.g., sets. The reason is that ideal objects do not combine with material ones, forming mixed systems. Material objects can combine with one another to produce material systems, and ideal objects can associate, forming ideal systems. There are no mixed entities, except in hylomorphic metaphysics, which Popper does not approve of explicitly. Moreover, if one holds that there are such mixed beings, then one is supposed to exhibit, if not empirical evidence for or against such a hypothesis, then at least a calculus containing the operations producing them out of material entities on the one hand and ideal objects on the other. (It won't do to say that a written sentence is the *embodiment* of a proposition in itself, and that a theorem is an *unembodied* object: these are vague words suggested by religion and psychophysical dualism. Give me a calculus of embodiment and unembodiment and I may begin to take you seriously.) In short, World 3 does not constitute a world proper, i.e., a system, as long as its ideal inmates are allowed to mix promiscuously with the material ones without obeying any laws.

(iii) Thirdly, Popper does not explain what he means by "the content" of a drawing, a musical phrase, or any other nonconceptual component of World 3. Are they "messages", and if so, how about abstract art? Neither he nor anyone else seems to have proposed a semantic theory equally applicable to works of art and scientific theories. Because of this, no less than (i) and (ii), the World 3 thesis is inexact to the point of meaninglessness.

(iv) In common with the idealist metaphysicians who preceded him, Popper starts by considering the intellectual and artistic activities of human beings, and ends up by abstracting from such concrete entities and their activities, to focus his attention on the "products" (and sometimes also their "embodiments", i.e., books, paintings, phonograph records, etc.) into a single "universe", namely World 3. That is, he assumes that this heterogeneous set constitutes a system. Finally, forgetting how this pseudosystem came about, he declares that it leads an autonomous existence, i.e., one independent of its creators—and this simply on the strength of two *Gedankenexperimente*. What is this but *reification* together with *systematization by decree*? What is it but to take literally the fire-ashes metaphor, or the model of the factory and its products?

(v) Popper asks us to attribute autonomy to World 3, i.e., to assign it an existence independent from its creators and probably also from everything else. From everything? If there were a Supreme Creator and Annihilator capable of annihilating every material thing to the last electron, photon, and neutrino, would World 3 subsist? Being an agnostic, Popper has not envisaged this third *Gedankenexperiment*, so we won't know for sure what the ontological status of World 3 is.

(vi) Popper does not justify or corroborate, let alone try to refute, the conjecture that his World 3 exists or is real. Why does he think that his readers, usually attracted by his critical rationalism, should be able to swallow this monster of traditional metaphysics—the monster that gave metaphysics its bad name among scientists?

5. KNOWLEDGE: SUBJECTIVE AND OBJECTIVE

It is well known that the Greeks distinguished between *doxa* (opinion or uncertain or subjective knowledge) and *episteme* (science or certain and objective knowledge). Popper keeps this distinction, though leaving certainty aside: to him all knowledge, even mathematical knowledge, is conjectural and therefore uncertain and subject to revision.

Still, though human knowledge is fallible and therefore subject to revision, at least it can be objective: it need not be subject-dependent. (Subjective knowledge, or mere belief, does not deserve to be called 'knowledge', Popper tells us in 1972, chapter 2.) Unfortunately, Popper does not define what he means by 'objective knowledge'. (In general he refuses to define his terms, alleging that definitions lead nowhere.) However, from the context it would seem that Popper calls 'objective knowledge' all knowledge that does not depend upon the knowing subject—although without saying whether the independence is referential or alethic. (A proposition can be said to be referentially subject-independent if, and only if, it does not concern any particular subject; and alethically subject-independent if its truth value is the same for all knowing subjects.) I shall argue that, if this were so, there would be no knowledge.

To the psychologist knowing is a mental (or brain) state or process of some animal. The same to an epistemologist other than Popper. When we say "p is known", where p designates a proposition or denotes a fact, we do nothing but abbreviate "There is at least one animal that knows p". (Actually, we intend to say that there are several animals that know p, or that whoever is not an ignoramus knows p, or that anyone may get to know *p* if only he intends to. But this is unimportant in our case.) Similarly, when we assert that *q* is unknown, we abbreviate "No animal knows *q*", or at least "None of the animals I know knows *q*".

All knowledge is then knowledge of *something by somebody*, whether human or not. In particular, that somebody may be you or I. If either of us asserts "I know *p*", and this statement proves to be true, e.g., as a result of a test, "it is" concluded—i.e., anyone can validly conclude—that there is at least one animal that knows *p*, i.e., that "*p* is known (by somebody)".

At first sight, whereas 'I know *p*' is a subjective expression, "*p* is known" is a statement of objective knowledge. But "I know *p*" implies "*p* is known", i.e., "There is at least one animal that knows *p*". Hence, the result would be that subjective knowledge not only originates objective knowledge but is also its foundation. However, this is not so, for "I know *p*" may be true or false quite aside from my certainty or uncertainty concerning the propositions "*p*" and "I know *p*". The same holds for "*p* is known": this "objective" proposition may be factually true or false (to some extent). And this is what matters most in both cases: the degree of truth of the proposition *p*.

Neither mathematicians nor physicists nor sociologists are in the habit of writing sentences of the type '*p* knows *q*', except when investigating empirically what people know, e.g., when grading exam papers. They do publish, by contrast, sentences of the forms '*p*', 'not-*p*', 'If *p* then *q*', etc., which do not refer explicitly to those who formulate or believe them. Even when social psychologists or historians of ideas make statements about beliefs, they do so without putting themselves in their sentences. Thus, they will write 'People of kind *X* tend to believe *p*', instead of 'I tend to believe that people of kind *X* tend to believe *p*'.

In this sense the data and the hypotheses of science, formal or factual, basic or applied, are objective, or referentially and alethically free from the knowing subject. When stating that the statements of science are objective we do not mean that they exist by themselves as inmates of World 3. All one means is (a) that what matters is the referent rather than the speaker or writer, and (b) that the proposition in question has to be judged according to canons accepted beforehand rather than by its agreement or disagreement with some authority.

In other words, that a cognitive statement is objective does not entail that, once made, it has a "life" of its own or that it enters World 3. At most, one feigns that the proposition exists by itself, and this simply because it can be thought and examined by others similarly (but not identically) to the way that the vegetables and fruits in an open-air market can be freely examined by the public. That fiction is indispensable if we wish to divert attention from the subject to what she asserts or the way of justifying (proving or confirming) or else refuting the proposition concerned. It is also indispensable to deal with infinite sets of propositions, such as theories, since no finite being could think up all of them.

6. TWO APPARENT EXCEPTIONS

There are, however, two fields of scientific research that are sometimes believed to have been invaded by subjectivism, namely the quantum theory and psychology. It can be shown that both beliefs are false. The subjective interpretation of the quantum theory can be shot down with heuristic arguments (e.g., Bunge 1955; Popper 1967) or rigorously (Bunge 1967c, 1973b). What one does in the latter case is to analyze the "variables" (functions and operators) of the theory and to organize the latter axiomatically. The analysis fails to exhibit any references to a knowing subject, such as an observer. And the axiomatization fails to reveal any assumption about experimenters. To put it in a positive fashion: every formula of the quantum theory refers exclusively to physical entities, none to knowing subjects.

What happens is that it is easy to "interpret" any proposition p as "There is an observer that verifies p". For example, instead of saying that properties P and Q are related by some function F such that $Q = F(P)$, one can (but one ought not to) say "The knowledge of P determines that of Q when the computation indicated by '$Q = F(P)$' is performed". This is just a didactic (but misleading) prop that does not prove that the function F is subjective, in the sense that it is left to the arbitrary decision of the knowing subject.

Conversely, every pragmatic expression, such as 'The values an observer may obtain when measuring the dynamical variable P are the eigenvalues of the operator representing P' may (and must) be translated into an expression free from any knowing subject, such as 'The P values that a physical entity may take on are the eigenvalues of the operator representative of P'. The reason for preferring the latter reading of the formulas of the quantum theory is that it deals with physical problems, not with psychophysical ones. So much for the pseudosubjectivity of the quantum theory.

What happens with psychology is this. Here the referents are (experimental) subjects, which are often capable of knowing. Moreover, the psychologist can be his own subject of observation or experiment. But those who study such subjects scientifically are supposed to proceed objectively: their results are supposed to refer to entire classes, e.g., species, occupations, or age groups, and they are supposed to be publicly scrutinizable. This does not exclude introspection (whatever that may be) but renders its data mere heuristic starting points. The psychologist does not use such data as an unshakable foundation, but as a source of hypotheses to be tested objectively.

The objectivity of psychology, then, does not consist in that it is not interested in subjective experience, but in that its conjectures, data, and conclusions are, it is hoped, true irrespective of who formulates them. (Thus, we do not believe in the hypothesis of the stages in biopsychological development just because it was formulated by Jean Piaget but because it has been confirmed by many other psychologists and agrees with what neurophysiology tells us about the maturation and plasticity of the central nervous system. By contrast, we do not believe in projective tests, such as the Rorschach, because the "interpretation" of the inkblots is left to the imagination of the psychologist—so that they rarely work.) In short, scientific psychology is no less objective than the other factual sciences.

7. EPISTEMOLOGY WITHOUT THE KNOWING SUBJECT?

The last two sections can be summarized as follows:

(i) *All knowledge is knowledge of something by somebody*. There is no knowledge of nothingness or knowledge without a knowing subject. (But of course, the object or referent of knowledge can be imaginary, as is the case with those scientific theories that postulate entities that are eventually recognized as unreal.)

(ii) *Objective knowledge is intersubjective and (partially) true knowledge*—i.e., it is invariant with regard to subject replacement, even though it must have been discovered or invented by some subject to begin with.

In other words, an item of knowledge is objective not because it exists or subsists in a separate "world", one above corruption (to employ Platonic language) or sheltered from nuclear bombs (to use contemporary language). It is objective because, and to the extent that, there are animals capable of acquiring it and putting it to the test with the help of criteria independent of personal factors such as authority or firmness of conviction. The degree of objectivity of a proposition may be estimated by virtue of the rules of the knowledge game, such as those of logic and empirical testing.

Now, epistemology is concerned with knowledge in general and, in particular, scientific knowledge. And scientific knowledge, unlike mere opinion, which can be subjective and groundless, is objective or invariant with respect to the knowing subject. In other words, the rules of science, though not eternal, are not arbitrary. (Hence, scientific research is not really a game.) And although such rules are proposed, discussed, applied, violated, or modified by living researchers, they are not proclaimed on the strength of such personal considerations, for they are supposed to lead to truth, in particular factual truth.

However, the objectivity of scientific knowledge does not imply that it is above all knowing subjects nor, in particular, that it constitutes a "world" independent of its creators (and destroyers). Getting to know something is, by definition, a process or activity and, like every other activity, it can be formalized as a relation that is at least binary. We think and say '(Subject) X knows (fact or proposition) p', neither 'X knows' nor just 'p', when we deal with the process of knowledge, i.e., when we consider knowledge as a biological, psychological, or cultural fact. It is only when we are concerned with the outcome or "product" of this process that we disregard the subject and focus on "the content of knowledge", such as a proposition or a theory. But when we do so, we engage in something other than epistemology.

Epistemology cannot dispense with the knowing subject because, by definition, it is concerned with what the subject can know, how she gets to know, and related questions. For example, genetics is concerned with genotypes and their relations to phenotypes. By contrast, the epistemologist is interested in the ways the geneticist investigates his genotypes and phenotypes, what motivates him, what guides or misguides him, what he succeeds in discovering or inventing, what are the philosophical presuppositions of his research, etc. To be sure, some epistemologists are not interested in real knowledge but speculate on abstract knowledge by an ideal knowing subject. They are antihistorical, antisociological, and antipsychological epistemologists who deal with the fiction of "objective knowledge without a knowing subject".

Whereas the classical epistemologist dealt with the knowing subject and his activities in general, some contemporary epistemologists have understood that the suprahistorical knowing subject hovering above society is a fiction. We have come to know that every knowing subject is a member of a given culture, and that membership in a culture opens up some horizons while it may close others. Contemporary epistemology does not ignore the history of knowledge: it takes the "context of discovery" as well as the "context of justification" into account. (Which is not to say that there is never objective truth, that truth value assignments are only seals of social approval.) In particular, the new epistemology that is in the making studies the factors of various kinds (cultural tradition, economic potential, political régime) that stimulate or inhibit the production and circulation of knowledge. In particular, epistemology must take into account that scientific research is just one cultural activity, hence the study of it cannot be isolated from the study of other branches, in particular philosophy and ideology. In short, epistemology, if it is to be realistic (not just realist), must be not only structural but also psychological, sociological, and historical.

Let us not forget the psychology of knowledge, a discipline which, like every other chapter of psychology, deals with one aspect of the brain activity of the higher vertebrates. Nor should we forget that there is an embryo evolutionary epistemology. Both are closely related to neurobiology and evolutionary biology. And the latter is about biopopulations, not about processes in themselves independent of the things that evolve. To be sure, Popper has written about evolutionary epistemology and has asserted that his is one (Popper 1972). But this claim contradicts his other claim that epistemology must ignore the knowing subject. Knowledge in itself, as conceived by Popper, i.e., as an "inmate of World 3", does not evolve: it is a product dwelling in the realm of Platonic ideas. Only an epistemology concerned with knowing subjects—and not only human ones—can be evolutionary. (See Vollmer 1975.)

8. CONCLUSION

Popper defends epistemological realism and criticizes philosophers who have held that epistemology is concerned with the beliefs of the knowing subject, instead of studying what he investigates (finds out, constructs, criticizes, etc.). He also insists that genuine knowledge is objective. In this eagerness to defend epistemological realism and the objectivity of science, Popper has also embraced metaphysical realism. In fact he has proposed two false theses.

The first thesis is that objective knowledge (and also the "contents" of works of art) constitutes a distinct and autonomous system: World 3 or the objective mind. The second thesis, or rather prescription, is that epistemology should study this "world" instead of the cognitive activities of live animals. The former thesis matches with psychophysical dualism and, by the same token, is inconsistent with biopsychology, or the study of the brain processes consisting in perceiving, imagining, thinking, remembering, and the like. And the second thesis is inconsistent with evolutionary epistemology—accepted by Popper himself—and at variance with the idea that most philosophers have about the task of epistemologists. Therefore, it is unlikely to become popular.

On the other hand, the many-worlds hypothesis, and in particular the thesis of the autonomy of World 3, is gaining in popularity. One reason is that it was rather popular even before being drummed up by Popper. Indeed, most of us are ready to distinguish a brain from a brain process, and the latter from its "products", e.g., constructs—which is all right. And most of us have a tendency to separate whatever we distinguish. (We may call this fallacy the 'ontological rule of detachment'.) This is what we do when we reify—and Popper's World 3 is nothing but an instance of reification. In short, the doctrine has popular appeal—and should therefore be suspect to philosophers.

Besides, Popper's trinity arrives at a time when modal logicians formalize and apply the old view that there are many, perhaps infinitely many, possible worlds. (Some of them hold that all these imaginary "worlds" are just as real as the real world. But if pressed, many of them acknowledge that such "worlds" are sets of formulas, hence even less real than the Queen and her court in Alice in Wonderland, which was no less than a dream "world".) This fantastic, escapist, and sterile metaphysics, which encourages unbridled speculation about physically impossible worlds instead of studying reality, may have facilitated the diffusion of Popper's trinitarian metaphysics. After all, the thesis that there are just three worlds comes as a refreshing breeze of rationality after the doctrine—worthy of J. L. Borges—that there are infinitely many worlds.

In short, the three worlds doctrine is false: there is but one immensely varied and forever changing world. The thesis of the autonomous and perennial world of culture is not only false. It is also noxious, for it fosters the illusion of immortality—for the works of intellectuals and artists—even after the nuclear Armageddon. The products of cultural activity are not perennial: they are being transformed or even destroyed all the time. We have at most the possibility of cultivating the arts or the humanities, the sciences or the technologies, during our own lifetime. And we have the duty to pass this possibility on to our children by doing something about the only world we have got.

II

METHODOLOGY AND PHILOSOPHY OF SCIENCE

■

8

ON METHOD IN THE PHILOSOPHY OF SCIENCE

(1973)

■

The philosophy of science is probably the fastest growing branch of philosophy nowadays. However, this growth is partly inflationary, in the sense that a very large part of the output is not concerned with real science. One therefore wonders what use that growth in bulk can possibly have for either philosophy or science.

When in doubt about the authenticity of an intellectual endeavor, the right thing to do is to perform a candorous reexamination of its three components: subject matter, method, and goal. In the case of the philosophy of science, we may generously assume that it has a clear goal regardless of the individual worker's orientation, namely to understand the presuppositions, means, products, and targets of scientific research in the light of philosophy. In order to explain why this goal is not always achieved, we must take a second look at the other two components of the enterprise: its subject matter or problem circle, and its method.

Problems and methods, though distinct, are mutually dependent.

Thus, it would be pointless to use advanced tools of formal logic, formal semantics, and mathematics to analyze and systematize an item of ordinary knowledge, e.g., "All ravens are black", which by its very nature is superficial, unsystematic, often imprecise, and in any case prescientific. Conversely, it would be impossible to handle a technical fragment of science with the poor resources of ordinary language analysis—which, though necessary, is insufficient. Subject matter and method determine each other to a large extent. So much so that the very choice of subject matter or problem circle may be regarded as a matter of method. At any rate, this is how we shall handle object and method in this paper.

A convenient way to try to discover what is wrong with present-day philosophy of science is to survey the main ways work is actually being done in this field. We shall review six such ways: apriorism, preface analysis, textbook analysis, historico-philosophical analysis, isolated item analysis, and systematic analysis. We shall thenceforward proceed to examine a typical item of a philosophy out of touch with science, namely the gruebleen paradox, and an authentic problem frustrated by the lack of adequate information about real science, namely the question of statistical explanation. We shall also touch on the exaggerated role allowed to probability theory in current philosophy of science as a product of an enthusiasm for exactness joined with disregard for the actual procedures of science. Finally, we shall sketch the main stations of the right method and shall propose some practical measures concerning the training of philosophers of science. The upshot of our study will be the slogan: *Primum cognoscere, deinde philosophari*.

1. APRIORISM

A philosophy of science may be said to be *a priori* if it consists in an attempt to force science into a prefabricated philosophical framework.

There are two genera of a priori philosophies of science: *apologetic* and *naive*. Those in the former genus are deliberate efforts to exhibit science as being in agreement with a given philosophy—usually a school philosophy—in order to add to the latter's prestige. Apologetic philosophies of science can be coarse or sophisticated: the former will just state that such and such theories and techniques of science endorse the given philosophical school, whereas the more sophisticated apologists will offer "interpretations" of both scientific items and articles of philosophic faith, so as to match them. (Actually there is a third kind of apologist, namely the one that beheads the more annoying heads of the hydra. But this type, though influenced by a philosophy, can hardly be regarded as a philosopher.) This militant type of a priori philosophy is by now utterly discredited: people have realized that it is sterile, its sole function being the defense of some establishment or other.

On the other hand, the a priori philosophies of science of the naive type are quite fashionable and may always remain in fashion, as it is far easier to discuss philosophical ideas that tradition associates with science than to handle philosophically a real piece of science. The most fashionable of all a priori philosophies of science is, of course, uncritical empiricism, according to which scientific knowledge is just an extension of ordinary knowledge, so that the findings of the general (and empiricist) philosophy of knowledge apply to the former. These findings, or rather tenets, include what may be called *Locke's principle*, namely "No sense impression, no worthwhile idea", or "One idea, one sense impression". This principle, or slogan, may be implemented in three possible ways.

A first way is to try and show that science *does in fact conform* to those principles: that every scientific concept is observational or can be reduced to observational concepts, e.g., by means of operational definitions or of reduction statements, and that every empirical procedure boils down to theory-free perception. This method comes close to apologetics but it is far less rigid: it is willing to surrender any number of subsidiary principles as long as the main tenets are kept. In any case, it solves the problem of philosophy of science in the simplest possible way, namely by obliterating the distinction between science and the empiricist version of ordinary knowledge. A second kind of enterprise in the same spirit is the criticism of certain scientific theories for failing to conform to Locke's principle or some variations of it, such as "No direct measurability, no basic concept" and "One region of experience, one theory". This sort of criticism has been two-edged: on the one hand it has weeded out some humbug in the sciences of man, but on the other hand it has effectively blocked theoretical science, thus getting dangerously close to the worst kind of apologetic a priori philosophy.

Finally, a third kind of enterprise in the spirit of naive a priori philosophy of science is the logical reconstruction of a fragment of science in such a way that it complies with a given set of philosophical requirements. A typical attempt of this kind is the construction of a purely observational (in particular phenomenal) language, allegedly for the use of some science. Another is the equally impossible definition of all basic scientific concepts in observational (in particular phenomenal) terms. These chimaeric projects go beyond the previously mentioned one (i.e., philosophical criticism), as they presuppose the latter's criticism of scientific theories and empirical procedures involving nonobservational items. Like logical reconstruction in general, it demands an impressive logical and mathematical apparatus. But, for all its scientific odor, it is a harmless *jeu d'esprit* with no relation whatever to real science, which takes off when observation ceases to suffice. Even a young science like linguistics is soaked with constructs that have no observational partners, such as those of structure and meaning. Older sciences, like physics and chemistry, employ no observational concepts at all in their theories. Moreover, they conduct every experiment in the light of a bunch of theories.

The conclusion is obvious: an a priori philosophy of science, one that sacrifices science to a set of preconceived ideas about science, is inauthentic: at its best it is a waste of time, at its worst it is an active enemy of science. The least a philosophy of science should be is a posteriori.

2. PREFACE ANALYSIS

A philosophy of science may be said to consist in *preface analysis* if it focuses on whatever some famous scientists may have said about their own fields, usually in the least significant parts of their writings, namely prefaces, introductions, and the like.

Preface analysis employs the argument from authority: if the father of a certain theory or empirical procedure says that its brain child has such and such virtues, then it must be so. For example, if Einstein said at one time that special relativity was the outcome of an attempt to square physics with the philosophies of Mach and Poincaré, this must be true: he should know best. If Heisenberg once said that quantum mechanics was a result of an effort to banish unobservables from atomic physics, why doubt his word? Preface analysis is thus a variety of late scholasticism, i.e., commentary on secondary sources without independent thinking.

Preface analysis is often used to bolster a given philosophy without ever coming to grips with the real thing, namely science: it thus becomes a technique for implementing the policy—and sometimes the police—of an a priori philosophy of science. Although it often passes for philosophy of science, it does not deserve to be counted as such, for it is not an analysis of science—just as a commentary on a history of philosophy does not count as original work in the history of philosophy. On the other hand, preface analysis, when performed by someone who happens to understand what the prefacer is talking about, may constitute a subsidiary source for the history and psychology of science, to the extent that it throws some light on the background and some of the motivations of the worker concerned. Still, it has mainly an anecdotal value. When Heisenberg told me that he has always been motivated by philosophical problems and hypotheses, hardly by experimental results, I found this revealing and encouraging, but I do not regard this as part of the philosophy of quantum mechanics.

Preface analysis is hardly a contribution to the philosophy of science, for three main reasons. Firstly, because a preface is no substitute for what it prefaces: it may at most label and advertise the contents of the can. Secondly, because scientists are seldom philosophically equipped to spin a reasonable philosophy out of their own scientific work. Thirdly, because given one authoritative opinion on the philosophical implications of a chunk of science, there exists at least one other such view that contradicts it.

3. TEXTBOOK ANALYSIS

Some philosophers venture beyond the preface, plunging into the text itself. Unfortunately, more often than not the text happens to be either a popularization work or a freshman textbook. Occasionally, impressive analytic tools are brought to bear on such texts, which is as though historians thrived exclusively on secondary sources. In fact, even the best science textbook is a faint imitation of some aspects of science: it selects, condenses, simplifies, and systematizes a few results of a complex process, disregarding purposefully most of the problems that prompted it as well as all, or nearly all, the unsolved difficulties those results present.

Every scientist is aware of the limitations of textbooks: he knows the difference between a piece of work, an original memoir on it, and a more or less journalistic report on it. He knows that even the original paper is an outcome of editing—selecting, condensing, simplifying, and systematizing—so much so that only a badly written memoir could give sufficient indications to duplicate the original work. A concise and well-organized paper has to leave out not only psychical and social circumstances, such as motivation, bias, and budget, but many technical details (kinds of material, presuppositions, detailed proofs, etc.) as well. All these gaps are greatly widened in the textbook—the supreme source of science for many of the most conscientious philosophers.

It may be rejoined that not even original investigators can dispense with textbooks for, after all, when they wish to be briefed on a new or half-forgotten subject, they have to resort to advanced textbooks and survey articles. True, but such material usually does not cover up the difficulties but, to the contrary, it is supposed to point them out, for while one goal of a high-level expository work is to show what has been done, another goal of it is to point to what remains to be done. Besides, the investigator is aware of the distance between such material and original work—something the layman can hardly fathom. True, all elementary textbooks ought to be more honest and give at least a glimpse of the shortcomings of the work they popularize: they need not be so dogmatic as in fact most of them are. But this is just the point: a dogmatic presentation of science—the undogmatic activity par excellence—cannot convey the gist of scientific research and is therefore of little use for the philosopher who intends to take science apart.

Granted, textbook commentary is superior to preface commentary, for at least the textbook sketches, however poorly, some end results and special techniques of research. True, some general characteristics of science can be gleaned from textbooks, e.g., something about the form, content, and use of elementary law statements. Surely, unlike apriorism and prefacism, textbookism is a first step in the right direction, namely in the direction of meeting the object of analysis. Granted it is a *first* step: not the whole walk.

4. HISTORICO-PHILOSOPHICAL ANALYSIS

Preface commentary and textbook analysis can bear on present-day cases or they can handle cases taken from the remote past. Moreover, a relatively well understood episode in the history of science may prove more valuable for the philosopher than an ill-digested result of recent research. Even if the documentary evidence is fragmentary, it is possible and rewarding to attempt rational reconstructions of important scientific achievements in a more or less distant past, even if such a reconstruction involves a lot of guesswork. Such a piece of work is a legitimate part of the philosophy of science, to be distinguished from a historical analysis sprinkled with philosophical remarks. For a historico-philosophical analysis will emphasize the historical status of a given item of scientific research: the inference, interpretation, and valuation problems faced by the scientist will tend to remain in the dark.

The great virtue of the historico-philosophical approach to science is, of course, that it brings out the input (background knowledge, paradigm, problem, and technique) and the output (result and new problem) in single cases of real science. But the inner mechanism of research, in particular the interplay of hypotheses and data, of background knowledge and new conjecture—all this will tend to remain out of focus. Yet this is precisely what the philosopher is interested in, in addition to building general theories concerning key metascientific concepts, such as those of scientific theory and measurement.

To the philosopher, the history of science is one of the two sources of raw material, the other being living science. This material, if analyzed with the proper tools, will serve various ends: it will pose problems, it will suggest metascientific hypotheses (e.g., on the role of mathematics in science), and it will help check such hypotheses. But, if the same material is subjected to a purely historical analysis, then little if any philosophical juice may be extracted. To put it in a slightly paradoxical way: in order to derive a philosophy of science from the history of science, the latter must be approached with a definite (but provisional) philosophy of science.

It is unfortunate, though psychologically unavoidable, that the historically oriented philosopher of science should care so little for the technical aspects of the foundations and philosophy of science, to the point of despising them as sheer pieces of formalism or even scholasticism. By adopting this romantic attitude he denies himself the pleasure of imagining general theories within which his episodes would make sense. However, we can afford to pass over such a romantic denial of theory, for the historically oriented philosopher of science can do much to enrich our stock of examples and thus correct the overwhelming tendency to dispense with real case histories.

In short, historico-philosophical analyses are complementary to, rather than a substitute for, cross-section analyses and, moreover, they may supply the very object of analysis. A thorough understanding of any product of intellectual culture is the outcome of a variety of approaches: historical, psychological, sociological, philosophical, and occasionally ethical and aesthetic. The understanding of science is the business of the joint sciences of science, of which philosophy is just the one in which we happen to be interested. Surely, a nonphilosophical history of science is dull and uninstructive reading, just as a philosophy of science sounds hollow unless it has a foothold on present science and on the history of science. But to grant the need for a closer cooperation between the historian and the philosopher of science does not entail letting a historico-philosophical examination of science pass for a philosophy of science.

5. ISOLATED ITEM ANALYSIS

A philosophy of science may be said to be concerned with isolated items of genuine research if it picks them in isolation from their context, both theoretical and experimental.

Voltaire and Kant performed isolated item analyses of Newtonian physics: they selected the law of gravity without, however, plugging it into the equation of motion and therefore missing the whole of Newtonian mechanics. During the nineteenth century whole philosophies were built on "the" law of conservation of energy without caring for its theoretical context, which is actually multiple. Something similar happened later with the second law of thermodynamics and, more recently, with Shannon's formula for the amount of information. Whole philosophies of space and time are sometimes built on (pseudo)measurements of distances and durations, in isolation from the theories that make such measurements possible. Whole philosophies of microphysics revolve exclusively on the alleged wave-particles duality or, even more especially, on Heisenberg's "uncertainty" formulas.

Surely, a particular item of science is a scientific item, provided its relations to the whole are brought out rather than cut off. Because, if these relations are ignored, then that item ceases to be itself, to become a tiny straw that can be manipulated at will, quite apart from its real significance in the whole. The case of $E = mc^2$ is as glaring as the cases recalled above: since the little formula can be written out with a few strokes, and since its component symbols can be given ordinary-sounding names ('energy', etc.) anyone can feel he masters it. (Recall essay 3.) However, this is a delusion, as shown by the multiplicity of interpretations. Only by taking the whole theory into account does one effectively reduce the number of possible interpretations to the point of hoping to be able to give the correct one. By itself, a formula has no interpretation whatever—whence it can be assigned any number of arbitrary (extrasystematic) interpretations. Mind, it is not that the whole (in this case a theory) has a meaning that transcends in a mysterious way the meaning of its components, but rather that the meanings of the basic symbols of a body of knowledge determine one another.

Beside this semantic argument against isolated item analysis, there is the logical one: unless a context is provided, one cannot know whether the given concept is primitive or definable, and whether the statement concerned is postulated or derivable—and this is surely essential if one is to evaluate the item as fundamental or as derivative. A famous case in point is the attempt to define "time" in terms of other concepts, e.g., "entropy increase", via law statements that could not even be written out unless the time concept had been available to begin with, and which anyhow are synthetic, not analytic, statements.

Similar criticisms may be raised against the philosophical examination of empirical procedures, such as measurement, unrelated to the substantive theories (e.g., mechanics) that guide the design and reading of measuring instruments. Likewise, a philosophy of measurement based on a fictitious theory of measurement, such as von Neumann's, i.e., one that bears on no empirical procedure, even though it claims that it does, is footless. Here again, isolation ensues in artificiality, if not in nonsense. (See essay 18.)

Granted, the analysis of single items of real science is superior to textbook commentaries, in that it does come to grips with the beast. But one does not overpower the lion by fighting its left upper canine alone. Surely, the systematic neglect of every single item will fail to give us an adequate picture of the whole, but only an analysis of the item concerned in its proper context can tell us something about both that item and the whole of which it is a part. But this deserves another section.

6. SYSTEMATIC ANALYSIS

In contrast to isolated item analysis, *systematic analysis* bears on whole units of scientific research, such as theories and complete cycles of experimental investigation. In this approach, a particular item is a part of the whole, so that its status and significance are clear, and the danger of misinterpretation is decreased.

Systematic analysis may be performed either in vivo or in vitro. The former takes a live analysandum, such as is found in the scientific literature, and dissects it with the best available tools, particularly logic, mathematics, and formal semantics, without, however, neglecting the epistemological and metaphysical associations. Systematic analysis is thus capable of shedding considerable light on science, if only because it handles problems emerging from real research, and because it does not sever their connections before examining them. However, to understand how a mechanism is built and how it works, there is nothing like dismounting and then reassembling it, preferably in an improved way. If systematic analysis in vivo is similar to observation and to description, systematic analysis in vitro is similar to experiment and to explanation.

If bearing on a theory, systematic analysis will be preceded by a rational reconstruction, or axiomatization, of the theory: it will start by classing the concepts of the theory into defining and defined, and by ordering the statements of the theory by means of the relation of deducibility. By so doing the philosopher not only subjects his philosophical hypotheses to the supreme test, but becomes a foundations research worker. This puts him in the best possible position for passing judgment on the technical (both substantive and formal) merits and shortcomings of the theory, as well as for disclosing its philosophical presuppositions and pointing to its possible philosophical significance. In any case, once the theory has been so reconstructed, any item of it— concept, definition, law statement, subsidiary hypothesis, or what not— can be hoped to be properly examined. Thus, any correct analysis of the role of the velocity of light in relativity theory requires the prior axiomatization of this theory, and a correct analysis of the meaning of Heisenberg's inequalities calls for a previous axiomatization of quantum mechanics. To proceed otherwise, i.e., along the indefinite lines of isolated item analysis, is as wasteful as trying to guess the ecology of a plant from a chromosome analysis of it alone.

Empirical procedures can be approached in a similar way, though they are seldom, if ever, so handled. Here we have to start by digging out the background knowledge, both theoretical and experimental, substantive and methodological; we must then proceed to describe and analyze the procedure itself, supplying the hypotheses and theories actually employed, if only fragmentarily, in the implementation of the design and execution of the experiment or observation. Most of these ingredients are taken for granted by the practitioners of the procedure and are therefore to be critically scrutinized by the philosopher of science. Such a scrutiny is bound to puncture a number of philosophical globes, such as the myth that pointer readings are ultimate data in no need of analysis.

In sum, systematic analysis, whether applied to theories or to cycles of empirical research, is the most promising of all the avenues of approach to science, for it gives us both the whole and its parts in their actual (or rather conjectured) relationships. It is superior to isolated item analysis, in turn better than historical analysis, which defeats textbook analysis, which is an improvement on preface analysis, which is finally better than the forcing into an a priori framework.

Heaving reached this conclusion we may now look back to the performance of the weaker approaches, which are still favorites among philosophers of science. First to apriorism.

7. A NONPROBLEM: GRUE-BLEEN

If real science is not made the object of analysis, no amount of sophisticated analytical means will produce a genuine result in the philosophy of science. I am not saying that an acquaintance with science is sufficient to discover genuine problems and propose sensible solutions to them: the naive homespun philosophies of most scientists bear witness to the contrary. All I am claiming is that such an acquaintance is necessary, and the failure to satisfy this condition is bound to result in problems irrelevant to science and in solutions at variance with science. The blind cannot guide the perplexed.

A paradigm of a piece of epistemology out of touch with science and thoroughly dominated by a certain philosophical tradition—in this case phenomenalism—is the so-called grue riddle. For Hume and his followers, there are no laws of nature but rather empirical regularities, and there is nothing necessary about the latter: from the repeated observation of an exceptionless joint occurrence of the properties E and G we cannot infer that they are intrinsically related. Moreover—here is the snag—since experience is the sole possible source of knowledge, it is impossible to go beyond that constant conjunction. Thus, we have observed countless times that emeralds are green, but this does not allow us to conclude that they cannot help being green: the *EG* association might be discontinued in the future. For example, emeralds might turn blue overnight, either spontaneously or as result of an unforeseeable cause.

Therefore, the probability that anything that is an emerald be also green, is not necessarily equal to unity: we should make room for the possibility of observing emeralds with different colors. In short, all we can state is that the probability of *G* given *E* seems, so far, to be larger than the probability of not-*G* given *E*: but the two probabilities are unknown, and therefore we cannot forecast with any certainty whether the next emerald we look at will be green or not. Thus far, the Humean philosopher. How about the scientist: does she face the same riddle?

During the past half century physics has learned to explain colors and other phenomenal properties in terms of deep-seated physical and chemical properties. In particular, the color of emeralds has proved not to be an *accidens* or removable label, but to be related in a physically necessary, i.e., lawful, way to their chemical composition. Likewise with most of the other macroproperties of emeralds, in particular their thermal and electrical conductivities. Furthermore, the ordinary knowledge concept of an emerald has been reduced to the corresponding scientific concept by way of a definition: we call a body an emerald just in case that body is a beryl with either of two definite chemical compositions, *C1* or *C2*. Otherwise, we shall give it a different name. In particular, if a gem looks like an emerald but fails to have the right chemical composition, we are allowed to call it a *simil-emerald*; calling it an emerald would be a fraud punishable by law.

In other words, as a result of a scientific development contrary to the grain of phenomenalism, we have adopted the following

DEFINITION. For every x, x is an emerald $=_{df}$ x has the chemical composition *C1* or the chemical composition *C2*. $\qquad$ (1)

In addition to this linguistic convention that matches ordinary knowledge with scientific knowledge, we have the following

LAW STATEMENT. For every x, if x has the composition *C1* or the composition *C2* then x is green under white light (or rather, the intensity of the reflected light has a pronounced maximum at the frequencies whose physiological correlate is green). $\qquad$ (2)

Now, by virtue of the previous definition (1), the chemical predicate "*C1* or *C2*" can be eliminated in favor of E in the preceding law statement, to yield the rewrite:

For every x, if x is an emerald, then x is green under white light.   (2')

If we now call $\mathcal{E}$ and $\mathcal{G}$ the extensions of the predicates E and G respectively, the preceding law statement may be rephrased simply thus: $\mathcal{E}$ is included in $\mathcal{G}$, or $\mathcal{E} \subset \mathcal{G}$ for short. Consequently, the intersection of the two sets yields the smaller of them: $\mathcal{G} \cap \mathcal{E} = \mathcal{E}$. Therefore, the conditional probability of $\mathcal{G}$ given $\mathcal{E}$ is unity:

$$Pr(\mathcal{G} \mid \mathcal{E}) =_{df} Pr(\mathcal{G} \cap \mathcal{E})/Pr(\mathcal{E}) = Pr(\mathcal{E})/Pr(\mathcal{E}) = 1, \tag{3}$$

provided that $Pr(\mathcal{E}) \neq 0$. Hence, the probability of not-$\mathcal{G}$ given $\mathcal{E}$ is strictly zero. In particular, grue (or bleen) has no chance. (Strictly speaking, since the predicates "is an emerald" and "is green" are not associated at random, the preceding formula should be read as follows: The probability of picking at random a green gem from an arbitrarily large heap of emeralds, at any time, equals unity. See essay 15.)

Note the following differences between the Humean and the scientific view of the question. Firstly, the former ignores the typically scientific predicate "$C1$ or $C2$". Moreover, it cannot accept it, for it is nonphenomenal. Accordingly, it cannot accept definition (1) either. Secondly, on the Humean view the odds for the anomalous association of E with not-G are not vanishing, but they are as unknown and unknowable as the odds for the observed association of E and G. Instead of a law statement we have an *ignorabimus*; instead of a scientific prediction we are left with a guess based only on numerous yet fallible observations unrelated to the whole body of science.

8. A MISCARRIED PROBLEM: STATISTICAL EXPLANATION

The good intention to deal with a genuine item of science is not enough, particularly if the real problem is missed. Thus, the analysis of explanations performed with the help of statistical laws is an authentic problem in the philosophy of science, and also a nearly virgin one, because the logically prior problem of elucidating the concept of a statistical law has not been solved in a satisfactory manner—mainly because hardly any actual scientific examples are ever examined.

It is widely believed that all statistical laws are inductions gotten by observing random samples: hence, that statistical laws are reasonable generalizations of empirical data, e.g., that N percent of the population of a given area are illiterate. This statement is not spatiotemporally universal, as it concerns a bounded spatiotemporal region. Moreover, it is not referentially universal either, as it does not apply to every one of the units into which the total sampled region can be divided: indeed the figure is an average that says nothing about any particular region (unless, of course, N is either 0 or 100). The statement concerns a collective, holistic, nonhereditary property of an ensemble, not any and every member of it. No wonder then, that it has hardly an explanatory power in reference to the constituents of the whole or statistical ensemble. If statistical *laws* were just empirical statistical generalizations of the type of demographic and economic data, then there would be no such thing as a statistical explanation. (To speak of statistical syllogisms and the like, in reference to the passage from populations to individuals, is just an abuse of language.) For this reason, some philosophers refuse to speak of statistical law and, a fortiori, of statistical explanation.

To be sure, some progress has been made in this field in more recent writings: it is now sometimes recognized that a statistical law statement is a proposition in which the probability concept occurs in an essential (or nonvacuous) way. But this is not enough, because the term 'probability' may be misunderstood to mean "frequency". What is characteristic of a statistical law is that (a) it belongs to a substantive theory (not just to a description of empirical material), and (b) that it contains some concept peculiar to probability theory, such as the ones of probability and distribution—not, however, those of frequency and average, which are shared by the descriptions of statistical material. The reason is simple enough: only by hypothesizing definite probability distributions (or a whole class of them) can one derive (hence explain and predict) observable statistical features, such as histograms (frequency distributions), averages, and average dispersions around averages—and even nonstochastic ("deterministic") laws. (See also essay 15.)

A statistical "law" gotten by sampling is not explanatory, as we saw before: it is not a law proper. Therefore, it does not count as an explanans premise but can at most occur as an explanandum: it is something to be explained. In particular, a statistical generalization of the form "f percent of the A's are B's" is something to be explained in terms of stochastic laws rather than a premise in the explanation of a particular fact, e.g., that a given individual of the kind A happens to be a B. The chief explanans in a statistical explanation is, as we saw before, a distribution function (such as the Maxwell-Boltzmann velocity distribution) or a stochastic model (such as the kinetic model of gases) formally described by that distribution. If the stochastic theory within which the statistical explanation takes place is of the phenomenological or black box type, it will supply just the distribution function, but no random model, e.g., a shuffling or urn mechanism. This is how generalized statistical mechanics can be developed, namely by trying out several partition functions restricted only by the condition that they yield the desired thermodynamic quantities. But a deeper theory will contain a stochastic model, e.g., a Markovian one, that will delve into some details of the process and thus will be able to supply a deeper explanation of the statistical regularities and irregularities (fluctuations) encountered in experiment.

Unfortunately, one misses these crucial points in the standard philosophical accounts of statistical explanation. This seems to be due to two main reasons. One is just *ignoratio elenchi*, sheer lack of acquaintance with typical stochastic theories of science, such as kinetic theory, quantum mechanics, genetic theory, and stochastic learning theory. As a consequence, those accounts are not even clear about what a statistical explanation is supposed to explain. Moreover, they typically mix up three distinct items: probability theory (a chapter of pure mathematics), mathematical statistics (an application of probability theory to definite stochastic models and statistical inference), and substantive statistical theories, such as the theory of Brownian motion.

A second reason for the failure to give an accurate account of statistical explanation is the compulsion to press everything new into a prefabricated philosophical mold. In particular, if our philosopher happens to be committed to empiricism, he will not speak of probability but rather of statistical probability, a mongrel concept recognized in no branch of mathematics or science, and one designed to filling the gap between theoretical probability and observed frequency, thus ignoring the important philosophical problem of the relation between the two. In addition, if our philosopher happens to be a subjectivist, he will try to couch every statistical statement, even one concerning objective random processes, in terms of subjective probability or degree of rational belief, thus hoping to obliterate the gap between objective facts and our expectations about them.

In conclusion, the choice of a genuine problem, though essential, is insufficient to produce adequate results, if the problem is handled with inadequate information and within a preconceived philosophical framework.

9. TOWARDS THE RIGHT METHOD

It is far easier to criticize erroneous procedures than to describe correct ones. Moreover, it is well known that even the fullest description of a correct method is no guarantee of success, since at least three crucial steps in original research are not rule-directed: the choice of problems, the invention of conjectures, and the evaluation of solutions. However, a few guidelines can be helpful and, in any case, they are worthwhile being discussed.

In philosophy, as in science, the first task is to get hold of a problem—an interesting and promising problem, not just an exercise that may instruct without adding to the body of common knowledge. Metascientific problems have two origins: one is a critical reflection about science itself, the other is a critical reflection about some philosopher, in particular oneself. These two sources may join: one may analyze a trait of science in the light of X's philosophy, or one may check X's philosophy against some trait of actual science. In original research other people's opinions act only as starters: the rest has to be supplied by oneself—otherwise one does metaphilosophy of science (e.g., this chapter) rather than philosophy of science.

The problem must concern real science and it must be a philosophical problem if it is to be a problem in the philosophy of science. Thus, the question 'How does one identify the referent of a scientific construct (concept or statement)?' complies with the two requirements: it is a semantic problem that comes up in real science whenever there are doubts about factual interpretation. Moreover, it is a problem that has received no satisfactory answer: there is no adequate semantics of scientific theories (unlike the semantics of mathematical theories) and, as a consequence, no adequate criteria for spotting the referents are generally accepted.

The next step will be to pick some typical examples of actual science, e.g., some typical scientific concepts (e.g., "mass", but not "red") and scientific statements (e.g., Mendel's laws, but not "Ravens are black"). Here we must be careful not to be misled by popular exposés or by previous philosophical treatments, too often inspired in such popular exposés. Thus, in the case of the question of the referent of a scientific construct, the main pitfall to be avoided is the confusion between the semantic and the pragmatic interpretations of the given symbol. A predicate letter 'P' occurring in a theoretical formula may be read as representing an objective property of a concrete system (semantic interpretation) and, sometimes, as representing a trait of certain operations, such as a measurement (pragmatic interpretation). Empiricists and, in general, subjectivists are bound to deny this distinction, which is, in contrast, essential to realism. Our problem is to find criteria for deciding which of the two interpretations is admissible in any given case; this will solve eo ipso the problem of ascertaining whether any of them is redundant. To solve this problem, we must look at original scientific memoirs, the least distorted by didactic concerns and philosophical commitments. But this won't be enough, for scientific writings, even if exacting in matters of form and procedure, are notoriously slack when it comes to pointing to their referent, which is usually either taken for granted or specified in a few imprecise extrasystematic remarks. We may be forced to do a bit of rational reconstruction, supplying the semantic assumptions and rules that suffice for a reasonably unambiguous specification of meaning. That is, we may find that systematic analysis in vivo is insufficient: that it has to be supplemented by the more active systematic analysis in vitro. But let us not anticipate: we are at the data-gathering stage.

Once we have built a stock of typical examples of a certain kind, we must turn them around in the hope of hitting upon a general conjecture that will fit them all. In other words, we shall proceed inductively even should our metascience allow induction only a modest place in science. (By contrast, the typical inductivist proceeds aprioristically, not inductively, in philosophy.) If we are familiar with the subject and happen to be in the right mood, the conjectures will flock quicker than we are able to formulate them explicitly and clearly. Before we can assay and work out these intuitions, we must state them, or rather some of them, in an explicit way. Actually, we are going to make a preliminary selection before the explicit formulation process is over: we feel impelled to formalize only those conjectures which intuitively look promising.

The third step is that of selecting a subset of metascientific hypotheses that look viable on the face of them, i.e., prior to a detailed investigation. We shall weed out those which are not quite clear, or which are inconsistent with firmer chunks of knowledge, or which are ad hoc, or which seem too superficial, or which have failed before. Our inductive policies should be as objective (or rather as little biased) as in science: we should not let our prejudices decide the issue, even though we cannot help our pet prejudices from suggesting possible solutions. It is not a question of checking the philosopher's cap at the door, for this is impossible, but a question of choosing between the metascientific hypotheses which best fit a given philosophy and those which best fit actual science.

Once an acceptable metascientific hypothesis has been chosen (adopted temporarily) we must proceed to sharpen it, improve its form, and enhance its clarity. Because the best way to polish a construct and bring out its meaning is to embed it into a full-fledged theory (hypothetico-deductive system), it should be our desideratum to build and work out metascientific theories—theories about the meaning and truth of scientific constructs, about the testability, explanatory power, and predictive performance of scientific theories, and so on.

However, just as in science, in metascience we do not seek theories for their own sake: we want metascientific theories that will account for scientific research, that will be true of it regardless of their philosophical loyalties. Hence, as soon as we have thought out a theory of that kind, we have got to check it for adequacy: we must hunt for further examples and counterexamples. In this process, our metascientific theory may have to be modified or even given up—just as if it were a scientific theory. In both cases we must require *adaequatio ad rem*, the *res* being a chunk of the world in the case of science, and a chunk of science in the case of metascience. Otherwise, our theory will be a ghost theory: a piece of science-fiction philosophy.

The critical scrutiny of a metascientific theory should include not only an examination of adequacy and of compatibility with other theories of its kind, but if possible also some of the standard metamathematical tests, chiefly a consistency test. All this may be harder to perform than the building of the theory. Fortunately, the task is greatly facilitated by first axiomatizing the theory, for in this case the troublemaking components may be spotted with more ease. Also, this is apt to cut our inevitable losses: usually some basic components of a nontrivial theory can be salvaged if they have been formulated in a clear-cut way. This, if none other, is the advantage of building with bricks over building with mud.

We started with particular cases which we tried to account for by hypotheses which we attempted to cohere into theories which we tried to assemble in the best possible way, and which we checked against fresh examples and possible counterexamples. But surely there is nothing new in this: it is the scientific method applied to the philosophical study of science.

10. THE TRAINING OF PHILOSOPHERS OF SCIENCE

Let us apply the foregoing to a practical problem: how best to train philosophers of science.

Up till a few years ago philosophers of science were mostly self-taught; they had been either scientists intrigued by some philosophical problems they encountered in their own work, or philosophers who wished to understand the means and goals of science, or even to learn from it the right way to philosophize. In either case the formation of philosophers of science was spontaneous and erratic, hence rather free from indoctrination, but also spotty and inefficient.

In the past few years a number of attempts have been made to correct this situation, by establishing special programs in the philosophy of science, mostly in the United States. Unfortunately, some of the existing programs do not regard a scientific training as compulsory—as if one could become a philosopher of science just by taking a number of courses in the philosophy of science. This policy ignores all we have been saying in the preceding pages, which comes to this: philosophers of real science are amphibious animals, at home both in scientific waters and in the philosophical air.

The training of such an amphibian should include work in—not just a study of—some science, for otherwise the student will not advance beyond a secondhand report on science, namely the one conveyed by textbooks. This does not exclude the need for a parallel training in philosophy. Moreover, an exclusive concentration on science for a lengthy period is likely to extinguish whatever philosophical flame there may have been in the beginning—not because science is unphilosophical but because the standard training of scientists is antiphilosophical, i.e., addressed to the quick mastering of techniques and committed to naive and obsolete philosophical tenets.

A judicious combination of the two kinds of food—science and philosophy—should constitute the diet of our amphibian apprentice. The precise dosing should be left to the student herself: after all, if she has chosen to displease both narrow-minded scientists and narrow-minded philosophers, she may be credited with seriousness of purpose. In other words, there is no ideal program for training philosophers of science: the only reasonable program is the one tailored to the skills, likes, and needs of the individual student. This is not a plea for the irresponsible choice and ordering of subjects: it is a plea for flexibility and against authoritarianism as expressed in uniform curricula and dull learning methods devised for the mass production of mediocre scholars. A responsible student, stimulated and helped by inspiring teachers, will understand that, in order to become a philosopher of science, she must get herself a training in some science (and thereby in mathematics) and must get hold of a number of philosophical tools. There should be no need to insist that a freely chosen discipline is far more effective than an externally imposed one.

The general conclusion is pretty obvious: if we wish to breed philosophers engaging in a philosophical scrutiny of real science, we had better face the fact that we must catch (1) young scientists with a philosophical motivation and willing to learn to handle certain philosophical tools, and (2) young philosophers with a scientific bent and willing to learn some science at firsthand—preferably some new science, as it can be mastered faster and will pose a carload of new philosophical problems. A scientific turn of mind and a feel for genuine scientific research are best gotten by doing some science, just as a poetic frame of mind and competence to discuss poetry are best cultivated by writing poetry.

Such a drastic, yet only commonsensical, reorientation of the training of philosophers of science might prove of consequence for the whole of philosophy, for it should quicken the constitution and diffusion of a scientific philosophy—one done in the spirit of science rather than along the literary and historical tradition. As Bertrand Russell (1914, p. 246) wrote over half a century ago: "The one and only condition, I believe, which is necessary in order to secure for philosophy in the near future an achievement surpassing all that has hitherto been accomplished by philosophers, is the creation of a school of men with scientific training and philosophical interests, unhampered by the traditions of the past, and not misled by the literary methods of those who copy the ancients in all except their merits".

11. AN ALARM SOUND

The philosophy of science is comparatively a newcomer to philosophy—at least by comparison with metaphysics, logic, epistemology, and ethics. As every other newcomer, it is still groping in search for its "identity": its proper subject matter, its method, its goal, and even its own standards of scholarship. Yet all this should be pretty obvious: The object should be real science (both natural and social), and the method should be essentially the same as the method of science, since in either case one tries to know something given. The goal should be to dismount and then to reassemble the mechanism of science in order to expose its structure, content, and functions. The standards of scholarship should be essentially those prevailing in more mature philosophical disciplines, such as logic and the history of philosophy: genuine problems, whether newly posed or handled in a novel way; clarity, cogency, use of the best available tools of analysis and synthesis, and concern for the adequacy of the results.

The consistent adoption of such standards should make it easier to avoid the five main pitfalls threatening the philosopher of science nowadays: amateurism, fashionableness, artificiality, hollow exactness, and scholasticism. By *amateurism* I mean the handling of philosophical problems without the technical apparatus of philosophy, chiefly formal logic: it is the typical shortcoming of scientists and historians of science interested in philosophy. By *fashionableness* I mean the tendency to work on whatever seems to be à la mode while neglecting subjects, old and new, that may be more rewarding. At one time many-valued logics and their alleged applications were hoped to constitute a breakthrough in the philosophy of science. Then information theory was expected to answer all sorts of questions, both in science and in philosophy. At all times during our century, probability has been issuing promissory notes that have seldom been cashed. This concern for fashion is a sign of immaturity and of a lack of a program. We should know by now that there is no master key—that not even logic will unlock philosophical treasures unless we first locate these treasures and then use the tool adroitly. To busy oneself only with fashionable problems in philosophy is worse than handling only classical problems, for at least the latter are entrenched in philosophy and probably none of them will ever receive a final solution, unless shown to be a pseudoproblem. If our concern is truth rather than popularity, then it is better to be "out" and right than "in" and wrong, and in any case it is more fun to wrestle with one's own genuine problems than to wear whatever happens to be in fashion.

The third danger mentioned above is the one of *artificiality*, in the sense of remoteness from real science. So far, inductive logic has been a paradigm of artificiality, as it is hardly concerned with studying the actual inductive reasonings made in forming, testing, and evaluating scientific hypotheses and theories. As a result, inductive logic has hitherto neither helped scientists reason better nor assisted philosophers in understanding scientific inference. Which is a pity, for the problem is genuine.

An allied danger is the one of *hollow exactness*: it affects all metascientific theories which, though exact, have no relevance to scientific research. Such theories may well be meticulous exercises in applied logic or applied mathematics, but they will be ghostly if they have no application to actual science. Most applications of probability theory to the elucidation of metascientific concepts are of this kind, if only because they all presuppose that there is a method for assigning numerical values to the probabilities of the propositions concerned (chiefly hypotheses and data). But this is not the case: unless a definite model (e.g., an urn model) can be set up (as in the case of classical genetics), one is at a loss. Such models do not concern propositions but possible *facts*, such as the chance encounter of two given atoms. Still, people will be found to speak with a straight face of the probability of scientific hypotheses such as Newton's laws, not just in the informal way peculiar to classical British empiricism, but pretending that they can "somehow" be assigned definite numerical values. Our probability fan will moreover be undeterred by the fact that his ghost theories have not the slightest contact with mathematical statistics, which is the science actually concerned with the problem of weighing inductive support—not, however, with evaluating the probability of Newton's laws. He feels sheltered in his calculi and confident that, since they are exact, they must also be relevant. This is a delusion. But at least this hollow clarity is preferable to any solid obscurity: it makes people fit for sterner enterprises. (See also essay 15.)

The fifth and last danger mentioned above is the one of *scholasticism*, by which I mean the exclusive concern with other people's views on science. Surely, to discuss them is part of the job, but our main task is to handle authentic metascientific problems regardless of their origin. Let the historian of the philosophy of science specialize in the contributions already made to the philosophy of science; let the philosopher of science make some contribution of his own.

The upshot is this. If we want the philosophy of science to be an authentic discipline with a definite "identity", and if we want it to be useful to both philosophy and science, then we must start by acknowledging that someone is a philosopher of science just in case he is a philosopher of science—not a philosopher of ordinary knowledge, nor a general epistemologist, nor someone committed to a fixed philosophical doctrine, nor a preface commentator, nor a textbook analyst, but somebody who philosophizes on chunks of real science, who analyzes them with the best available tools and tries to conceive general views about them, doing his best to check them for adequacy.

9

INDUCTION IN SCIENCE
(1963)

■

Laws, the vertebrae of science, are sometimes believed to be established by induction (empiricist tradition, as represented by Bacon), and at other times to be the product of reason and free imagination (rationalist tradition, as exemplified by Einstein). The first belief is frequent among field and laboratory workers, the second among theoreticians. When the *establishment* of a law statement is mentioned, either of two entirely different inferential procedures may be meant: the *inception* or introduction of the statement, or its *test*. In either case it is accepted that inferences are involved, rather than direct and immediate apprehensions. The question is whether the inferences are inductive, deductive, or perhaps neither exclusively inductive nor exclusively deductive, but a combination of the two with the addition of analogy and of some kind of invention or creation.

In this paper we shall investigate the question of whether scientific inference is predominantly inductive, as claimed by inductivist meta-science (e.g., Keynes 1921; Reichenbach 1949; Carnap 1950; Jeffreys

1957; von Wright 1957), or predominantly deductive, as maintained by deductivism (e.g., Duhem 1914; Wisdom 1952; Popper 1959)—or, finally, whether it actually goes along a third way of its own. The discussion will be confined to factual statements, usually called 'empirical sentences', without thereby denying the great heuristic value that case examination also has in mathematical invention and problem solving (Pólya 1954).

1. INDUCTION PROPER

Before approaching the problem, let us clear the ground. By *induction stricto sensu* I shall understand the type of nondemonstrative reasoning consisting in *obtaining or validating general propositions on the basis of the examination of cases*. Or, as Whewell put it long ago (1858), "by *Induction* is to be understood that process of collecting general truths from the examination of particular facts". This linguistic convention makes no appeal to epistemological categories such as "new knowledge", which are often used in the characterization of inductive inference, although the enlargement of knowledge is the purpose of both inductive and deductive inference.

The proposed equation of induction and generalization on the basis of case examination leaves the following kinds of inference *out* of the domain of inductive inference: (1) *analogy*, which is a certain reasoning from particular to particular, or from general to general, and which probably underlies inductive inference; (2) *generalization* involving the introduction of *new* concepts, i.e., of concepts absent in the evidential basis; (3) the so-called *induction by elimination*, which is nothing but the refutation of hypotheses found unfit because their observable consequences, derived by deduction, do not match with the empirical evidence at hand; (4) scientific *prediction*, which is clearly deductive, since it consists in the derivation of singular or existential propositions from the conjunction of law statements and specific information; (5) *interpolation* in the strict sense (not, however, curve fitting), which is deductive as well, since it amounts to specification; (6) *reduction*, or assertion of the antecedent of a conditional on the ground of the repeated verification of the consequent.

With the above definition of induction in mind, let us inquire into the role of induction in the formation and testing of the hypotheses that are dignified with the name of *laws of nature* or *of culture*.

2. INDUCTION IN THE FRAMING OF HYPOTHESES

The premises of induction may be singular or general. Let us distinguish the two cases by calling *first-degree induction* the inference leading from the examination of observed instances to general statements of the lowest level (e.g., "All humans are mortal"), and *second-degree induction* the inference consisting in the widening of such empirical generalizations (leading, e.g., from such statements as "All humans are mortal", "All lobsters are mortal", "All snakes are mortal" to "All metazoans are mortal"). First-degree induction starts from singular propositions, whereas second-degree induction is the generalization of generalizations.

Empirical generalizations of the type of "Owls eat mice" are often reached by first-degree induction. Necessary, though not sufficient, conditions for performing a first-degree induction are: (a) the facts referred to by the singular propositions that are to be generalized must have been observed, must be actual facts, never merely possible ones like the burning of this book or the establishment of a democratic government in Argentina; (b) (the referents of) the predicates contained in the generalization must be observable *stricto sensu*, such as predicates designating the color and size of perceptible bodies. Hence, the "observables" of atomic theory, such as the variables representing the instantaneous position or angular momentum of an electron, will not do for this purpose, since they are actually theoretical predicates (constructs).

Condition (a) excludes from the range of induction all inventions, and countless elementary generalizations, such as those involving dispositions or potential properties. Condition (b) excludes from the domain of induction all the more important scientific hypotheses: those which have been called *transcendent* (Kneale 1949) or *noninstantial* (Wisdom 1952), because they contain nonobservable, or theoretical, predicates, such as "attraction", "energy", "stable", "adaptation", or "mental". Transcendent hypotheses, i.e., assumptions going beyond experience, are most important in science because, far from merely enabling us to colligate or summarize empirical data, they enter into the explanation of data.

The hypothesis "Copper is a good conductor" is a second-degree inductive generalization. It contains the class terms 'copper' and 'conductor' (a dispositional term). Its generalization "All metals are good conductors" is, a fortiori, another second-degree induction: it refers not only to the class of metals known at the moment it was framed, but to the conceptually open class of metals known and knowable. We do not accept the latter generalization just because of its inductive support, weighty as it is, but also—perhaps mainly—because the theoretical study of the crystal structure of metals and the electron gas inside them shows us that the property denoted by the predicate "metal" or, if pre-ferred, "solid" is functionally associated with the property of being a con-ductor. This association, which transcends the Humean juxtaposition of properties, is expressed in law statements belonging to the theory of solid state. We accept the generalization with some confidence, because we have succeeded in understanding it by subsuming it under a theory. Similarly, we know since Harvey that "There are no heartless verte-brates" is true, not because this statement has been found and verified inductively, but because we understand the function of the heart in the maintenance of life.

Compare the above examples with the low-level generalization "All ravens are black", the stock-in-trade example of inductivists. Ornithology has not yet accounted for the constant conjunction of the two properties occurring in this first-degree induction. The day animal physiology hits upon an explanation of it, we shall presumably be told something like this: "All birds having the biological properties $P, Q, R, \ldots$ are black". And then some ornithologist may inquire whether ravens do possess the properties $P, Q, R, \ldots$, in order to ascertain whether the old generalization fits in the new systematic body of knowledge.

In sum, enumerative induction does play a role in the framing of general hypotheses, though certainly not as big a role as the one imag-ined by inductivism. Induction, important as it is in daily life and in the more backward stages of empirical science, has not led to finding a single important scientific law, incapable as it is of creating new and transem-pirical (transcendent) concepts, which are typical of theoretical science. In other words: induction may lead to framing *low-level, pretheoretical, ad hoc*, and *ex post facto* general hypotheses; the introduction of com-prehensive and deep hypotheses requires a leap beyond induction.

3. INDUCTION IN THE TEST OF HYPOTHESES

Scientific hypotheses are empirically tested by seeking *both* positive instances (according to the inductivist injunction) and unfavorable ones (deductivist rule). In other words, the empirical test of hypotheses includes both confirmations and unsuccessful attempts at refutation. But only first-degree inductive generalizations *have* instances; hence they are the only ones that can be directly checked against empirical evi-dence. Statements expressing empirical evidence, i.e., basic statements, do not contain theoretical predicates such as "mass", "recessive char-acter", or "population pressure". Hence, case examination by itself is irrelevant both to the framing and to the testing of transcendent hypotheses.

However, we do perform inductive inferences when stating plausible "conclusions", i.e., guesses, from the examination of observed consequences of our theories. Granted, we cannot examine instances of transcendent hypotheses such as "The intensity of the electric current is proportional to the potential difference", because they are noninstantial. But hypotheses of this kind, which are the most numerous in the advanced chapters of science, do have observable consequents when conjoined with lower-level hypotheses containing both unobservable and observable predicates, such as "Electric currents deflect the magnetic needle". (The deflections can literally be observed, even though electricity and magnetism are unobservable.) And, if we wish to validate transcendent hypotheses, we must examine instances of such end points of the piece of theory to which they belong.

To sum up, in the factual sciences the following rule of method seems to be accepted, at least tacitly: "All hypotheses, even the epistemologically most complex ones, must entail through inferential chains as long and twisted as is necessary instantial hypotheses, so that they can be inductively confirmed". This rule assigns to induction a place in scientific method, the overall pattern of which is admittedly hypothetico-deductive.

Inductivism rejects the deductivist thesis that what is put to the test is always some (often remote) observable consequence of theories, and that we never test isolated hypotheses but always some potpourri of fragments of various theories—eventually including those involved in the building and reading of instruments and in the performing of computations. Inductivism maintains that this description of scientific procedure might square only with very high-level hypotheses, such as the postulates of quantum mechanics. However, an analysis of elementary scientific hypotheses, even of existential ones such as "There is an air layer around the Earth" confirms the deductivist description, with the sole, though important, exception of the contact line between the lowest-level theorems and the empirical evidence.

Consider, for instance, the process that led to the establishment of the existence of the atmosphere. An analysis of this process (Bunge 1959b) will show that Torricelli's basic hypotheses, "We live at the bottom of a sea of elemental air" and "Air is a fluid obeying the laws of hydrostatics", were framed by analogy, not by induction, and that the remaining process of reasoning was almost entirely deductive. Induction occurred neither in the formulation nor in the elaboration of the hypotheses: it was equally absent in the design of the experiments that put them to the test. Nobody felt the need of repeating the simple experiments imagined by Torricelli and Pascal, nor of increasing their poor precision. Rather on the contrary, Torricelli's hypotheses were employed to explain further known facts and were instrumental in suggesting a number of new spectacular experiments, such as Guericke's and Boyle's. Induction did appear in the process, but only in the final estimate of the whole set of hypotheses and experimental results, namely when it was concluded that the former had been confirmed by a large number and, particularly, by a great variety of experiments, whereas the rival peripatetic hypothesis of the abhorrence of void had been conclusively refuted.

To sum up, enumerative induction plays a role in the test of scientific hypotheses, but only in their *empirical* checking, which is not the sole test to which they are subjected.

4. INDUCTIVE CONFIRMATION AND DEDUCTIVE REFUTATION

Deductivists may object to the above concessions to induction, by stating that confirming instances have no value as compared with negative ones, since the rule of *modus tollens* ("If p, then q; now, not-q; hence, not-p") shows that a single definitely unfavorable case is conclusive, whereas no theorem of inductive logic could warrant a hypothesis through the mere accumulation of favorable instances. But this objection does not render the examination of cases worthless and does not invalidate our "concluding" something about them; hence, it does not dispose of induction by enumeration.

Consider, in fact, a frequent laboratory situation, such as the one described by the following sentence: 'The results of n measurements of the property P of system S by means of the experimental set-up E agree, to within the experimental error ε, with the values x_i predicted by the theory T'. Certainly, ninety favorable instances will have little value in the face of ten definitely unfavorable measured values, at least if high precision is sought. (On the other hand, a single unfavorable case against ninety-nine favorable ones would pose the question of the reliability of the anomalous measurement value itself rather than rendering the theory suspect.) But how do we know that an instance is definitely unfavorable to the central hypothesis of the theory we are examining, and not to some of the background hypotheses, among which the usual assumption may occur, that no external perturbations are acting upon our system? Moreover, do we not call 'negative' or 'unfavorable' precisely those instances which, if relevant at all, *fail to confirm* the theory under examination?

Confirmation and refutation are asymmetrical to each other, and the latter is weightier than the former; moreover, a theory that can only be confirmed, because no conceivable counterexample would ruin it, is not a scientific theory. But confirmation and refutation cannot be separated, because the very concept of negative instance is meaningful only in connection with the notion of favorable case, just as "abnormality" is meaningless apart from "normality". To say that hypotheses such as natural laws (or, rather, the corresponding statements) are only refutable, but not confirmable by experiment (Russell 1948; Popper 1959), is as misleading as to maintain that all humans are abnormal.

How do we know that a skilled and sincere attempt to refute a hypothesis has failed, if not because the attempt has *confirmed* some of the lowest-level consequences of the theory to which the given hypothesis belongs? How do we know that an attempt has succeeded—thereby forcing us to abandon the hypothesis concerned, provided we are able to isolate it from the piece of theory to which it belongs, and provided better ones are in sight—if not because we have obtained no positive instances of its low-level consequences, or even because the percentage of positive instances is too poor?

The falsifiabilist rule enables us to discard certain hypotheses even *before* testing them; in fact, it commands us to reject as nonscientific all those conjectures that admit of no possible refutation, as is the case with "All dreams are wish fulfillments, even though in some cases the wishes are repressed and consequently do not show up". But refutability, a necessary condition for a hypothesis to be *scientific*, is not a criterion of truth: to establish a proposition as at least partially true, we must confirm it. Confirmation is insufficient, but it is necessary.

The falsifiabilist rule *supplements* the characterization of the difficult notion of positive instance, or favorable case, but provides no substitute for it. Refutation enables us to (provisionally) eliminate the less fitted assumptions, which are those that fit the data less adequately, but it does not enable us to justify alternative hypotheses. And, if we wish to resist irrationalism, if we believe that science and scientific philosophy constitute bulwarks against obscurantism, we cannot admit that scientific hypotheses are altogether unfounded but lucky guesses, as deductivism claims. Law statements do not hang in the air: they are both *grounded* on previous knowledge and successfully *tested* by fresh evidence, both empirical and theoretical.

The attitude of attempting to refute a theory by subjecting it to severe empirical tests belongs to the pragmatic and methodological level, and pertains even to the ethical code of the modern scientist. The problem of confirmation and, consequently, the problem of the degree of validation and, hence, of acceptability of factual theories belong both to the methodological and the epistemological levels. There is no conflict between the procedure that aims at refuting a theory, and the assignment to it of a degree of validation or corroboration on the basis of an examination of positive instances: they are complementary, not incompatible operations. Yet none of them is sufficient: pure experience has never been the supreme court of science.

5. THEORIFICATION

Neither unsuccessful attempts to refute a hypothesis nor heaps of positive instances of its observable consequents are enough to establish the hypothesis for the time being. We usually do not accept a conjecture as a full member of the body of scientific knowledge unless it has passed a further test which is as exacting as the empirical one or perhaps even more so: to wit, the rational test of *theorification*, an ugly neologism that is supposed to suggest the transformation of an isolated proposition into a statement belonging to a hypothetico-deductive system. We make this requirement, among other reasons, because the hypothesis to be validated acquires in this way the support of allied hypotheses in the same or in contiguous fields.

Consider the hypothesis "All humans live less than two hundred years". In order to test it, a confirmationist would accumulate positive instances, whereas a refutationist would presumably establish an enrolling office for bicentenaries, the simplest and cheapest but not the most enlightening procedure. Old age medicine does not seem to pay much attention to either procedure, but tends in contrast to explain or deduce the given statement from higher-level propositions, such as "The arteries of all humans harden in time", "All cells accumulate noxious residues", "Neurons decrease in number after a certain age", and so on.

The day physiology, histology, and cytology succeed in explaining the empirical generalization "All humans live less than two hundred years" in terms of higher-level laws, we shall judge it as established in a much better way than by the addition of another billion deaths fitting the low-level law. At the same time, the hypothesis will, after theorification, offer a larger target to refutation—which is, after all, a desideratum of geriatry—since it will become connected with a host of basic laws and may consequently contact with a number of new contiguous domains of experience.

The degree of support or sustenance of scientific hypotheses—which is not a quantitative but a comparative concept (among other reasons because hypotheses have philosophical supports besides empirical ones)—increases enormously upon their insertion into nomological systems, i.e., upon their inclusion in a theory or development into a theory.

No inference can even provisionally be justified outside the context of some theory, including of course one or more chapters of formal logic. Factual hypotheses can be justified up to a certain point if they are grounded on deep (nonphenomenological) laws that, far from being just summaries of phenomenal regularities, enable us to explain them by some "mechanism" (often nonmechanical). Thus, the age-long recorded succession of days and nights does not warrant the inference that the sun will "rise" tomorrow, as Hume rightly saw. But a study of the dynamic stability of the solar system and of the thermonuclear stability of the sun, as well as a knowledge of the present positions and velocities of other neighboring celestial bodies, renders our expectation highly plausible. Theory affords the validation refused by plain experience: not *any* theory but a theory including deep laws transcending first-degree inductive generalizations. In this way inductivism is inverted: we may *trust inductions to the extent that they are justified by noninductive theories*.

In sum, empirical confirmation is but one phase, though an indispensable one, of the complex and unending process of inventing, checking, mending, and replacing scientific hypotheses.

6. INDUCTIVIST METHODOLOGY AND THE PROBLEM OF INDUCTION

According to inductivism, empirical knowledge (a) is obtained by inductive inference alone, (b) is tested only by enumerative induction, (c) is more reliable, as it is closer to experience (epistemologically simpler), (d) is more acceptable, as it is more probable, and consequently (e) its logic, inductive logic, is an application or an interpretation of the calculus of probability. Deductivists, especially Popper (1957a, 1959, 1960), have shown that these claims are untenable, particularly in connection with theoretical laws, which are neither obtained nor directly tested by induction, and which have exactly zero probability in any universe that is infinite in some respect. They and a few others (e.g., Kneale 1949) have also conclusively shown that the theory of probability does not solve the riddles of induction and does not provide a warrant for inductive leaps.

All this, however, does not prove the vanity of the cluster of problems concerning induction, conceived as the set of questions connected with both the inductive inception and, particularly, the inductive confirmation of hypotheses; hence, those arguments do not establish the impossibility of *every* logic of induction, even though they considerably deflate the claims of available systems of inductive logic. It is, indeed, a fact that induction is employed in the formulation of some hypotheses both in formal and in factual science, even though it is true that such hypotheses are rarely impressive and deep. And it is a fact too (or rather a metascientific induction!) that induction is employed in the validation of all factual theories. The mere mention of statistical inference should suffice. Now, if a subject exists, scientific philosophy suggests that the corresponding scientific (or metascientific) approach should be attempted. And why should induction be left in the hands of inductivists?

Granted, there is no inductive *method*, either in the context of invention or in the context of validation; at least, there is no inductive method in the sense of a set of secure rules or recipes guaranteeing once and for ever the jump to true general conclusions out of case examination. Nor is there an intuitive method or a hypnotic method. Yet induction, intuition, and hypnosis do exist and deserve to be studied scientifically. An analysis of scientific research shows the current employment of various patterns of plausible inference, such as analogy, reduction, weakened reduction, and weakened *modus tollens* (Keynes 1921; Pólya 1954; von Wright 1957; Czerwinski 1958); it also shows the operation of inductive policies, such as those connected with sampling, and which are after all designed to provide the best possible inductions. Why should we disregard these various kinds of nondemonstrative inference, especially knowing as we do that successful patterns tend to be accepted as rules admitted uncritically unless they are critically examined?

The rules of deductive inference, to which we all pay at least lip service, were not arbitrarily posited by some inspired genius in the late Neolithic: they were first *recognized* in sound discourse and then explicitly adopted because they lead from accepted statements to accepted statements—and statements are accepted, in turn, if they are deemed to be at least partially true. Conversely, statements that are not postulated by convention are regarded as true if they are obtained by procedures respecting accepted rules of inference. Such a *mutual and progressive adjustment* of statements and rules is apparently the sole ultimate justification of either (Bôcher 1905; Goodman 1955). Analogously, the belief in the possibility of a logic of plausible (nondemonstrative) reasoning rests not only on a false theory of knowledge which minimizes the role of constructs, and on a history of science biased against the theoretical, but also on the plain observation that some nondemonstrative inferences are crowned with success. (Usually, this is the case with recorded inferences, because humans, as Bacon pointed out, mark when they hit.) This is what entitles us to adopt as (fallible) rules of inference, and as inductive policies, those patterns that in good research lead from accepted propositions to accepted propositions.

Of course, the theory of plausible inference should not restrict itself to a *description* of the types of argument found in everyday life and in science: it should also refine them, devising *ideal* (least dirty) patterns of inference (Barker 1957). However, such a rational reconstruction should be preceded by a realistically oriented investigation into patterns of *actual* scientific inference, rather than by another study of the opinions of distinguished philosophers concerning the nature and role of induction.

Furthermore, ideal patterns of plausible reasoning should be regarded neither as binding rules nor as inference tickets, but rather as more or less successful, hence advisable, patterns. This, at least in the constructive stage, when the greatest freedom to imagine is needed, since creative imagination alone is able to bridge the gap separating precepts from concepts (Einstein 1944; Bunge 1962), first-degree inductions from transcendent hypotheses, and isolated generalizations from theoretical systems. Logic, whether formal or informal, deductive or inductive, is not supposed to concoct recipes for jumping to lucky conclusions—jumps without which there is as little science as there is without careful test—but it may show which are the best patterns that can be discerned in the test of hypotheses framed in whatever way.

7. CONCLUSION

As must have been suspected by many, scientific research seems to follow a *via media* between the extremes of inductivism and deductivism. In this middle course, induction is instrumental both heuristically and methodologically, by taking part in the framing of some hypotheses and in the empirical validation of all sorts of hypotheses. Induction is certainly powerless without the invention of audacious transcendent hypotheses, which could not possibly be suggested by the mere examination of experiential data. But the deepest hypotheses are idle speculation unless their lower-level consequents receive instantial confirmation. Induction plays scarcely a role in the design of experiments, which involves theories and demands creative imagination; but experiment is useless unless its results are interpreted in terms of theories that are partly validated by the inductive processing of their empirically testable consequences.

To sum up, induction—which is but one of the kinds of plausible reasoning—contributes modestly to the framing of scientific hypotheses, but is indispensable for their test, or rather at the empirical stage of their test. Hence, a noninductivist logic of induction should be welcome.

10

THE GST CHALLENGE TO THE CLASSICAL PHILOSOPHIES OF SCIENCE

(1977)

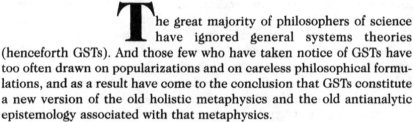

T he great majority of philosophers of science have ignored general systems theories (henceforth GSTs). And those few who have taken notice of GSTs have too often drawn on popularizations and on careless philosophical formulations, and as a result have come to the conclusion that GSTs constitute a new version of the old holistic metaphysics and the old antianalytic epistemology associated with that metaphysics.

This neglect, on the part of philosophers, of the technical literature in the various GSTs, is deplorable for a number of reasons. The main one, however, is that GSTs present a serious challenge to the two most popular philosophies of science, namely empiricism (or inductivism or confirmationism), as represented by Rudolf Carnap, and rationalism (or deductivism or refutationism), as championed by Karl Popper (1959). Indeed, none of these philosophers ever had GSTs in mind and, as a consequence, there is no room for GSTs in their philosophies. Worse, according to either of these philosophies, GSTs are nonscientific, for they yield no precise predictions that can be checked by observation or experiment. This sounds odd and even insulting to the practitioners of GSTs, none of whom seem to doubt that what they do is science.

This is a serious situation—not for GSTs but for philosophy. However, the situation is not new: actually, not even the great scientific theories, praised but not analyzed by the devotees of the standard philosophies of science, conform to the latter. Indeed, those great theories are far too general to be able to yield predictions without further ado, so by themselves they are untestable. The birth of GSTs has simply highlighted a philosophical crisis that existed before, albeit that it was unnoticed except occasionally.

In the face of a serious crisis in a discipline there is but one possible course of action: to examine the presuppositions, to reexamine the object of study, and to make a fresh start no matter which idols may fall. The present paper is addressed to the crisis outlined above.

1. THE CLASSICAL PARADIGM
OF A SCIENTIFIC HYPOTHESIS

Our problem is to find out what kind of animals GSTs are and, in particular, whether they are scientific theories. But this task requires that we first answer the question 'What constitutes a scientific theory?' or, equivalently, 'What are the necessary and sufficient conditions for a theory to qualify as a scientific theory?', or again 'What is the criterion of scientificity in the case of theories?'.

Now, theories are systems of statements (propositions, formulas) closed under deduction, i.e., sets such that every statement that can be deduced from any statements in the theory is already contained in the latter. And scientific theories are hypothetico-deductive systems, i.e., self-contained sets of hypotheses, or statements going beyond observation. Think of classical electrodynamics, or of the selective theory of evolution, or of the theory of social mobility. We may therefore start by asking the more restricted question 'What is a scientific hypothesis?'.

We have all been taught that a hypothesis is scientific to the extent to which it entails empirically testable consequences. But whether it passes the test of observation or of experiment is essential to determining not its scientific character but its truth value. A hypothesis may be false yet scientific, or true yet unscientific. Archimedes' law of the lever, a true scientific hypothesis, can be used to predict that a grandfather or his grandchild will lose balance if they ride a seesaw. By contrast, the hypothesis of telepathy does not allow one to predict anything, whence it is unscientific. And the hypothesis that the reader has at least one enemy is testable and, moreover, probably true but hardly of interest to science.

In sum, we have been conditioned to accept:

DEFINITION 1. A theory is *scientific* if, and only if, it entails empirically testable consequences.

Once formulated, all the schools seemed to agree on this definition. The only difficulty was determining what is to be meant by "testable". While the empiricists equated "testable" with "confirmable" (or having possible examples), rationalists equated it with "refutable" (or having possible counterexamples). Thus, whereas empiricists like Carnap and Feigl regarded psychoanalytic hypotheses as scientific because of the many cases that seem to confirm them, rationalists like Popper (1962) held that they are not, precisely because it is so easy to confirm them and so hard to refute them. But aside from this divergence, both empiricists and rationalists agreed that direct empirical testability is the trademark of scientific hypotheses.

2. A FIRST CRACK IN THE CLASSICAL PARADIGM

On the basis of Definition 1, inductivists posed the ambitious inverse problem: finding out, given a bit e of empirical evidence, which among the hypotheses entailing e is the most probable; that is, finding the h which maximizes the conditional probability $Pr(h \mid e)$. Rationalists pointed out that this was the wrong problem to pose, for the likeliest hypotheses are the least bold ones, namely those framed precisely to account for e and nothing else, i.e., the *ad hoc* hypotheses, incapable of undergoing generalization. As a reaction they demanded instead that the least probable hypotheses be preferred. And, just as the inductivists built systems of inductive logic (Carnap 1950) based on the assumption that the more probable a hypothesis the better, so the deductivists constructed theories of corroboration based on the negation of that assumption.

Both empiricists and rationalists turned out to be wrong on two counts. Firstly, they assigned probabilities to statements. It is impossible to do so—except of course by fiat, i.e., conventionally. Facts can be assigned probabilities on condition that they be handled by some stochastic theory involving a random mechanism such as blind shuffling. But propositions are not facts: they cannot be shuffled, particularly if they belong to orderly wholes such as hypothetico-deductive systems. (See essay 15.) Hence, both inductive logic (pace von Wright 1957) and the theories of corroboration are empty formalisms. None of them is in a position to make meaningful probability assignments to a formula, say, in quantum mechanics or in the genetics of populations. Both sets of theories are wholly artificial. Scientific research is not after maximally probable (or maximally improbable) hypotheses, but after maximally true, deep, and systematizable ones.

That was not the only flaw in the classical paradigm. The second flaw is even more apparent and it consists in inferring empirical (observational or experimental) consequences from hypotheses alone, which is impossible. It is false that every scientific h implies some e. For example, the gas law does not predict by itself what the exact value of the volume is unless the exact value of the pressure is given. Thus, from $pV = k$ and $p = 1$ atmosphere we infer that $V = k$ liters. No empirical information, no prediction. Hence Definition 1 should be replaced by:

DEFINITION 2. A theory is *scientific* if, and only if, jointly with empirical data, it entails empirically testable consequences.

Such was no fatal objection to the rationalists. On the contrary, it bolstered their view that scientific hypotheses are no mere inductive syntheses: they stated explicitly that scientific hypotheses are *noninstantial*, i.e., do not entail observational cases by themselves. As for the empiricists, they still have to learn that no theory t entails e's, so that the search for the best t that covers a given e is as illusory as the search for the promised land.

Finally, both the rationalists and inductivists have still to face the fact that statements, unlike facts, cannot be assigned probabilities, or that, if they are assigned probabilities by some conventional rule, then such probabilities are subjective (because of the arbitrariness or the assignment rule), hence just as repugnant to the empiricist as to the rationalist.

The upshot so far in our discussion is this. Empiricism is not a viable philosophy of scientific experience, which it largely ignores. And rationalism can be upheld only on condition that it give up all attempts at building probabilistic theories of scientific evidence. However, there is still worse to come: the next round will knock out rationalism as well.

3. THE CHALLENGE PRESENTED BY GENERAL SCIENTIFIC THEORIES

The specific hypotheses of science, such as Snell's law of refraction, certainly qualify as scientific according to Definition 2. How about the extremely high-level hypotheses, such as the general principles of classical mechanics, of electrodynamics, or of the selective theory or evolution? They are seldom studied by philosophers, so it should come as no surprise if we find that they do not fit Definition 2 either. A single counterexample to the latter will suffice to ruin it.

Consider the basic equation of motion of classical (i.e., continuum) mechanics, namely $\rho(d^2X/dt^2) = K + \text{div } T$, where X and t are the position and time coordinates respectively, ρ the mass density, K the body force density, and T the stress tensor. For putting this equation to empirical tests no data will suffice, not even in the simplest of cases, i.e., that in which $\rho = \text{const}$ and div $T = \text{const}$. Indeed, in this simple case we would still need to know the initial position and the initial velocity of every one of the infinitely many point particles. And in the general case we would have to know, in addition, the exact values of ρ and T at each point in the region of the manifold occupied by the body.

Hence, as it stands, the equation of motion cannot be solved even with the adjunction of a million empirical data. What one does is not to add data to it, but to make *further hypotheses*, namely about initial streamlines, about the mass distribution, and about the stress distribution. In other words, one builds a conceptual and hypothetical *model* of the body of interest. (Sometimes such a model may be suggested by data, as is the case with photoelasticity. But such suggestions are always imprecise and usually difficult to express mathematically.) Once the equation of motion has been supplemented with such subsidiary hypotheses, it can be integrated to yield the lines of flow. (Even so, chances are that no integration in closed form is possible unless one makes additional assumptions of a simplifying nature.)

The solutions obtained at the end of the process we have just summarized can finally be contrasted with the empirical evidence, provided the latter has been couched in the language of the theory that is being tested. The same holds, mutatis mutandis, for every other scientific theory of the generic as opposed to the specific kind. In all such cases the testing process, far from conforming to Definition 2, comes closer to satisfying:

DEFINITION 3. A theory is *scientific* if, and only if, jointly with the subsidiary assumptions and empirical data, it entails empirically testable consequences.

Will this refinement do? Not yet: there is still the matter of the compatibility of t with the body of background knowledge. Indeed, nobody but a crackpot wastes his time conducting empirical tests on conjectures that collide head-on with the bulk of scientific knowledge, as is the case with, e.g., the hypotheses of teletransportation and of communication with the dead. That is, before anything else, one subjects the hypothesis of interest to the test of compatibility with standard scientific theories and rejects it out of hand if it is inconsistent with the best knowledge at hand. More-over, one requires compatibility with equally firm philosophical hypotheses, such as that no thing is isolated and utterly unknowable.

The upshot is that one more reform in our definition of a scientific conjecture is called for.

DEFINITION 4. A theory is *scientific* if, and only if,
(i) it is compatible with the bulk of scientific knowledge, and
(ii) jointly with subsidiary hypotheses and empirical data, it entails empirically testable consequences.

Actually, even this refinement is insufficient for coping with standard scientific theories of the generic type, for it makes no provision for the theories and procedures underlying the production of the data d and the fresh empirical evidence e—bits of information which are anything but straightforward. However, Definition 4 will do as a good approximation for our ultimate goal, which is to see how GSTs fare. (For details on the prior theoretical and philosophical tests, as well as for the theoretical underpinnings of high-precision empirical tests, see Bunge 1967b, chs. 14 and 15, and Bunge 1973b, ch. 10.)

4. THE CHALLENGE PRESENTED BY GSTs

It would seem that all of the standard scientific theories, in particular the high-grade ones such as quantum mechanics, conform to Definition 4—if not exactly, at least reasonably well. By contrast, the pseudoscientific conjectures, such as those of telekinesis and of one's fond memories from intrauterine times, do not conform to the definition because they are incompatible with physics, as in the first case, and with neurophysiology as in the second. (Telekinesis violates energy conservation, and intrauterine memories are impossible because of the lack of maturity of the fetus's nervous system.) So, it seems that Definition 4 allows one to keep in science all that is worth keeping and to shed all that is not normally regarded as scientific. (The partisans of absolute and permanent scientific revolutions will of course object to the condition (i) of compatibility with the bulk of scientific knowledge. Never mind, for only crackpot theories claim to effect such radical and total revolutions. The compatibility condition, which amounts indirectly to a battery of instant empirical tests, does not prevent genuine scientific revolutions. On the contrary, it makes them testable and credible.)

Have we reached a satisfactory solution to the problem of determining the necessary and sufficient conditions for qualifying a theory as scientific? If this question had been asked before the GST revolution, the answer would have been an unqualified "yes". But the situation has altered radically since World War II, for since then we have reaped a rich crop of theories even more general than the general frameworks of classical electrodynamics, or quantum mechanics: theories not just general like the preceding ones but *hypergeneral*. I refer of course to GSTs, such as the statistical theory of information, game theory, control theory, automata theory, and even the general lagrangian framework, the general classical theory of fields, and the general quantum theory of fields (misnamed 'axiomatic field theory').

All of these theories are phenomenological, i.e., mechanism-free: they are black box or gray box theories, but not translucid box theories. Hence, they can describe the behavior of certain systems but not explain how they work. Moreover, they are stuff-free, i.e., they make no detailed assumptions concerning the nature of the components of the systems concerned; whence their extreme generality. They are in fact so general as to be unable to yield any predictions, not even when enriched with data. To convince yourself that this is so, try to predict the behavior of an information system or of an automaton, without resorting to any extra knowledge concerning their composition, the materials they are made of, and so forth. Clearly, such hypergeneral theories defy the standard philosophies of science even more openly than the standard generic scientific theories, such as quantum electrodynamics or the theory of selection.

Take, for example, the so-called law of requisite variety, a cybernetic analog of Shannon's Theorem 10. A possible formulation of that law might be this: The information-theoretic entropy in the output is at least as great as the excess of the entropy of the external disturbance over the entropy of the control device. Ashby (1956, pp. 208–209) made the shrewd comment that, although this formula does exclude certain events, it has *nothing to fear from experience* for it is independent of the properties of matter. In this case the function of experiment is not to check the theory but to feed it with data. Moreover, if a given set of empirical data seems to refute this formula of the theory or any other formula of it, then the indicated course is to redraw the boundary between the system and its environment until a system is circumscribed that conforms to the given formula. For example, if the above law of requisite variety seems to fail because the system is too noisy, then we include noise among the external disturbances and so agreement between theory and fact is restored (pp. 216–17).

Likewise with the theories of information and communication: if a system fails to conform to them, the system is flunked, not the theory, for an information-processing device is, *by definition*, one that fits these theories. In particular, "any device, be it human or electrical or mechanical, must conform to the theory [of communication] if it is to perform the function of communication" (Miller 1967, p. 14). Consequently, the concepts of degree of confirmation, corroboration, and testability are pointless with regard to this theory: the whole of inductive logic is irrelevant to it and so is the refutationist methodology.

Similar considerations apply to automata theory. (See, e.g., Harrison 1965, and Bunge 1973a, ch. 8, for an axiomatization of the theory.) This theory supplies a precise function of a sequential machine and enables one to study machine homomorphisms, behavioral equivalence among machines, the composition of machines, and even the entire lattice of machines. It is not a theory in abstract mathematics, because it concerns a certain genus of concrete system interacting with its environment, although it is totally uncommitted as to the precise nature of either. Any real system that happens to conform to this theory, regardless of its physics and chemistry, will qualify as an automaton. And those concrete systems that do not fit the description just do not qualify.

Such concrete systems may be *forced* to behave as automata, thus providing a cheap confirmation of the theory. For example, a pigeon may be trained to behave like a two-state automaton, and, if it fails to learn the trick, the theory is unscathed and the experimenter may have pigeon pie for supper. If no real system is found or built or even thought to be technically feasible, the automata theorist won't be deterred, provided he can show that his theoretical automaton is a good model for possible machines. (This is the case with Turing machines, which, being equipped with infinitely long tapes, are strictly speaking unrealizable.)

In sum, while automata theory is *applicable* (by specification) and, moreover, guides much of advanced engineering design and even some psychological research (Suppes 1969), it is *irrefutable*. It is not even confirmable in the traditional manner of predicting and checking: the theory makes no specific prediction, it prohibits hardly any event, and it suggests no experiments other than *gedankenexperimente*. In short, automata theory, like information theory and every other member of the variegated set of GSTs, is *empirically untestable* in any of the traditional ways discussed in section 1. The same holds even for some of the applications of GSTs, such as the cybernetic model of the reticular formation or RF: "there is no experiment that could invalidate our claims; our concept [actually a theoretical model] has not yet produced any risky predictions; it does not forbid any measurable RF event; and we have not yet proposed any real alternatives" (Kilmer et al. 1968, p. 321).

Now, if GSTs are unable to yield precise predictions even when enriched with subsidiary assumptions and empirical data, then they are not covered by Definition 4 in section 3. Why, then, should GSTs be regarded as scientific?

5. THE METHODOLOGICAL STATUS OF GSTs

GSTs, as we have just seen, are neither confirmable nor refutable the way standard scientific theories are. However, they are not dogmas above criticism. In fact GSTs are *corrigible* if not exactly refutable in the light of empirical evidence. To begin with, they can be improved upon formally, i.e., logically or mathematically. For example, they can be overhauled and made mathematically more powerful with the explicit use of new mathematical theories. (See, e.g., Klir 1972 as well as Padulo and Arbib 1974.) Or they can be made more complex in an attempt to fit better their intended referents. Thus, if the goal is to model a learning system then, since learning is largely a stochastic process, the theory of probabilistic automata may be found more useful than the theory of deterministic automata. In a sense, then, GSTs are *confirmable*.

Surely, GSTs are not confirmed in the classical way, i.e., through prediction and empirical checking. Nevertheless they are confirmed, although in a special way, namely by being shown either to fit a whole family of specific theories (i.e., theories concerning specific systems) or to take part in the design of viable systems. The former may be called *conceptual confirmation*, the latter *practical confirmation*, and either kind differs from the usual *empirical confirmation*. Actually, all GSTs are confirmed both conceptually and practically without ever being confirmed empirically, i.e., by contrasting their predictions with empirical evidence. Thus, general network theory is confirmed conceptually by being shown to capture the traits common to all nets, whether physical or informational. And it is confirmed practically by being used in the design of nets of some kind.

In other words, GSTs are not just generalizations of specific theories, for they can be applied. And, to the extent that they are applied successfully, they are shown to be suitable and fruitful without being true, let alone false. (A false dogma, such as racism, can be used for certain purposes.) Moreover, GSTs are doubly confirmable, conceptually and practically, but they can never be falsified. At most they can be shown to be irrelevant to the problem at hand. They either "apply" or they don't.

This does not entail that GSTs can be applied to particular cases, hence tested for usefulness, without further ado. Even though they are nonspecific and therefore empirically untestable, their application to specific situations requires some substantive knowledge of the latter: only this can provide a suitable interpretation of the theory in question. In other words, GSTs are so many *general frameworks that must be adjoined specific models* of the system of interest before they can be of any use. For example, in the case of information theory we must be able to identify at least the sources of information and of noise, the channel, and the receiver; and we must have a code for the system of signals. (Incidentally, because none of these conditions is fulfilled in the case of molecules, the use of information theory in molecular biology is purely metaphorical.) Likewise, in the case of cybernetics we must be able to identify the system, its regulator, the environment, and the disturbances originating in the latter; besides, we must know what the goal (or the set of final states) is, for otherwise we won't even suspect what is to be regulated, let alone how to achieve the regulation. In fact, a cybernetic problem looks like this: Given a system —which includes its inputs and outputs—together with its environment and the desired subset of output values (i.e., the goal), design a control system that, coupled to the main system, will keep its output within preassigned bounds. Without all these items of specific information—unobtainable without the help of observations or experiments supported by specific theories—one would be unable to pose the problem, hence to solve it.

In sum, every application of a GST calls for the building of a specific model of the system of interest—the model being built, of course, with the concepts of the theory if it is to be coupled to the latter. In other words, a member of the GST class *becomes a specific theory* of the standard type, perhaps a scientific theory, when enriched with specific information concerning the system to which it is applied. And *this* specific theory is of course subject to the strict canons of empirical testability. That is, Definition 4 refers not to any GST, but to specifications or applications of GSTs. (And we know well enough that a large number of such applications do not live up to those standards of scientificity. But this is another story that will have to be told some day.)

The upshot of this section is: GSTs, though neither confirmable nor refutable in the usual way, i.e., through the empirical checking of predictions, are confirmable in a *sui generis* way. Indeed, GSTs are confirmed (a) by fitting whole families of specific theories (*conceptual confirmation*) and (b) by helping in the building of specific theories that are tested the classical way (indirect confirmation). Yet this does not solve the problem of the scientific status of the GSTs. Let us now turn to this problem.

6. THE SCIENTIFIC STATUS OF GSTs

No matter what philosophers with their *a priori* criteria of scientificity may decide, GSTs are usually regarded as scientific. They are so classed because (a) they are precise (by virtue of being formulated mathematically); (b) they are not at variance with our antecedent scientific knowledge; and (c) when applied (by specification) they often yield either scientific knowledge, in particular scientific theories proper, or guides for efficient action, e.g., in the field of management. Surely, GSTs do not provide detailed accounts of any real systems. In particular, they do not explain and do not predict the behavior of any real systems. But they are no less useful for that reason. Indeed, GSTs are *generic frameworks* helping one to think of entire genera of entities in a variety of domains, from biology and psychology through hardware and human engineering to city planning and politics. True, GSTs solve no particular problems without further ado, but on the other hand they help in the discovery and formulation of new problems and they clarify basic ideas in all fields of inquiry (Rapoport 1972, p. 74). In short, GSTs are respectable members of the body of scientific knowledge, even though some scientists have misgivings because GSTs have not delivered all the goods promised by their most enthusiastic proponents.

But if GSTs are declared scientific, then we must modify our previous canons of scientificity: *we must change our methodology of science.* To begin with, we must replace Definition 4 by a more tolerant definition making room for conceptual and vicarious confirmability as an alternative to strict empirical testability. Let us try the following:

DEFINITION 5. A theory is *scientific* if, and only if,
(i) it is compatible with the bulk of scientific knowledge and either
(ii) jointly with subsidiary hypotheses and empirical data, it entails empirically testable consequences, or
(iii) jointly with subsidiary hypotheses and empirical data, it entails theories that in turn entail testable consequences as in (ii).

Shorter: *t* is scientific iff, when enriched with suitable subsidiary assumptions and empirical data, it becomes *empirically testable either directly or vicariously*, i.e., through some (specific) theory.

If we now look back on our previous definitions, except for thoroughly inadequate Definition 1, we realize that we have distinguished three levels of exigency. While we demand of all theories that they be compatible with the bulk of scientific knowledge, in some cases we require that they entail testable predictions when enriched with data, in others when they are adjoined data and subsidiary assumptions, and in still others we require that, when enriched with both data and subsidiary assumptions, they entail theories that are empirically testable.

In short, if we wish to keep GSTs within science, then we must adopt an enlarged criterion of testability allowing for vicarious testability. But if we do so, then we must give up the standard philosophies of science, none of which tolerates such an extension of the scientificity criterion. Would anything be gained by sticking to any of these philosophies? Let us see.

7. GSTs SANDWICHED BETWEEN SCIENCE AND METAPHYSICS

When finding out that GSTs do not conform to the classical standards of scientificity, as embodied in Definition 4, we had two options. One was to give up the definition and adopt a broader one making room for GSTs. This is what we accomplished through Definition 5. The other option was to stick to conventional wisdom and declare GSTs to be nonscientific. In this case, since all of the genuine GSTs are mathematical in form, they must belong either in pure mathematics or in exact, i.e., mathematical, philosophy. (Exact philosophy is philosophy tamed by mathematics. It can still be wild with regard to having scientific evidence or it can be utterly irrelevant to science, but it is formally correct because it makes explicit use of logical or mathematical tools. For a recent sample of exact philosophy see Bunge 1973 ed.) Let us explore these latter two possibilities.

That GSTs are mathematical theories is sometimes held by mathematicians. However, every one of the GSTs is concerned with concrete, though rather faceless, entities. Moreover, these theories are used in designing concrete systems such as communication networks or learning systems. And they are not on the same footing with mathematical theories, such as linear algebra or probability theory, which have a much higher degree of cross-disciplinarity. Instead GSTs, when applied, are employed as broad schemata or models of the things to be designed or controlled. In short, although GSTs are hypergeneral, they are not nearly as universal as mathematical theories. They are mathematical in form, not in content.

Hence, if we insist on regarding GSTs as nonscientific, we are left with the second possibility, namely that they belong in exact philosophy, in particular in mathematical ontology, or the formal theory of extremely wide genera of concrete things. Pause and ponder before smiling. GSTs share three basic traits with theories in exact ontology: (a) they are mathematical in form, (b) they concern genera (not just species) of concrete, material, real things, and (c) they are *empirically untestable* except in a devious way. Why, then, not accept GSTs in the fold of ontology? There is but one reason for refraining from doing so, namely the belief that there must be an unsurpassable frontier between science and metaphysics—a belief shared of course by empiricists and rationalists alike, even though they differ about the demarcation criterion.

Here again we have the choice of either following tradition or facing the facts and making our own decision. The fact, as I see it, is that GSTs are *both* scientific *and* ontological, precisely because they share the above-mentioned defining traits: exactness, hypergenerality, and empirical irrefutability. It may be rejoined that GSTs are, in addition, compatible with the bulk of scientific knowledge, whereas a number of theories in exact ontology are totally alien to science. (Thus the most fashionable among them consist in speculation about possible worlds, without caring for investigating the basic traits of the real world.) Agreed. But this shows only that there are two sorts of theory in present-day exact ontology: those which are and those which are not contiguous with the bulk of scientific knowledge.

We may then speak of *scientific ontology* when referring to theories that concern the most pervasive traits of reality, that are systematic (i.e., theories proper rather than bags of opinions), that make explicit use of mathematical logic or abstract algebra or any other branch of mathematics, that are congenial (not just compatible) with the science of the day, and that elucidate some of the key concepts occurring in philosophy or in the foundations of scientific theories, such as those of system, process, interaction, life, mind, or society.

The theories in scientific ontology are of course those satisfying Definition 5. For example, an ontological theory of space that explains in exact (mathematical) terms spatial relations as certain relations among physical things, is entitled to be called *scientific* if it ends up with a space that is metricizable, so that the physicist can add to it any metric function she needs for her theories. Another example: an ontological theory of society that defines the latter as a concrete system endowed with a structure consisting of a family of equivalence classes (one for each social group) is entitled to be called *scientific* because the sociologist can use it as a framework or matrix for working out his more specialized theories, such as the theories of stratification, of mobility, etc.

In sum, our enlarged criterion of scientificity (Definition 5) houses not only GSTs, but also theories in scientific ontology. There is a large overlap between the two areas, the extent of which depends upon the distinction between them, which is largely a matter of convention. If terms were to help we might say that, whereas GSTs are hypergeneral, theories in scientific ontology, or SO, are superhypergeneral. The two are so many rungs in a ladder of generality that goes like this:

Hyperspecific scientific theories, e.g., the theory of the simple pendulum.
Specific scientific theories, e.g., classical particle mechanics.
Generic scientific theories, e.g., continuum mechanics.
Hypergeneral scientific theories, e.g., lagrangian dynamics in GST.
Superhypergeneral scientific theories, e.g., an SO theory of change.

Since all five categories are scientific, and the last two are philosophic as well, the border between science and philosophy has disappeared in our perspective. And, since there is no frontier left, there is no occasion for frontier skirmishes and no point in looking further for a demarcation criterion. So, one more old philosophical debate has been bypassed in the advancement of science.

8. CONCLUSION

GSTs have defied the scientificity criteria upheld by the most influential philosophies of science of our time, namely empiricism and rationalism. These hypergeneral theories found philosophy unprepared for their degree of generality. Actually, within the classical sciences there already existed extremely general theories that challenged the accepted slogan *Deduce-and-check*. Indeed, the generic scientific theories, such as quantum mechanics and the theory of evolution are untestable without further ado: we must enrich them not only with data, but also with extra assumptions before we can deduce testable consequences. GSTs have, then, just given the coup de grâce to the standard epistemologies and methodologies.

The collapse of the standard philosophies of science has forced us to revise the testability criteria. This revision has led to proposing a broader definition of the concept of scientific theory, namely Definition 5. This definition makes room for vicarious empirical testability, i.e., testability through the intermediary of specific theories. GSTs are testable this way. But so are theories in scientific ontology. So this places GSTs in philosophy as well as in science.

This is not to say that GSTs and scientific ontology coincide: they have an appreciable overlap, but they may be distinguished if one wishes. For one thing, GSTs, as conceived here, take for granted a number of notions that ontology makes it its business to elucidate, such as those of thing and property, space and time, causality and chance, natural law and history. For another, some GSTs are a bit too specific for philosophical purposes. For example, ontology is not particularly interested in the properties of linear macrosystems as opposed to nonlinear ones. Moreover, ontology cannot restrict its attention as GSTs have done so far to classical, i.e., nonquantal and nonrelativistic, systems. This is why we call SO *superhypergeneral*. But the differences between GSTs and theories in SO are perhaps less interesting than their similarities in subject matter and in method. Likewise, although there are certainly differences between a hypergeneral theory, such as the general classical field theory, and a generic theory, such as classical electrodynamics, both categories are species of the genus science (Edelen 1962).

Should any GST practitioner feel uncomfortable with this reclassification of his field because of the bad name so many philosophical schools have called for themselves, he should take into account that being called a philosopher is the price he has to pay for calling himself a scientist.

11

THE POWER AND LIMITS OF REDUCTION

(1991)

■

1. THE REDUCTION OPERATION

In this paper we shall be concerned with reduction as an epistemic operation and, more precisely, a kind of analysis. We take it that to reduce *A* to *B* is to *identify A* with *B*, or to *include A* in *B*, or to assert that every *A* is either an *aggregate*, a combination or an *average* of *B*'s, or else a *manifestation* or an *image* of *B*. It is to assert that, although *A* and *B* may appear to be very different from one another, they are actually the same, or that *A* is a species of the genus *B*, or that every *A* results somehow from *B*'s—or, put more vaguely, that *A* "boils down" to *B*, or that "in the last analysis" all *A*'s are *B*'s.

For example, the following hypotheses are instances of reduction, whether genuine or illegitimate. The heavenly bodies are ordinary bodies satisfying the laws of mechanics; heat is random molecular motion; light beams are packets of electromagnetic waves; chemical reactions are atomic or molecular combinations, dissociations, or substitutions; life processes are combinations of chemical processes; humans are animals; mental processes are brain processes; and social facts result from individual actions, or conversely.

At least four of the above hypotheses exemplify a special kind of reduction, namely microreduction. A *microreduction* is an operation whereby things on a higher level (macrolevel) are assumed to be either aggregates or combinations of entities belonging to some lower level(s) (microentities); macroproperties are assumed to result either from the mere aggregation of microproperties, or from a combination of the latter; and macroprocesses are shown to result from microprocesses. In short, microreduction is the accounting of wholes by their parts. The converse operation, whereby the behavior of individuals is explained by their place or function in a whole, may be called *macroreduction*. Of the two kinds of reduction the former is so much more common, successful, and prestigious than the latter that the term 'reduction' is usually taken to mean "microreduction".

The following are examples of microreduction: The magnetic field of a magnet results from the alignment of the magnetic moments of the component atoms; water is composed of molecules resulting from the combination of hydrogen and oxygen atoms; and all medium- and large-size plants and animals are systems of cells. Note the difference between aggregation and combination, on which more will be said below.

Whereas reduction is an epistemic operation, *reductionism* is a research strategy, namely the methodological principle according to which (micro)reduction is in all cases necessary and sufficient to account for wholes and their properties. On the other hand, macroreductionism is usually called *antireductionism*. The ontological partner of (micro)reductionism is atomism, while that of antireductionism is holism. In recent years antireductionism has become one of the battle cries of the "postmodernist" reaction against science and rationality generally.

There is no gainsaying the sensational successes of reduction in all natural and social sciences ever since Descartes formulated explicitly the program of reducing everything except the mind to mechanical entities and processes—or, as he put it, to *figures et mouvements*. (No wonder Descartes is the *bête noire* of the postmoderns, whether on the right or on the left.) One only needs to recall the spectacular achievements of nuclear, atomic, and molecular physics, as well as of molecular biology, to understand the popularity of reductionism among scientists, and the reluctance they feel to acknowledge that reduction may be limited after all.

True, mechanism declined since the birth of field physics in the mid-nineteenth century. (See, e.g., D'Abro 1939.) It is now generally understood that mechanics is only a part of physics, so that it is impossible to reduce everything to mechanics, even to quantum mechanics. Moreover, not even the whole of physics suffices for biology—which is not to say that vitalism has been resurrected. Physicalism, one of the tenets of the Vienna Circle and the *Encyclopedia of Unified Science*, died with the latter three decades ago—which is not to say that a richer version of materialism is dead.

However, the sharp decline of physicalism has not been the end of reductionism. Quite the contrary, reduction is still extremely successful, and it will continue to be so because all real things happen to be either systems or components of such. To reject reduction altogether is to deprive oneself of the joy of understanding many things and processes, and of the power that this knowledge confers.

Yet, as will be seen below, reduction is not omnipotent: it has limitations. In general, analysis is not enough: eventually it must be supplemented with synthesis. This is because the world and our knowledge of it happen to be systems rather than mere aggregates of independent units. In particular, it is often necessary to combine two or more theories or even entire research fields rather than reducing one to the other. Witness the very existence of physical chemistry, biochemistry, physiological and social psychology, bioeconomics, economic sociology, and hundreds of other interdisciplines.

In the following we shall begin by attempting to identify the ontological roots of the limitations of the reduction operation. We shall subsequently examine reduction at work in physics, chemistry, biology, psychology, and social science. This examination will exhibit the limits of reduction and the ontological roots of such limits.

2. WHOLES: AGGREGATION AND COMBINATION

There are two ways a whole may come into being: by aggregation or by combination. The accretion of dust particles or of sand grains exemplify aggregation; so do garbage dumps, water pools, clouds, unruly mobs, and columns of refugees fleeing from a disaster. What characterizes all of these wholes is the lack of a specific structure composed by strong bonds: such wholes are not cohesive. As a consequence, once formed they may break up rather easily under the action of external forces.

When two or more things get together by interacting strongly in a specific way they constitute a system, i.e., a complex thing possessing a definite structure. Atomic nuclei, atoms, molecules, crystals, organelles, cells, organs, multicellular organisms, biopopulations (reproductive communities), human families, business enterprises, and other formal and nonformal organizations are systems. They may all be said to emerge through combination or self-organization rather than aggregation, even though, once formed, some of them may grow by accretion or decline by attrition. Unlike mere aggregates, systems are more or less cohesive. However, they may break down either as a result of conflicting relations among their parts or in response to external forces.

The relevance of the distinction between aggregates and systems to the problem of reduction should be obvious. In accounting for the emergence and dismantling of aggregates, we focus on their composition and environment, in particular the external stimuli that favor the aggregation (or the disgregation) process. In this case structure matters little: a heap does not cease to be a heap if its parts exchange places. Therefore, basically we explain aggregates (and their disgregation) in terms of their composition and environment.

On the other hand, structure, in particular internal structure, is essential in the case of systems. Indeed, to account for the emergence of a system, we must uncover the corresponding combination or assembly process and, in particular, the bonds or links resulting in the formation of the whole. The same holds, mutatis mutandis, for any account of its breakdown. In other words, we explain the emergence, behavior, and dismantling of systems in terms not only of their composition and environment, but also of their total (internal and external) structure.

In short, we may model any system s by a triple $\langle C(s), E(s), S(s) \rangle$, the coordinates of which designate respectively the composition (collection of parts), environment, and structure (collection of links, couplings, or bonds) of s. The structure is partly internal (links among the system components) and partly external (links among system components and environmental items). I call the above the *CES model*. As noted previously, aggregates lack a definite internal structure; however, they do have an external structure, which is composed by their inputs and outputs. Hence, they too can be described by *CES* models.

One of the uses of *CES*-type modeling is that it prevents one from falling for what may be called *pseudoreduction*. This is the identification of a system with its composition, as exemplified by the assertions that an ice cube is "nothing but" a bunch of H_2O molecules, a living being "nothing but" a bag of atoms and molecules, and a society "nothing but" a collection of individuals. This is mistaken because it involves overlooking structure, i.e., the way the components of a system combine with one another. Shorter: Nothing-but-ism is wrong because it denies the very existence of systems.

As with things so with their properties: in this case too we must distinguish between aggregation and combination. The aggregation of things is accompanied by the aggregation of their properties. When no qualitative novelty is involved, the properties of the whole are said to be *resultants*. For instance, joining two liquid bodies, or two heaps of sand, results in a body the mass of which equals the sum of the partial masses. On the other hand, although a system is bound to have resultant or hereditary properties, it also has *emergent* ones. When a system breaks down those new properties may be said to *submerge*. The concepts of emergence and submergence being ontological categories, they are consistent with explanation, in particular of the reductive kind. Thus, contrary to a widespread prejudice, emergentism and reductionism are mutually compatible (Bunge 1977b). (Recall essay 5.)

Chemical reactions and biological evolution are paragons of emergence. The keys to the former are not only the atomic composition, but also the chemical bonds of various kinds and the environment. And the keys to biological evolution are genic variations, which are molecular changes, and the environmental forces causing natural selection. In both cases the peak of the process is speciation, or the emergence of systems of new kinds. (See, however, Mahner and Bunge 1997.)

In both cases the environment is usually described in a schematic macroscopic way. In the case of chemical reactions, more often than not, only the external pressure, temperature, and humidity are noted along with the kind of enclosure or the lack of it. Likewise, in the case of the emergence of new biospecies, only some of the global features of the environment, in particular climate and geographical accidents, are listed. In neither case is the environment analyzed into its elementary components.

This is one of the limitations of microreduction: that it treats the environment as a whole. It might be argued that this is only a practical limitation which an omniscient being would not face. This is doubtful: certain environmental features, such as the shape of a container and the climate of a region, are macroproperties that cannot be usefully analyzed in microphysical terms because they are properties of systems acting as wholes.

As with things and their properties so with processes, i.e., changes in properties. Two processes may run parallel courses or they may interact. For example, two chemical reactions may or may not interfere with one another. If they do, they may either compete for a given chemical or, on the contrary, they may cooperate with one another producing what each "needs" for its completion. Likewise, two children growing up together influence one another's development. In both cases it is insufficient to study each process component: we must also study how these components act upon one another and how they modify or are modified by third processes. And, while some of the items (things, properties, processes) may be described in microscopic terms, others will be described in macroscopic ones.

3. MICROLEVELS, MACROLEVELS, AND THEIR RELATIONS

For any kind of system we may distinguish at least two levels: the macrolevel and the microlevel. The *macrolevel* is the kind itself, i.e., the collection of all systems sharing certain peculiar properties. The corresponding *microlevel* is the collection of all the components of the systems in question. (In a moment it will be seen that there may be more than one microlevel.) For example, the atomic level is the set of atoms, while the molecular level is the set of molecules. (The fact that the molecular level is composed of a number of sublevels is beside the point.) In general, a level L system is composed of things of level $L - 1$.

The distinction between levels need not be arbitrary: it often has a real counterpart in the qualitative differences between systems and their components. However, levels are collections of things, hence concepts, not concrete things. Therefore, levels cannot act upon one another. In particular, the expression 'micro-macro interaction' does not denote an interaction between micro- and macrolevels, but an interaction between individual things belonging to a microlevel and things belonging to a macrolevel. (See also essay 5.)

Actually, it is only in particle physics and electrodynamics that a single microlevel need be distinguished. All the other sciences study systems or even supersystems composed of systems, so that they involve the distinction between a number of microlevels. In other words, most sciences tackle nested systems, sometimes called 'hierarchies'. Think, e.g., of the human brain, with its many subsystems, such as the hippocampus and the primary visual cortex, every one of which is composed by further systems, namely neurons and glial cells.

The complexity of the real systems studied in most sciences forces us to analyze the concept of composition into as many levels as needed. In other words, given the composition $C(s)$ of a system s, and a level L of organization (or integration), we define the *L-composition* of s, or $C_L(s)$, as the overlap or intersection of $C(s)$ and L. Hence, when speaking of *the* microlevel, we should specify exactly which one we have in mind.

All of the natural, social, and mixed sciences are faced with micro/macro gaps because all of them study systems of some kind or other, and all systems have components (the micro-aspect) as well as macroproperties of their own (the macro-aspect). In many cases one knows how to solve problems concerning the microlevel or the macrolevel in question, but one does not know how to relate them. In particular, one seldom knows how to account for macrofeatures in terms of microentities and their properties and changes thereof. Consequently, the microspecialists (e.g., microeconomists) and macrospecialists (e.g., macroeconomists) outnumber the experts in gap bridging.

Every problem concerning a micro-macro relation is intrinsically difficult. This difficulty is compounded by the dearth of careful philosophical analyses of the micro-macro relations. We shall proceed to sketch such an analysis. The very first task we must perform is that of distinguishing two basic types of micro-macro relations: *De re*, or ontological, and *de dicto*, or epistemological. Ontological micro-macro relations are a particular case of the part-whole relation, whereas epistemological micro-macro relations are conceptualizations of relations between microlevels and macrolevels. Let me explain.

The assembly of two or more atoms (or molecules, or cells, or animals) into a higher-level entity is a case of ontological micro-to-macro relation. Likewise, the converse process of system dismantling exemplifies ontological macro-to-micro relation. The water condensation effect of a table salt molecule, the effect of the pacemaker cells on the heart, and of a leader on an organization, are further examples of the ontological micro-to-macro relation. When a limb is cut off, its cells die; when "the sun goes down", the average kinetic energy of the air molecules decreases; and when an organization is banned, all of its members suffer. These are instances of the ontological macro-to-micro relation. In all of these cases, links (or bonds or couplings) between micro and macro things or processes are involved.

No such bonding relations are involved in the relations among levels of organization, because—as noted above—levels are sets, hence concepts, not concrete things or processes. For example, the formula for the entropy of a thermodynamic system in a given macrostate, in terms of the number of atomic or molecular states or configurations compatible with the given macrostate, i.e., the famous formula "$S = k \ln W$", is a micro-to-macro relation of the epistemic type. So are the formulas for the specific heat, the conductivity, and the refractive index of a body in terms of properties of its atomic constituents. Ditto Hebb's theory of learning in terms of the reinforcement of interneuronal connections: here the macrolevel is composed by brain subsystems ("psychons") capable of performing mental functions, and the microlevel by neurons. There is no action of neurons on the brain or on the mind: there is only a conceptual relation between two levels of organization.

4. MM, Mm, mM, AND mm RELATIONS

Combining the micro-macro distinction with the ontologico-epistemological one, we get a total of eight interlevel relations:

1. MICRO-MICRO (mm)

(a) *Ontological*, e.g. atomic collisions and the love bond.

(b) *Epistemological*, e.g., quantum theories of atoms, and psychological theories of interpersonal relations.

2. MICRO-MACRO (mM)

(a) *Ontological*, e.g., the interaction between an electron and an atom as a whole; a landslide caused by a small atmospheric perturbation; a social movement triggered by a charismatic leader.

(b) *Epistemological*, e.g., statistical mechanics, astrophysics, and a theory of animal behavior triggered by microstimuli, such as a handful of photons striking a retina.

3. MACRO-MICRO (Mm)

(a) *Ontological*, e.g., the action of a flood or of an earthquake on an animal, and the effect of governments on individuals.

(b) *Epistemological*, e.g., a theory of an intrusive measurement on a microphysical entity, and a model of the course of a ship adrift in an oceanic current.

4. MACRO-MACRO (MM)

(a) *Ontological*, e.g., Sun-Earth interaction, and rivalry between groups of animals (e.g., families).

(b) *Epistemological*, e.g., planetary astronomy, plate tectonics theory, and international relations models.

These distinctions are relevant to the theories of definition and of explanation. They are also relevant to the old dispute between reductionists and antireductionists. This philosophical controversy occurs in every science. For instance, molecular biologists fight with organismic biologists, and in the social sciences individualists fight with collectivists. Whereas reductionists claim that only *mm* and *mM* relations have explanatory power, their opponents write these off and hold that only *Mm* and *MM* relations can explain.

In our view, both contenders are partially right, hence partially wrong: since all four relations exist, all of them pose problems. In particular, we need to investigate how individuals of all kinds interact (*mm*), and how they assemble forming systems (*mM*). But we also need to know how being part of a system affects the individual (*Mm*), and how systems affect one another (*MM*). The need for such broader research projects shows that radical reductionists are just as wrong as radical antireductionists. Hence, we had better adopt the systemic approach, which embraces all four relations—and more when required.

Evidently, whenever we distinguish more than two levels, we are faced with several further relations. For example, interpolating a mesolevel between a micro- and a macrolevel, and adding a megalevel on top of the latter, we get four same-level relations plus three interlevel relations, thus a total of seven without skipping levels—actually fourteen when the ontological-epistemological distinction is drawn. This happens, for instance, when relations between genome, cell, whole organism, and ecosystem are investigated. Another well-known four-level distinction is that between individual, firm, cartel, and whole economy.

We wind up this section by rewording the above in terms of the kinds of proposition that can be constructed when distinguishing just two levels, a microlevel (m) and a macrolevel (M). By combining the corresponding concepts we may form propositions of four different kinds, two same-level and two interlevel ones:

1. mm, e.g., hypotheses concerning specific nuclear forces, interneuronal connections, and face-to-face personal relations.

2. mM, e.g., the formulas of statistical mechanics, genotype-phenotype relations, and the macrosocial outcomes of individual actions, such as voting.

3. Mm, e.g., Lorentz's formula for the force exerted by an external magnetic field on an electron, and the hypotheses on the influence of social structure upon individual behavior.

4. MM, e.g., Newton's law of gravity, the rate equations of chemical kinetics, the ecological equations on interspecific competition, and data on international conflicts.

As with propositions so with explanations. That is, in principle an explanation may contain same-level or interlevel explanatory premises or conclusions (Bunge 1967a, b). Moreover, explanations of all four kinds are needed because ours is a world of systems, and systems must be understood on their own level as well as resulting from the assembly of smaller units and constraining the behavior of their components (Bunge 1979a).

In other words, ontology must be realistic and it must guide epistemology if the latter is to be useful. (This view of the dependence of epistemology upon ontology is in sharp contrast with the positivist thesis that the "logic of science", and in particular the analysis of reduction, must be free from ontological assumptions: see, e.g., Carnap 1938.) One-sided ontologies, such as individualism (atomism) and collectivism (holism), outlaw each other's explanation types and thus curtail the power of science and technology. Only a systemic ontology encourages the search for explanations of all four kinds.

In the following we shall examine a few typical examples drawn from five branches of contemporary science: physics, chemistry, biology, psychology, and sociology.

5. MICROREDUCTION IN ACTION—AND AT BAY

The quantum theory is usually regarded as the nucleus of microphysics and, moreover, as the key to the reduction of macrophysical to microphysical things and processes. This view is largely correct, though only with some qualifications. Indeed, the quantum theory contains a number of concepts borrowed from macrophysics, such as those of space, time, mass, electric charge, classical linear momentum, and classical energy. Thus, four of these concepts, namely those designated by x, t, p, and E, occur in the most basic of all state functions, namely the plane "wave" $\Psi = A\ exp\ [i(px - Et)/h]$. Besides, the boundary conditions, which are part of the definition of every state function, constitute schematic macrophysical representations of the environment. (Example: the condition that the state function vanishes on the surface of a container.) A more dramatic and popular case is that of inseparability: a quantum system remains a system even if its components become spatially separated. (See essay 17.) Finally, in every measurement the measuring apparatus is treated as a macrophysical system described in classical or at most semiclassical terms (see, e.g., Bohr 1958).

What holds for quantum physics holds, a fortiori, for quantum chemistry. This discipline contains not only the above-mentioned macrophysical concepts, but also certain macrochemical ones. Thus, one of the accomplishments of quantum chemistry is the *ab initio* calculation of the equilibrium constants of chemical reactions. In classical chemistry these constants are treated as empirical parameters. In quantum chemistry they are part of the theory of chemical reactions construed as inelastic scattering (collision) processes. However, this theory presupposes the rate equations of chemical kinetics, a part of classical chemistry. Indeed, consider the problem of calculating the rate or equilibrium constant k of a chemical reaction of the type "$A + B \rightarrow C$". The phenomenological (macrochemical) equation for the rate of formation of the reaction product C is $dn_C/dt = kn_A n_B$. This classical equation is not deduced but postulated when the rate constant k is calculated in quantum-theoretical terms. Hence, quantum chemistry does not follow from quantum mechanics without further ado. In other words, chemistry is not fully reducible to physics. The epistemological reduction is only partial, even though the ontological one is total. (For details see Bunge 1982a; Lévy 1979.)

Our third case is that of biology. Undoubtedly, the most sensational advances in contemporary biology have been inspired by the thesis that organisms are nothing but bags of chemicals, whence biology is nothing but extremely complex chemistry. (See, e.g., Bernard 1865.) But this thesis, though heuristically enormously powerful, is not completely true, as we shall attempt to show by considering the cases of genetics and of the very definition of the concept of life.

At first sight, the discovery that the genetic material is composed of DNA molecules proves that genetics has been reduced to chemistry (see, e.g., Schaffner 1969). However, chemistry only accounts for DNA chemistry: it tells us nothing about the biological functions of DNA, e.g., that it controls morphogenesis and protein synthesis. In other words, DNA does not perform any such functions when outside a cell. Analog: an airline pilot does not function as such at home.

The reason is, of course, that the very concept of a living cell is alien to chemistry. True, the cell components are physical and chemical entities, but in the cell these components are organized in a peculiarly biological way. It is also true that every single property of a cell, except for that of being alive, is shared by some physical or chemical systems. But only living cells possess jointly the dozen or so properties that characterize living systems. Consequently, biology, though based on physics and chemistry, is not fully reducible to the latter. For example, any system of biological classification based exclusively on the degree of difference in RNA is bound to fail for missing the supramolecular features of organisms.

What about psychology: is it reducible to biology? Assume, for the sake of the argument, that all mental processes are brain processes. Does this entail that psychology is a branch of biology and, in particular, of neuroscience? Not quite, and this for the following reasons. First, because brain processes are influenced by social stimuli, such as words and encounters with friends or foes. Now, such psychosocial processes are studied by social psychology, which employs sociological categories, such as those of social group and occupation, which are not reducible to neuroscience. A second reason is that psychology employs concepts of its own, such as those of emotion, consciousness, and personality, as well as techniques of its own, such as interrogation and deceit, which go beyond biology.

We conclude, then, that even though the psychoneural identity hypothesis is a clear case of ontological reduction, and tremendously fertile to boot, psychology is not reducible to neuroscience, even though it has a large overlap with it. (For details see Bunge 1990; see also essay 19.) Shorter: ontological reduction does not imply epistemological reduction. The parallel with quantum chemistry is striking.

Our final example is that of social science. Scholars otherwise as different as the idealist philosopher Wilhelm Dilthey and the behaviorist and utilitarian sociologist George Homans have attempted to reduce social science to psychology. According to psychological reductionism, every social fact is the outcome of individual actions steered by the actors' beliefs, values, goals, and intentions. In this perspective neither the natural nor the social environment would play any role in constraining individual actions: all persons would be free rational agents acting so as to maximize their utilities (see, e.g., Homans 1974).

However, even the most extreme reductionists admit that an individual is likely to act differently in different situations or circumstances. But they do not bother to analyze such situations in terms of individual thoughts and actions. Instead, they describe them in molar terms: there had been a drought, a new law prescribed this or that, the country was at war, the transportation system was overloaded, and so on. In the end, then, sociological individualists cannot possibly carry out their methodological prescription.

Social scientists are bound to formulate propositions of the forms "Individual x in situation y performs action z", and "Forced by circumstance (or institution) x, individual y did z", where the concepts of situation (or circumstance) and institution are not analyzed in psychological terms. In conclusion, although social science needs psychology, it is not reducible to it. Shorter: psychological reductionism does not work in social science.

What often does work is relating two levels without attempting to reduce one to the other, as suggested by the following elementary example. It is well known that, when the interest rates rise above a certain level, the construction industry declines. In obvious symbols, $R \Rightarrow D$. This relation between two macroeconomic variables can be explained as follows. If the interests rates rise beyond a certain level, poor people cannot afford to buy or build houses, as a consequence of which the construction industry declines. In symbols,

$$R \Rightarrow \forall x(Px \Rightarrow \neg Bx) \qquad M\text{-}m$$
$$\forall x(Px \Rightarrow \neg Bx) \Rightarrow D \qquad m\text{-}M$$
$$\therefore R \Rightarrow D \qquad M\text{-}M.$$

In conclusion, the microreduction preached by the methodological individualists cannot work because every person is a component of several social systems that constrain individual freedom in several ways. (Shorter: structure constrains agency, which reacts back on structure.) By the same token, mM and Mm relations are bound to be key components of social science. The fact that eminent social scientists, such as

Boudon (1981) and Coleman (1990), call themselves 'methodological individualists' while investigating such mM and Mm social relations, should not fool the philosopher. (See also Bunge 1996, 1998.)

6. KINDS AND LIMITS OF REDUCTION

Reduction may bear on concepts, propositions, explanations, or hypo-thetico-deductive systems (Bunge 1983b). To reduce a concept A to a concept B is to define A in terms of B, where B refers to a thing, property, or process on either the same or a different (lower or higher) level than that of the referent(s) of A. Such a definition will be called a reductive def-inition. (In the philosophical literature *reductive definitions* are usually called 'bridge hypotheses', presumably because they are often originally proposed as hypotheses. History without analysis can be misleading.) Example: "Heat $=_{df}$ Random atomic or molecular motion". Downward reductive, i.e., microreductive, definitions may also be called *bottom-up* definitions. By contrast, upward reductive, i.e., macroreductive, ones may be called *top-down* definitions. Example: "Conformism $=_{df}$ An indi-vidual's bowing to the ruling norms or habits". But there are also same-level definitions, such as "Light is electromagnetic radiation".

The reduction of a *proposition* results from replacing at least one of the predicates occurring in it with the definiens of a reductive definition. For example, the psychological proposition "X was forming a linguistic expression" is reducible to the neurophysiological proposition "X's Wer-nicke's area was active", by virtue of the reductive definition "Formation of linguistic expressions $=_{df}$ Specific activity of the Wernicke area".

An *explanation* can be said to be reductive if, and only if, at least one of its explanans premises is a reductive definition or a reduced proposi-tion. For example, the explanation of the existence of a concrete system in terms of the links among its parts is of the microreductive (or bottom-up) kind. By contrast, the explanation of the behavior of a component in terms of the place or function it holds in a system is of the macroreductive (or top-down) type. Work on a car assembly line (or on the origin of life) induces explanations of the first kind, while the car mechanic (and the medic) typically resorts to explanations of the second type.

The analysis of theory reduction is somewhat more complex. Call T_1 and T_2 two theories (hypothetico-deductive systems) sharing some ref-erents, R a set of reductive definitions, and S a set of subsidiary hypotheses not contained in either T_1 or T_2. (However, these hypotheses must be couched in the language resulting from the union of the lan-guages of T_1 and T_2 if they are to blend with the latter.) We stipulate that

(a) T_2 is fully (or strongly) reducible to $T_1 =_{df} T_2$ follows logically from the union of T_1 and R;

(b) T_2 is partially (or weakly) reducible to $T_1 =_{df} T_2$ follows logically from the union of T_1, R, and S.

Ray optics is strongly reducible to wave optics by way of the reductive definition "Ray $=_{df}$ Normal to wave front". In turn, wave optics is strongly reducible to electromagnetism by virtue of the reductive definition of "light" as "electromagnetic radiation of wavelengths comprised within a certain interval". On the other hand, the kinetic theory of gases is only weakly reducible to particle mechanics because, in addition to the reductive definitions of the concepts of pressure and temperature, the former includes the subsidiary hypothesis of molecular chaos (or random initial distribution of positions and velocities).

Likewise, as we saw in section 5, quantum chemistry, cell biology, psychology, and social science are only weakly (partially) reducible to the corresponding lower-level disciplines. We also saw that even the quantum theory contains some classical concepts as well as subsidiary hypotheses, e.g., about macrophysical boundaries, so that it does not effect a complete reduction to microphysical concepts. (More on the various kinds and aspects of reduction in Bunge 1983b, 1989b.)

7. CONCLUSIONS

We make bold and generalize the preceding results by stating that, while partial microreduction is often successful, full microreduction seldom is. Moreover, we hazard to explain the failures of full microreduction by the hypothesis that every real thing, except for the universe as a whole, is embedded in some higher-level system or other. Hence, every mM relation is accompanied by some Mm relation, and both have often mm or MM concomitants. For this reason, same-level definitions and explanations must be supplemented with interlevel, in particular bottom-up and top-down, definitions or explanations. Which goes to show that, to be of any use to science and technology, epistemology must match ontology. More precisely, a realistic epistemology must be paired off to a systemic ontology.

In short, although reduction should be pushed as far as possible, its limits should be realized. Consequently, moderate reductionism is a more realistic research strategy than radical reductionism, and either is more powerful than antireductionism.

12

THINKING IN METAPHORS

(1999)

■

Poetry is metaphorical. When not, as is the
case of Robert Frost's writing, one suspects
that it is not poetic. The postmodernists too are long on metaphor, which
is not surprising, since most of them have a literary background—and
none of them is interested in truth, the exclusive property of some literal
statements. However, love of metaphor is not restricted to poets and
postmodern scribblers. Every teacher uses analogies that, rightly or
wrongly, are expected to draw attention and awaken understanding.

Metaphor is also conspicuous in the sciences. For example, during
the long reign of the mechanistic worldview (ca.1600– ca.1900) every
new scientific discovery was expected to be presented in mechanical
terms, which were often metaphorical. This was the case with Maxwell's
reluctant and unsuccessful attempt to clad electromagnetic field theory
in mechanical garb. However, outside the field of mechanics, mechanical
models are now just historical curiosities.

Likewise, the information revolution that started nearly half a cen-

tury ago led molecular biologists and psychologists to crafting information-processing models of all the processes they could not fully understand in chemical or biological terms. If you do not understand how
thing *A* activates thing *B*, say that there is a flow of information from *A*
to *B*, and abstain from defining 'information'. In this way you can buy the
illusion of understanding, though at the high price of voiding chemistry
from chemical bonds, and of turning biology into a branch of communication engineering.

Philosophers too have actively participated in promoting metaphors.
A first systematic endeavor in this direction was Hans Vaihinger's fictionism (1920). According to fictionism, science replaces "that"
(description) and "because" (explanation) with "as if". For example,
bodies are said to behave as if they satisfied the laws of mechanics, and
individuals to behave as if they were selfish. All knowledge is allegedly
fictitious and none is true, though some is useful. In particular, some
myths are useful social control tools. Half a century later, Max Black
(1962) and Mary Hesse (1966) reinvented fictionism. They claimed that
scientific theories are metaphors. Their pet example was Bohr's model of
the atom as a miniature solar system—a model that was soon superseded
by the quantum theory, which disallows pictures. (Hesse is often fondly
cited as authority by some postmoderns. See also Ricoeur 1975.)

No doubt, we often think in metaphors in all fields, just as we often
think in intuitive or pre-analytic terms. The question is: When is
metaphor justifiable? To answer this question, let us start by recalling
half a dozen well-known statements, half of them literal (L) and the
others metaphorical (M).

L1 $E = mc^2$.

L2 Criminality increases with unemployment.

L3 Nothing comes out of nothing, and nothing turns into nothing
(Lucretius).

M1 The Lord is my shepherd.

M2 The genome is like a language. It has a vocabulary (the genes), a
grammar (the way in which information is arranged), and a literature
(the instructions needed to make an organism). (Summary from any of
hundreds of textbooks in molecular biology.)

M3 Society is like a text to be interpreted.

To begin with, note that the above literal statements are generally
regarded as testable. Surely, L3 is far too general to be empirically

testable: it is a piece of ontology. But, far from being an arbitrary claim, it is a presupposition of modern science. Indeed, if something seems to pop out of nothing, or to disappear without leaving traces, scientists are driven to look for the missing item or to hypothesize it. For example, when in 1930 a certain nuclear reaction seemed to violate the principle of conservation of energy, Wolfgang Pauli proposed the hypothesis that the missing energy was carried off by a neutrino. This thing was eventually discovered a quarter century later.

In short, literal statements about matters of fact are true or false, at least potentially. (Only some mathematical statements are undecidable, but they have not stood in the way of mathematical progress.)

Note the following features of metaphors in contrast to literal statements. First, they are expected to be evocative or suggestive rather than true. Hence, they can be tested for fertility but not for truth. This suffices to disqualify them from being part of scientific theories. In other words, if a body of knowledge contains a metaphor, then that body is not a scientific theory proper, but at most an embryo of such.

Second, many a metaphor is culture-bound rather than cross-cultural. For example, the Judeo-Christian metaphor "The Lord is my shepherd" may mean nothing to the modern city dweller, who is unlikely to have ever seen real shepherds or live lambs. (He or she is likely to find "The CEO is my boss" more familiar and persuasive. But this statement must be interpreted literally: as a metaphor it might be blasphemous.) Or take the famous verse in the Song of Solomon (4:5): "Your two breasts are like fawns, twins of a gazelle, which feed among the lilies." To our modern ear this metaphor sounds far-fetched, even ridiculous. Obviously, both metaphors are local or culture-bound, whereas literal statements are supposed to be universally understandable.

Third, some metaphors, though heuristically fertile at the beginning, end up by being obstacles to understanding. For instance, the ancient sperm-seed metaphor is likely to have slowed down the development of embryology by centuries. Indeed, the mammalian egg was found only in 1827.

Another example: in the early days of quantum mechanics, electrons and the like were pictured as either particles or waves. People wondered about the so-called wave-particle duality, and some despaired of ever getting over this apparent contradiction. It took nearly three decades to realize that electrons and their kin are neither waves nor particles, but quanta (units) of the fields described by quantum field theory. (This is why I coined the term *quanton* to denote them: see Bunge 1985a.) In other words, the ideas of wave and particle had been used as classical analogs. Mature quantum theory has no need for such metaphors, just as mature electrodynamics has no need for mechanical analogs.

Fourth, some metaphors block scientific research from the very beginning. For example, the ideas that people interpret themselves and others, whence social facts are "texts or like texts", invites hermeneutic (textual) analysis. By the same token, it discourages objective observation, measurement, and mathematical modeling in social studies. Moreover, this metaphor is pathetically inadequate, since social groups have neither syntactical nor semantical nor phonological nor literary properties. Its popularity is only due to the fact that it demands neither empirical investigation or mathematical modeling.

Fifth, some metaphors slow down research because they produce the illusion of understanding. Thus, the students who parrot the creed that protein synthesis is the result of the transfer of some of the genetic information or instructions encoded in DNA, are unlikely to bother asking for the chemical mechanisms involved in the process. They will not realize that, so far, no theory (hypothetico-deductive system) of protein synthesis has been built and corroborated. Much the same holds for the information flowcharts rampant in cognitive psychology: the blocks and arrows occurring in them are only question marks. In short, in these fields the word 'information' serves only as a cloak to cover naked ignorance. (See also Mahner and Bunge 1997.)

All this is not to deny that metaphor has a legitimate function: that of conveying part of the meaning of a novel idea in familiar terms. This is so not only in science but also, and primarily, in teaching. Thus, when telling a young child that an electric current is like a flow of water, that an electron is like a tiny marble, that the sun is like a huge bonfire, that the president of the nation is like the principal of a big school, or that digits are like letters, we draw coarse sketches in the expectation that they will quench momentarily the child's thirst for knowledge. And we dread or cherish the thought that, sooner or later, the child will feel dissatisfied with our effort, and will demand to know more. In short, we know full well that our metaphors are just crude cartoons.

To conclude. Metaphors may be useful in teaching as well as in the beginning of a scientific project, in suggesting similarities with known things or processes, as in the case of the pair light wave-water wave. But a time is bound to come when the peculiarity of the new thing proves to be more important than its similarity with the old one. This is the time to get rid of the crutches and interpret the new idea in its own terms— for instance, to speak of the electromagnetic field rather than of the elastic ether, of quantons rather than of particles or waves, and of the structure of the genetic material rather than of genetic information.

Metaphor may be the seal of poetry, but in science it is a sign of immaturity. Beware, then, the information-processing metaphor of protein synthesis, the computer metaphor of the brain, and the textualist metaphor of society, for they are no better than "you are my sunshine".

III

PHILOSOPHY
OF MATHEMATICS

■

13

MODERATE MATHEMATICAL FICTIONISM

(1997)

■

1. INTRODUCTION

Radical fictionism (or fictionalism) is of course the doctrine that all discourse is fictive, so that there is no truth of any kind—mathematical, factual, or other. Like other epistemological doctrines, fictionism has old roots. One of them is radical skepticism or Pyrrhonism ("Nothing can be known with certainty"); another is nominalism ("There are no concepts: there are only things and names of things"). However, fictionism only attained adulthood in pragmatism ("Ultimately only action counts"). And it flowered in Vaihinger's monumental book *Die Philosophie des Als-ob* of 1911 (1920), which owed much to Kant, Lange, and Nietzsche.

I submit that fictionism is utterly false of factual science, which seeks to account for the real world, but quite true of pure mathematics. More precisely, I suggest that the sciences of reality, such as physics, biology, and sociology, fit epistemological realism, for they presuppose the reality of the external world, which they explore in order to know it. That there are stars out there is fact, not fiction. I also suggest that the modern (or science-based) technologies, from engineering to biotechnology to management science, aim at helping change the world—rather than any fictions—on the basis of its scientific understanding. Hence they, too, presuppose that the world exists by itself and can be known, though not necessarily fully or at one go (Bunge 1985b).

By contrast, pure mathematics is not about the real world or about experience, and mathematical proofs are not empirical. The mathematical objects, such as sets, functions, categories, groups, lattices, Boolean algebras, topological spaces, number systems, vector spaces, differential equations, manifolds, and functional spaces, are not only *entia rationis*: they are *ficta*. Consequently, the concept of existence occurring in mathematical existence theorems is radically different from that of real or material existence. Therefore, mathematical existence proofs—and all other mathematical proofs for that matter—are purely conceptual procedures. In short, mathematicians, like abstract painters, writers of fantastic literature, and creators of the animated cartoons featuring talking animals deal in fictions.

This, in a nutshell, is the kind of mathematical fictionism to be sketched and argued for in this paper. As will be seen below, it is fictionism of a moderate rather than a radical kind, for it regards mathematics as a science, not as a game, or a convention, let alone an arbitrary fantasy. Furthermore, it distinguishes between mathematical fictions on the one hand and myths, fairy tales, parapsychological and psychoanalytic fictions, as well as many-worlds philosophical theories, on the other.

2. THREE KINDS OF TRUTH: FACTUAL, FORMAL, AND ARTISTIC

I follow Leibniz and a few others, such as Grassmann, in distinguishing *propositions de raison* from *propositions de fait*. The former refer exclusively to *entia rationis*, and they are proved or refuted by purely conceptual means, namely argument (deduction and criticism) or counterexample. By contrast, the *propositions de fait* refer at least partly to real (concrete) entities and they are confirmed or infirmed with the help of direct or indirect empirical operations, such as observation, counting, measurement, or experiment.

The formal/factual distinction calls for distinguishing formal from factual truths (or falsities). In particular, we distinguish mathematical theorems on the one hand, and scientific (e.g., biological) data and hypotheses on the other. The difference between the two kinds of truth is so pronounced, that a factual theory, such as classical electrodynamics, contains some mathematically true formulas (e.g., those for advanced potentials) that fail to match the facts, i.e., that are factually false. Likewise, mathematics is full of theorems that have yet to be employed in factual science.

We have just smuggled in the distinction between the formal and the factual sciences. We define formal science as a science that contains exclusively formal propositions, or *propositions de raison*. By contrast, at least some of the propositions of a factual science must be factual: they must describe, explain, or predict things or processes belonging to the real (natural or social) world. Logic, philosophical semantics, and mathematics are formal sciences. On the other hand, the natural, social, and socionatural sciences are factual. So are all the technologies, from mechanical engineering to knowledge engineering to management science.

The formal/factual distinction leaves out all the propositions and fields that are neither formal nor factual; hence it is not a dichotomy. Among them the artistic fictions stand out. When reading about Don Quixote we may feign, and may actually feel, that he exists along with the figments of his own sick imagination. And when attending a performance of *Othello* we may believe for a moment that in fact Othello kills Desdemona. But when reflecting critically upon these and other works of fiction, we do not mistake them for factual accounts, unless of course we happen to be mad. We group them together under artistic fiction. Moreover, occasionally, we are justified in talking about artistic truth and falsity, as when we say that Don Quixote was generous, and Othello's suspicion false. In order to establish the artistic truth or falsity of an artistic fiction, we only resort to the work of art in question. That is, artistic truth, like mathematical truth, is internal and therefore context-dependent. That is, it only holds in some context and it need bear no relation to the external world.

Allow me to repeat a platitude: Mathematical truth is essentially relative or context-dependent. For example, the Pythagorean theorem holds for plane triangles, but not for spherical ones; and not all algebras are commutative, or even associative. By contrast, a factual statement, such as "There are photons", and "The computer cult is threatening pure mathematics", are absolute or context-free.

Finally, let us admit that the problem of truth, though central in factual science and philosophy, is peripheral in mathematics. As Mac Lane (1986) writes, it is not appropriate to ask of a piece of mathematics whether it is true. The appropriate questions are whether a piece of mathematics is correct, "responsive" (i.e., solves a problem or carries further some line of research), illuminating, promising, or relevant (to science or to some human activities).

In sum, we distinguish between formal, factual, and artistic truths and falsities. (We may even add moral truths, such as "Racial discrimi-

nation is unfair", and "Poverty is morally degrading", but these are irrelevant to our present subject. See essay 27.) Moreover, from the above
discussion it is clear that, in our view, mathematics is closer to art than
to science as regards its objects, methods, and relations to the real world,
as well as regards the role of truth. However, as will be argued in section
7, there are important differences between mathematics and art.

3. MATHEMATICS IS ONTOLOGICALLY NEUTRAL

To say that logic and mathematics are formal sciences is to say that they
have no ontological commitment, i.e., that they do not assume the existence of any real entities. In other words, logic and mathematics, and a fortiori metalogic and metamathematics, are not about concrete things but
about constructs: predicates, propositions, and theories. For example,
predicate logic is about predicates and propositions; category theory is
about abstract mathematical systems; set theory is about sets; number
theory is about integers; trigonometry is about triangles; analysis is about
functions; topology and geometry are about spaces; and so on. (Warning:
Quine and others misuse the word 'ontology' when they equate it to the
universe of discourse, or the reference class of a construct. An ontology is
not a set of items but a theory about the world.)

The thesis that mathematics is about mathematical objects seems
self-evident, but it may not be proved in a general way. What we can do
is to support it in two ways: by methodological and semantical considerations. The methodological consideration is in two stages. A first stage
consists in recalling that all known mathematical objects are defined,
explicitly or implicitly, in purely conceptual ways, without resorting to
any factual or empirical means, except occasionally as heuristic devices.
The second step consists in recalling that mathematical proofs (and refutations) too are strictly conceptual processes making no reference to
empirical data. (In this regard computer-assisted proofs are no different
from pencil-assisted ones.)

As for the semantical consideration, it consists in identifying the referents of mathematical constructs, i.e., in finding what they are about.
(For example, set theory is about nondescript sets, whereas number
theory refers to natural numbers.) This task requires a theory of reference, as distinct from a theory of extensions. (Unlike the latter, the
former makes no use of any truth concept. For example, the predicate
"even and not-even" refers to integers while its extension is empty, for it
is not true of any number.)

I shall presently use my own axiomatic theory of reference, which is
couched in elementary set-theoretic terms (Bunge 1974a). Let us begin

by elucidating the general notion of a predicate. Unlike Frege and his followers, I define a predicate as a function from individuals, or n-tuples of individuals, to the set S of propositions containing the predicate in question. That is, an n-ary predicate P is to be analyzed as a function $P\colon A_1 \times A_2 \times \ldots \times A_n \to S$, with domain equal to the Cartesian product of the n sets of individuals concerned, such that the value of P at $\langle a_1, a_2, \ldots, a_n \rangle$ in that domain is the atomic statement $Pa_1a_2 \ldots a_n$ in S. For example, the predicate "prime" is a function from the natural numbers to the set of propositions containing "prime", i.e., $P\colon \mathbb{N} \to S$.

I next define two reference functions, $\mathcal{R}_p$ and $\mathcal{R}_s$, the first for predicates and the second for statements, through one postulate each:

AXIOM 1. The reference class of an n-ary predicate P equals the union of the sets occurring in the domain of P, i.e., $\mathcal{R}_p(P) = \cup_{1 \le i \le n} A_i$.

AXIOM 2.

(a) The referents of an atomic proposition are the arguments of the predicate(s) occurring in the proposition. That is, for every atomic formula $Pa_1a_2 \ldots an$ in the set S of statements, $\mathcal{R}_s(Pa_1a_2 \ldots a_n) = \{a_1, a_2, \ldots, a_n\}$.

(b) The reference class of an arbitrary propositional compound, such as a negation, a disjunction, or an implication, equals the union of the reference classes of its components. (Corollary: A proposition and its negation have the same referents. By the way, the insensitivity of reference to the differences among the logical connectives is one of the differences between reference and extension, two concepts that are usually conflated in the philosophical literature.)

(c) The reference class of a quantified formula (i.e., one with the prefix "some" or "all") equals the union of the reference classes of the predicates occurring in the formula.

With the help of this theory we can identify the referents of any construct (predicate, proposition, or theory). For example, since the logical operations (such as conjunction) and logical relations (such as that of entailment) relate propositions, they are about the latter and nothing else. The formal proof for disjunction goes like this. Disjunction can be analyzed as a function from pairs of propositions to propositions, i.e., $\vee\colon S \times S \to S$. By Axiom 1 above, $\mathcal{R}_p(\vee) = S \cup S = S$. That is, disjunction refers to arbitrary propositions—formal, factual, moral, artistic, or what have you: it is ontologically noncommmittal. This proof dispenses with the possible-worlds metaphor.

It is also easily seen that our own theory of reference is about arbitrary predicates and propositions. It is then a formal theory just like logic. The same applies to our theories of sense and meaning and, indeed, to our entire philosophical semantics (Bunge 1974a, b, d).

To be sure, no theory of reference can prove that every single mathematical formula refers exclusively to constructs. But ours can test any *particular* claim concerning reference.

4. THE ALLEGED ONTOLOGICAL COMMITMENT OF THE "EXISTENTIAL" QUANTIFIER

What about Quine's well-known claim that the so-called existential quantifier involves an ontological commitment? This claim is false, as one realizes upon recalling that, unless the context is indicated, an expression of the form '$(\exists x)Px$' does not tell us whether the individuals in question are real or imaginary, i.e., whether we are talking about real or ideal existence. In other words, the symbol $\exists$ is ambiguous, hence incomplete. Therefore, the "existential" quantifier should always be completed by indicating the set over which the bound variable in question ranges. The once-standard notation "$(\exists x)_D Px$", where D names the universe of discourse, will do. (The bounded "existential" quantifier can be defined thus: $(\exists x)_D Px =_{df} (\exists x)(x \in D \& Px)$. See also essay 6.)

A modicum of conceptual analysis, clarifying the ambiguity of the word 'existence', would have spared us the false but influential thesis of the ontological commitment of the "existential" quantifier, hence of predicate logic and everything built upon the latter. For the same reason such analysis would have spared us Field's (1980) revival of the dual nominalist thesis, that "there are" no mathematical entities.

Such analysis would also have shown that, outside mathematics, the proper interpretation of that quantifier is not "existence" but "someness". That is, in mathematics "$(\exists x)Px$" can be read as "There are P's", with the understanding that such individuals are conceptual, i.e., exist in some mathematical universe. But outside mathematics it is best to avoid the ambiguity and read "$(\exists x)Px$" as "Some individuals are P's". In other words, only in mathematics "there are" amounts to "some". In alternative contexts, e.g., in physics and in ontology, when speaking of existence we may have to make two distinct statements: one of real or physical existence, and the other of someness. (See essay 6.)

In conclusion, pure mathematics is ontologically neutral and, more precisely, a gigantic (though not arbitrary) fiction. This explains why pure mathematics (including logic) is the universal language of science, technology, and even philosophy: why it is the most portable and serviceable of all sciences. It also explains why the validity or invalidity of mathematical ideas is independent of material circumstances, such as the state of the brain and the state of the nation. More on this anon.

5. MATHEMATICS, BRAINS, AND SOCIETY

The reader who has come this far may wonder how mathematical fictionism differs from Platonism. The difference is that the Platonic philosophy of mathematics is part and parcel of an objective idealist metaphysics, one that postulates the autonomous existence of ideas, i.e., the real existence of ideas out of brains, and their ontological priority. In contrast, mathematical fictionism is not included in any ontology, because it does not regard mathematical objects as self-existing but as fictions.

When introducing or developing an original mathematical idea, the mathematician creates something that did not exist before. As long as she keeps the idea to herself, it remains locked in her brain, for, as a physiological psychologist would say, the idea is a process occurring in the mathematician's brain. However, the mathematician does not assign to her idea any neurophysiological properties. She *feigns* that the idea in question has only formal properties. For example, she may pretend that the theorem she has just proved holds even while she is asleep and that, if made known to others, it will continue to hold long after she is gone.

This is of course a fiction, since only wakeful brains can do (correct) mathematics. But it is a necessary fiction because, although a proof process is neurophysiological, the proof itself does not contain any neurophysiological data or assumptions. In short, although we hold that theorems are human creations, we pretend that their validity is independent of any human circumstances. We are justified in adopting this fiction because mathematical ideas are not about the real world: every mathematical idea refers to some other mathematical idea(s). Mathematics (including logic) is the self-reliant science.

What holds for brains holds, mutatis mutandis, for societies too. While it is undeniable that there is no (sustained) mathematical activity in a social vacuum, but only in a community, it is equally true that pure mathematics has no social content. If it did have any, then mathematical theories would include social science predicates such as "commodity", "competition", "social cohesiveness", and "political conflict". Furthermore, they would double as social science theories, perhaps to the point of rendering the latter redundant.

Our view about the social neutrality of mathematics is at odds with the fashionable social constructivist-relativist sociology of science (Bloor 1976; Restivo 1992). The latter contends that "mathematics is through and through social", not just because all mathematical research is conducted in a scholarly community, but because all the mathematical formulas would have a social content. Needless to say, no evidence has ever been offered for this thesis: it is just an opinion. However, we can put it to the test in every particular case with the help of the theory of reference sketched in section 3. Take for instance the recursive definition of the factorial function "!", namely "0! = 1 and (n + 1)! = (n + 1)n!". Obviously, it refers to natural numbers: there is no trace in it of the social circumstances surrounding its origin. Likewise, the Pythagorean theorem is about plane Euclidean triangles, not about ancient Greece. And the Taylor series concerns functions, not the rise of capitalism. Because neither of the three formulas describes any social circumstances, they will hold for as long as there remain people interested in mathematics. (For a detailed criticism of the social constructivist-relativist sociology of science see Bunge 1991a, 1992, as well as essay 23.)

6. HOW TO MAKE ONTOLOGICAL COMMITMENTS

Pure mathematics, then, is not about concrete or material things such as photons or societies. It is about conceptual or ideal objects. If preferred, mathematics is about changeless or timeless objects, not about events or processes. The ontological neutrality of mathematics explains why this discipline is the universal language of science, technology, and even philosophy, i.e., why it is portable from one intellectual field to the other.

Yet when looking at any work in theoretical physics, chemistry, biology, or economics, a mathematician may be tempted to regard it as a piece of mathematics. There is a grain of truth in this belief. After all, rate equations, equations of motion, field equations, and more, are mathematical formulas, and they are solved using mathematical techniques, such as those of separation of variables, series expansion, and numerical integration. This is why Pierre Duhem, a conventionalist, claimed that classical electromagnetic theory is identical with Maxwell's equations. It is the same reason that Clifford Truesdell invokes to hold that rational mechanics is part of mathematics.

However, this is only a partial truth, as can be seen in the case of any formula of elementary logic. Thus, the formula "$(\forall x)(Px \Rightarrow (Qx \Rightarrow Rx))$" does not state anything about the real world or even about mathematical objects: it is an empty shell. In order to "say" something definite—true, false, or half true—the formula must be interpreted. A possible interpretation is this: $Int\ (P)$ = human, $Int\ (Q)$ = thinks, $Int\ (R)$ = is alive. This semantic assumption turns the above formula into "Any human who thinks is alive"—a mere generalization of Descartes's famous *Cogito, ergo sum*. Any change in the interpretation yields a formula with a different content.

What holds for logic holds for the whole of mathematics. Indeed, a mathematical formula does not become part of physics, or of any other factual science, unless enriched with a factual content. Such enrichment is achieved by pairing the formula to one or more semantic assumptions. Such assumptions state that the formula refers to such and such concrete things, and that at least some of the symbols occurring in it denote certain properties of such things. For example, the equation "$(\exists x)(d^2x/dt^2 + ax = 0)$" is the equation of motion of a linear oscillator in classical mechanics, provided x is interpreted as the instantaneous value of the elongation, t as time, and a as the ratio of the elastic constant to the mass of the oscillator. The concepts of linear oscillator, elongation, elastic constant, mass, and time, which confer a physical interpretation upon the mathematical symbols, are extramathematical.

In sum, mathematics does not suffice to describe or explain the real world. But it is necessary to account for it in a precise and deep manner. Indeed, mathematics supplies one of the two components of any theory in advanced theoretical factual science or technology, namely the mathematical formalism. The other component is the set of semantic assumptions, or "correspondence rules", that "flesh out" the mathematical skeleton.

More precisely, a mathematical theory or model of a domain of factual items is a triple: domain-formalism-interpretation, or $\mathcal{M} = \langle D, F, Int \rangle$ for short. Here D is the factual domain or reference class, F the union of some theories in pure mathematics, and Int a partial function from the formalism F to the power set of D, which assigns some predicates and formulas in F to sets of factual items in D. One and the same F may be paired to any number of different factual domains and interpretations. (Think of the multiple uses of the infinitesimal calculus.) A mathematician can check the formal correctness of F, but only empirical tests can tell whether any given theoretical model $\mathcal{M}$ matches the domain D, i.e., is factually true. In other words, the mathematical truth of the theorems in F does not guarantee the factual truth of $\mathcal{M}$. (On the other hand, any important mathematical flaw guarantees factual falsity.)

And yet it is easy to fall into verbal traps, taking literally such expressions as 'dynamic logic' and 'dynamical systems theory', suggesting to the unwary that some mathematics deals with time and change after all. But actually those theories are just as timeless as arithmetic and geometry. What happens is that, when applied, some of the concepts occurring in them get interpreted in factual terms. (For instance, the independent variable is routinely interpreted as time—a nonmathematical concept.) This is how dynamic logic is applicable to computer programs, and dynamical systems theory is applicable to the analysis or design of concrete systems of many kinds, even though it is often but an excuse for working on systems of ordinary differential equations. Likewise, elementary logic can be applied to analyze reasoning processes such as the arguments we construct in real life, by assuming that the steps in a logical sequence match the temporal sequence of our thoughts.

In sum, mathematics supplies ready-made formal and timeless skeletons, some of which scientists and technologists see fit to flesh out (interpret) in alternative ways in order to map concrete changing things. To change the metaphor: from a purely utilitarian viewpoint, mathematics is a huge warehouse of ready-to-wear clothes that scientists, technologists, and humanists can help themselves to. When none of those clothes fits, the user has got to do the tailoring himself, thus becoming a mathematician for a while. (Ptolemy and Newton come to mind. So do Einstein and Heisenberg but only up to a point, because the former only reinvented Riemannian geometry and the latter matrix algebra.)

7. RESPONDING TO SOME POSSIBLE OBJECTIONS

Let us now address some of the possible objections to mathematical fictionism. One of them is that mathematicians do not invent but discover—or at least they prefer to say that they have discovered something rather than invented it. (Some of them may do this out of modesty, others for being either Platonists or empiricists.) If this were true, either Platonism or empiricism would be true, and fictionism false.

I espouse the commonsensical view that there are mathematical inventions as well as discoveries. That is, the original mathematician sometimes posits and at other times he finds. He posits definitions, assumptions (in particular axiomatic definitions), and generalizations; and he discovers logical relations between previously introduced constructs. In particular, mathematical "entities", such as categories, algebras, number systems, and functional spaces, are invented: there are no mathematical quarries where one can find them ready-made. Even theorems are usually first conjectured, then proved—or disproved. But the proof process consists in discovering that the new theorem follows from previously known assumptions: axioms, lemmas, or definitions. (However, sometimes such discovery necessitates inventing further items, such as the auxiliary constructions of elementary geometry.)

For example, by using a convergence test, one discovers that a certain infinite series is divergent. But all convergence tests have been invented, not discovered. Likewise, by expanding a function in series and integrating term by term, one finds the integral of the function, provided the series converges uniformly. But the given function, the concepts of integral, and much more, had to be invented before they could be handled.

In recent years chaos theory and fractals have highlighted the "fact" that a radically new invention may have unexpected or even scandalously queer properties. For example, a tiny alteration in the value of a parameter in a nonlinear equation may have explosive effects. This "fact" may be discovered by trial—trial and wonder, though, rather than trial and error.

The general rule is then: *First invent, then discover.* And if the discovery is negative—e.g., that a bunch of assumptions is inconsistent, that they do not entail a given theorem, or that a series is divergent—then alter some of the assumptions, i.e., revise the invention process.

Since in mathematics there is invention as well as discovery, Platonism is false. But evidence for mathematical invention is not enough to substantiate fictionism. Yet, to do so we only need to recall that we handle an infinite totality, such as a line or a hypothetico-deductive system, *as if* it were an individual and *as if* it were "there" all in one piece, without of course saying where "there" is. (Mathematicians are in space and time, but they assume that their own creations are out of space and time.) For example, we prove the consistency of an abstract (or formalized) theory by exhibiting a model of it.

A second possible objection goes as follows. If mathematics is a work of fiction, why don't librarians group mathematical works together with novels and, in particular, with fantastic literature? In other words, what if any are the differences between mathematical fictions and artistic ones? For example, how does the fundamental theorem of algebra differ from the claim that Superman can fly, or Donald Duck can speak?

I submit that the crucial differences between mathematical fictions and all the others are the following (Bunge 1985a, pp. 39–40):

(a) far from being totally free inventions, mathematical objects are constrained by laws (axioms, definitions, theorems); consequently, they cannot behave "out of character"; e.g., there can be no such thing as a triangular circle, whereas even mad Don Quixote is occasionally lucid;

(b) mathematical objects exist (ideally) either by postulate or by proof, never by arbitrary fiat;

(c) mathematical objects are either theories, components of theories, or referents of theories, whether in the making or full-fledged; by contrast, myths, fables, stories, poems, sonatas, paintings, cartoons, and films are atheoretical;

(d) mathematical objects and theories are fully rational, not intuitive, let alone irrational (even though there is such thing as mathematical intuition);

(e) mathematical statements must be justified in a rational manner, either by their fruits or by their premises, not by intuition, revelation, or experience;

(f) far from being dogmas, mathematical theories are based on hypotheses that must be repaired or given up if shown to lead to contradiction, triviality, or redundancy;

(g) there are no strays in mathematics: every formula belongs to some theory, and theories are linked together forming supersystems, or they are shown to be alternative models of one and the same abstract theory; thus logic employs algebraic methods, and number theory resorts to analysis; by contrast, artistic or mythological fictions are self-sufficient: they need not belong to any coherent system;

(h) mathematics is neither subjective nor objective: it is ontologically noncommittal; but the process of mathematical invention is subjective, and that of proof (or disproof) intersubjective; what is real (concrete) about mathematics is only living mathematicians and active mathematical communities;

(i) some mathematical objects and theories find application in science, technology, and the humanities;

(j) mathematical objects and theories are socially neutral, whereas myth and art are sometimes used to support or undermine the powers that be; and

(k) because it deals in timeless objects, correct mathematics does not age, even though some of it may go out of fashion.

The practical upshot of the preceding is that librarians have good reasons for placing mathematical fiction and literary fiction in different sections. After all and above all, mathematics is a science—nay, the old queen of the sciences.

A third possible objection is this: if mathematics is fictive and made up exclusively of timeless objects, how can it represent real things and processes? The answer lies of course in the concept of symbolic representation. Ordinary language allows us to form sentences designating propositions which may represent ordinary things and processes, even though such sentences do not resemble their denotata:

$$\text{Sentences} \xrightarrow{\mathcal{D}} \text{Factual propositions} \xrightarrow{\mathcal{R}} \text{Facts}$$

The key is, of course, a system of linguistic (in particular semantic) conventions, mostly tacit. In the case of scientific theories, the semantic components are not conventional and therefore irrefutable, but hypothetical. (This is why the name 'semantic assumption' is more adequate than the traditional 'correspondence rule'.) Thus, the semantic assumption that the value of the metric tensor at a given point in spacetime represents the intensity of the gravitational field at that point is not a convention, or even a rule, but a hypothesis that can be put to the experimental test—and one, moreover, that makes no sense in action-at-a-distance theories of gravitation. However, the matter of convention deserves a separate section.

8. MODERATE MATHEMATICAL FICTIONISM AND CONVENTIONALISM

At first sight moderate mathematical fictionism is identical with conventionalism. On closer inspection one realizes that it is not. Indeed, mathematical fictionism retains the concept of (mathematical) truth, as well as the distinction between assumption and convention (in particular definition), which conventionalism rejects.

The mathematician uses the concept of mathematical truth when claiming that a certain formula (other than an axiom) is true for following validly from some set of assumptions in accordance with the inference rules of the underlying logic. Likewise, we use its dual, namely the concept of mathematical falsity, when disproving a conjecture. In other words, theoremhood equals provability, which is anything but conventional. If theorems were conventions, they would be posited, not proved.

The dual of the concept of truth, namely that of error, is equally important in mathematics. It occurs whenever a mathematical idea is criticized as erroneous for some reason, and it is central in approximation theory and numerical analysis. Think of approximating an infinite convergent series by the sum of its first few terms. (In particular, one can make true statements about quantitative errors, e.g., that they are bounded, or that they fit some distribution.) A moderate fictionist is enough of a realist to admit error and thus truth. Conventionalists have no use for the concept of error for the same reason that they have no use for the concept of truth.

What about the axioms of a mathematical theory: can one say that they are true? One might be tempted to claim that they are "true by convention". But this expression strikes me as an oxymoron. Indeed, conventions have no content, and they are checked for well-formedness and convenience, not truth. In any event, there is no need to assign truth values to the axioms of a mathematical theory. (On the other hand, we need to know whether the axioms of a factual theory are true or false to some extent.) In practice, the most important piece of knowledge about a postulate system is that it entails the standard theorems in the given field and, preferably, some new interesting ones as well. (To be sure, it would also be important to be able to prove the consistency of the system. But here again Gödel has taught us some humility.)

As noted above, a second difference between moderate mathematical fictionism and conventionalism is that the former keeps the difference between assumption and definition, whereas conventionalism holds, in Poincaré's famous thesis, that axioms are disguised definitions. That this view is false is best seen in the light of the theory of definition and, in particular, in the light of Peano's thesis that definitions are identities. Indeed, an axiom may be an equality, such as a differential equation, but not an identity, such as "1 = the successor of 0". Identities are symmetrical, equalities are not. The difference is sometimes indicated by the symbols ≡ and := respectively. (The latter occurs in Pascal and other computer languages.) Examples: "For any real number x: $(x + 1)(x - 1) \equiv x^2 - 1$", and "$x := 5$", or "let x equal 5", or "assign x the value 5".

A particularly shallow version of mathematical conventionalism is Carnap's view that mathematical statements are "empty linguistic conventions", such as "In every Romance language adjectives follow names". Actually, the only linguistic conventions occurring in mathematics are notational conventions, such as "Let n designate an arbitrary natural number". The well-formed mathematical statements are meaningful and testable. They refer to definite mathematical objects, such as abstract sets, numbers, figures, functions, spaces, or what have you. If the (well-formed) mathematical formulas were meaningless, there would be no way of checking them and no point in doing so. But they are checked in several ways: definitions for noncircularity, axioms for fertility, conjectures for theoremhood, etc. (Gödel refuted Carnap's conventionalism in an unpublished manuscript that Rodríguez-Consuegra [1992] located, translated into Spanish, and commented on.)

In conclusion, mathematical conventionalism won't do. (See further objections in Quine 1936.) In any event, mathematical fictionism differs from it and has no use for it.

9. MODERATE MATHEMATICAL FICTIONISM AND THE IDEALISM/MATERIALISM CONTROVERSY

Vulgar materialists (physicalists) are bound to reject mathematical fictionism for regarding it as a variety of idealism. (Nominalists, who are the crassest of all vulgar materialists, reject all concepts: they only admit inscriptions.) Actually, mathematical fictionism is neutral in the debate between materialism and idealism. In fact, mathematical fictionism is only about mathematical objects: it makes no assertions about the nature of the world. As far as mathematical fictionism is concerned, one may hold the world to be material, spiritual, or either a mixture or a combination of material and ideal objects.

A materialist should not feel uneasy about the thesis that mathematical objects are ideal and therefore timeless, as long as she subscribes to the thesis that mathematics is a human creation. All the reassurance the materialist needs is that mathematical ideas do not exist by themselves in some Platonic Realm of Ideas both immaterial and eternal. (If she so wishes, she may regard any construct as an equivalence class of brain processes occurring in different brains or in the same brain at different times: see Bunge 1983a.)

Yet it might be argued that mathematics is "ultimately" about the real world for, after all, arithmetic originated in counting concrete things like shells and people, and geometry originated in land surveying and, in particular, in the need to allot land lots to farmers. No doubt, such were the humble origins of the parents of mathematics. But mathematics proper is not about such empirical operations as counting and surveying. In fact, mathematics freed itself from its empirical origins about twenty-five centuries ago, when the first general mathematical propositions were stated and the first mathematical proofs were constructed in classical ancient Greece.

A cognate argument for both vulgar materialism and empiricism is that mathematicians often make use of ordinary (incomplete) induction or of analogy to find patterns. (See Pólya 1954.) True, but the result of any such plausible reasoning is a conjecture that has got to be proved (or disproved) by purely mathematical means, e.g., by *reductio ad absurdum*, or using the principle of complete induction, neither of which is suggested by ordinary experience. In short, induction by enumeration and analogy have at most a heuristic value: they prove nothing, and proving happens to be the main job and only privilege of the mathematician. Moreover, incomplete induction and analogy may lead us to error unless we check their outcome.

True, factual science has sometimes stimulated mathematics by posing new problems. For example, dynamics encouraged, nay required, the development of the infinitesimal calculus and the theory of differential equations; gravitation theory stimulated the growth of differential geometry; and quantum mechanics that of functional analysis and group theory. But all of this fails to prove that mathematics is about the real world, because a mathematical formula can be given alternative factual interpretations—or none. Besides, nowadays mathematics is far more useful to science and technology than the latter are to the former. In the modern intellectual production line the main arrow goes from the abstract to the concrete.

Finally, another familiar objection to the autonomy of mathematics is the idea that some mathematical theories are more "natural" than others, being closer to human experience. For instance, according to mathematical intuitionism in its radical version, no mathematical formula is meaningful unless it is somehow related to the natural numbers (Dummett 1976). But in fact there are plenty of nonnumerical mathematical fields, such as logic, category theory, set theory, much of abstract algebra, and topology. Worse, intuitionists have not produced a suitable semantic theory elucidating the notion of meaning and thus helping us test formulas for meaningfulness (in their sense).

In conclusion, mathematics is semantically and methodologically self-sufficient. But at the same time it feeds all the other sciences and it is occasionally stimulated by them. So much so that mathematics is at the very center of the system of human knowledge, which may be pictured as a rosette of partially overlapping petals. In other words, far from being independent from the rest of knowledge, mathematics is at the very center of it. But this does not entail the thesis, held consistently by Quine, and at times by Putnam and others, that there are no important differences between mathematics and the rest, and that in principle mathematics might be refuted by experiment. Like mates, mathematics and science are neither equal nor separate.

10. CONCLUSIONS

We have outlined and defended one kind of mathematical fictionism. This philosophy of mathematics is fictionist, for it holds that all mathematical objects are fictions, so that they are neither empirically compelling nor logically necessary (even though they are constrained by logic). They are fictions because, although they are human creations, they are deliberately detached from personal and social circumstances. We pretend that these timeless ideal objects exist in a "world" of their own, alongside other fictions, such as myths and fables.

However, our mathematical fictionism is moderate because it is at variance with conventionalism, and it regards mathematics as a science, not as a grammar, much less as a game or pastime on the same footing with chess. It differs from conventionalism in assigning conventions a rather modest role in mathematics in comparison with hypothesis, proof, and computation. And it regards mathematics as a serious activity that enriches our stock of ideas and helps factual science, technology, and even philosophy. In particular, mathematics helps us discover structures and patterns, as well as pose and solve problems of all kinds in all fields of knowledge and rational action. Thus, far from being escapist, mathematical fictions are necessary to understand and control reality.

If moderate mathematical fictionism is true, i.e., if it fits mathematical research, then both mathematical Platonism and mathematical empiricism are false, even though they are the most popular of all the philosophies of mathematics. Platonism is false because fictions are human creations, not self-existing eternal objects. And mathematical empiricism is false because most mathematical fictions go far beyond experience and none are tested by it. The same holds, a fortiori, for mathematical pragmatism, the philosophy of mathematics tacitly espoused by those who demand that mathematics should be responsive to the market.

As for mathematical intuitionism, it is inadequate for restricting severely the invention of mathematical fictions on doubtful philosophical grounds, such as constructivism and the Pythagorean cult of natural numbers. (Caution: This criticism concerns mathematical intuitionism, not intuitionist logic or mathematics.) Nominalism is false because, as Frege pointed out, it confuses signs with their designata, e.g., numerals with numbers. Also, because it cannot account for the (mathematical) existence of nameless objects, such as the overwhelming majority of real numbers. Finally, conventionalism is false for a different reason, namely because it conflates assumption with definition, and it jettisons the concept of mathematical truth.

Furthermore, if mathematics (including logic) is the most exact and comprehensive of fictions, then we need no special logic of fictions, such as Woods's (1974) and Routley's (1980). In particular, the free logics (i.e., logics with empty domains), deliberately invented to handle fictions, turn out to be unnecessary. This remark may be useful to curb the inflationary growth of deviant logics, the vast majority of which are mere academic pastimes "nur gut zum Promovieren" ("suitable only to get a Ph.D."), as Hilbert might have said.

On the other hand, moderate mathematical fictionism has little to say about any of the foundational strategies. In particular, one may couple mathematical fictionism to either logicism, formalism, or moderate (post-Brouwerian) intuitionism, as well as to the combination of all three, which seems to be tacitly used nowadays by most working mathematicians.

Moderate mathematical fictionism ought to exert a liberating influence on the mathematical researcher, in reminding him that he is not out to discover a ready-made universe of ideas, to explore the real world, or even to latch on to natural numbers. His tasks are to create (invent or discover) mathematical concepts, propositions, theories, or methods, and to discover their mutual relations, subject only to the conditions of consistency and conceptual fruitfulness.

As philosophers, our duties are to defend the freedom of mathematical creation from philosophical and ideological strictures; to see to it that mathematical fictions be not reified; to use some of them to elucidate, refine, or systematize key philosophical ideas, i.e., to do some exact philosophy; and to build a philosophy of mathematics matching actual mathematical research as well as our own philosophy.

14

THE GAP BETWEEN MATHEMATICS AND REALITY

(1994)

■

1. INTRODUCTION

Mathematicians have often marveled at what they have seen as the perfect fit of mathematics to reality. For example, Euclid believed that his geometry mirrored the structure of the physical world; Galileo held that nature is a book written in mathematical symbols; and, until the 1920s, curves were often defined as trajectories of mass points, and rational mechanics was taught mainly by mathematicians and as a branch of mathematics—hence as impregnable to experiment.

These were not, of course, silly mistakes. Indeed, when suitably interpreted in physical terms, and applied to the regions of space and the bodies we encounter in ordinary experience, Euclidean geometry and Newtonian mechanics have proved to be excellent first approximations. Nor are those the only examples. In fact, the huge success of Galileo's program to mathematize physics led some to adopt more or less tacitly the methodological postulate according to which, given any new mathematical idea, it will eventually find its "embodiment" in the real world. The heuristic power of this optimistic principle is undeniable.

So, there was some ground to believe in Leibniz's preestablished harmony between mathematics and the real world, or in Kant's secular version, the synthetic a priori judgment. However, as we all know, Einstein's two relativities refuted this belief and taught us that pure mathematics is necessary but not sufficient to account for facts. Moreover, relativistic physics only highlighted and dramatized the gap between pure mathematics and factual science. A proper epistemological analysis of prerelativistic physics should-have given the same result, as should become clear from the arguments to follow.

My thesis is that the gap between pure mathematics and reality has five main sources: the idealizations and approximations involved in every factual theory, the dimensionality of most magnitudes, the existence of redundant solutions, and the ontological neutrality of pure mathematics. Let me explain and substantiate.

2. IDEALIZATIONS AND APPROXIMATIONS

Every theorist knows that the modeling of any domain of facts involves schematizations or idealizations. Here are some elementary examples from physics: point masses, frictionless motions, nonviscous liquids, perfectly elastic collisions, instantaneous forces, plane waves, light rays, completely isolated systems, reversible macroprocesses, and perfect thermal equilibrium.

There are of course huge gains in feigning that an extended body can be treated as a point, a wave as an infinite plane, an atom as dwelling in a perfect vacuum, and so on. The main gains from such idealizations are relative ease of formulation of the assumptions, and computational convenience. However, such gains are often offset by severe losses. Thus, point charges and plane electromagnetic waves must be assigned infinite energies, which are of course physically meaningless. Thus, one is sometimes forced to swallow half-rotten eggs.

Approximations are parallel. They are convenient, and sometimes unavoidable, when the original problem involves equations that cannot be solved in closed form. Such cases call for either simplification, e.g., linearization in the case of nonlinear equations, or the use of approximation methods, such as expanding a function in infinite power series and keeping only the first few terms.

Idealization is involved in theory construction, and approximation in theory application, i.e., in the handling of special cases. The two procedures involve simplification, but they are quite dissimilar in other respects. For one thing, whereas approximations can be improved upon without touching the initial assumptions (axioms), the latter cannot be substantially refined: if they are, a different theory results.

Furthermore, there is a philosophically interesting difference between idealization and approximation, namely that the former is unavoidable—in that all theorizing overlooks details, particularly when ugly—whereas the latter is avoidable in principle though not in practice. For example, one can increase without limit the number of terms in a series expansion, but one cannot make changes in an equation of motion, or a field equation, and hope to get all the standard theorems, among them those which have been experimentally confirmed. Thus, the addition of a term representing magnetic monopoles utterly spoils Maxwell's equations: for one thing, it violates the conservation of electric charge, one of the most robust pieces of physics. Likewise, an inhomogeneous Schrödinger equation would belong to some theory other than standard quantum mechanics, if only because it would violate the superposition principle.

In short, there is tension as well as trade-off between mathematical rigor and the matching with reality. If we want rigor we must forsake much detail and departure from mathematical beauty. Likewise, if we are reluctant to part with detail and rough edges and noise, we postpone indefinitely theory building and, consequently, must grope our way blindly in the laboratory or in the field. We must try and find a wise compromise between the two extremes. Physicists, chemists, and more recently biologists have learned the trick. Most social scientists are yet to learn that there is a problem here.

3. DIMENSIONS AND UNITS

Unlike the functions studied in pure mathematics, most physical "quantities", or magnitudes, "come with" dimensions and units. Thus, time, the most common independent variable in physics, is assigned dimension $[T]$, and is accompanied by the family of time units, such as the second and the day. The pure mathematician who studies a function, operator, or equation does not bother to inquire into the physical interpretation of his independent variable(s) and, consequently, he is not faced with the problem of finding out its dimension(s), let alone with the problem of choosing units. He proceeds as if all "quantities" were dimensionless, and he could not care less about units.

In contrast, the physicist is supposed to know the dimension(s) of the magnitudes she handles, and she uses dimensional analysis in order to discover them. She must do this for two reasons. One is to observe Fourier's principle of dimensional homogeneity or, to put it negatively, to avoid writing ill-formed and therefore meaningless expressions such as 'Distance plus time equals mass'. The second reason is to have an inkling as to the kind of instrument(s) she needs to measure the magnitude in question.

Except in the infrequent case of dimensionless quantities, such as relative increments and the fine structure constant, dimensions and units enter in the very construction of magnitudes. For example, the various concepts of velocity occurring in classical physics are construed as functions from triples like ⟨physical entity, reference frame, unit⟩ to triples of real numbers. That is, $V: P \times F \times U_V \to \mathbb{R}^3$, where P names the collection of physical entities or parts thereof, F the collection of reference frames, U_V that of velocity units, and $\mathbb{R}$ the real line. The general concept of a function, as well as the system of real numbers, belong in pure mathematics. On the other hand, the particular function V and the three factors in its domain are physical concepts, not mathematical ones. Likewise, the dimensional equation "$[V] = L \cdot T^{-1}$" belongs in physics not in pure mathematics.

When stating that a certain concept is physical (in general factual) rather than mathematical, one means that it has an extramathematical content, not that it lacks a mathematical form. For example, the concept F of a reference frame, such as the sun or a laboratory, can be represented by a coordinate system, but coordinate systems are purely mathematical objects. Indeed, unlike reference frames, coordinate systems—Cartesian, spherical, cylindrical, or other—are massless, timeless, and motionless. Likewise, the concepts of dimension and unit do not belong in pure mathematics but they can be mathematized. (Every dimension is a value of a function [ ] satisfying, among others, the obvious conditions $[A + B] = [A] + [B]$ and $[AB] = [BA] = [A] [B]$ for any magnitudes A and B. And a system of units is an algebra similar to a group.)

In sum, pure mathematics has no use for dimensions and units, without which most magnitudes could not be constructed, let alone measured.

4. FACTUALLY MEANINGLESS MATHEMATICAL CONSTRUCTS

We can and should assume the optimistic methodological postulate that every minimally clear construct can be exactified, i.e., mathematized. This holds not only for science and technology, but also for philosophy. (In fact, there is such a thing as exact philosophy, underdeveloped and often irrelevant as it is.) But not every mathematical (or exact) construct has a counterpart in the real world. It is impossible to prove the above methodological postulate: we cannot exclude the possibility that some important concepts be inherently inexact. On the other hand, to prove that not every mathematical concept can be interpreted in factual (or empirical) terms, one example suffices. Let us exhibit a few examples.

Consider a declining human population. Curve fitting with the help of any interpolation method is likely to yield, to a good approximation, a formula of the form $P(t) = at - bt^2$, with a and b positive constants. If we extrapolate after $t = a/b$, we get a negative population value—which, of course, makes no demographic sense. Hence, we must drop the negative part of the curve. Likewise, some solutions to certain equations of mathematical economics correspond to negative quantities or prices. We drop these too, for they make no economic sense.

Another example is provided by classical electromagnetic theory. Maxwell's equations have two (infinite) sets of solutions: one corresponds to retarded potentials, the other to advanced ones. The second set is usually eliminated, because it violates the principle of antecedence (an ingredient of the causality principle), according to which the effect cannot precede its cause. (However, some physicists have managed to superpose solutions of two kinds in a physical meaningful if controversial fashion.) In relativistic physics there is the similar case of the spacelike (*raumartige*) lines or trajectories, which must be excluded for representing physically impossible processes such as superluminal signals.

So much for the easy examples. Other cases are much tougher. Suppose, for example, that in a given theory certain important physical magnitudes, such as velocities and energies, are given by series or integrals that happen to diverge. Since we know that in the real world velocities and energies are finite (and sometimes measurable), we have two choices: to give up the theory, or to correct its false theorems. Obviously, we opt for the former strategy if there are better alternatives in hand or if we hope to be able to construct one. This is the case considered in the standard philosophies of science. However, a better alternative may not be in sight. In this case one opts for the second strategy. For example, we force a divergent series to converge by using some summation procedure such as Cesàro's. In the case of a divergent integral, we postulate a cutoff point or we take Hadamard's finite part. I submit that these and similar procedures are legitimate, provided they are postulated from the beginning (at the axiom level) rather than being applied in extremis. Otherwise, they are only ad hoc repairs that may introduce inconsistencies.

To sum up, some scientific theories are mathematically too rich: they contain factually meaningless constructs. We do not tolerate them. In some cases we prune them, in others we discipline them.

5. PURE MATHEMATICS IS ONTOLOGICALLY NEUTRAL

The foregoing suggests that pure mathematics never faces facts in the real (material) world. What can and must face facts are mathematical theories interpreted in factual (physical, biological, sociological, etc.) terms. For example, any monotonously increasing function of one variable may serve to exactify the loose hypothesis (or the empirical finding) that an increase in the independent variable of interest is accompanied by an increase in the corresponding dependent variable. But since the hypothesis (or the finding) concerns some real thing, the variables in question will represent (actual or putative) properties of the thing.

In other words, a mathematically well-defined function F from a set A to a set B is a piece of pure mathematics, uncommitted to any feature of reality. This is why the same mathematical functions occur in so many different research fields. A commitment to reality is made the moment we add that the domain A of F is a collection of concrete things, or of states of things, or of events happening in things, and that F itself represents a given property of the A's. Think of the people-age and people-weight functions.

What holds for functions holds, mutatis mutandis, for all the other mathematical constructs occurring in science and technology, except for the purely computational devices. (For example, if a function F is an infinitely differentiable function representing a physical property, then it is impossible to assign a physical interpretation to every term in a series expansion of F. This is why, incidentally, most Feynman diagrams are physically meaningless: they are only computational devices.)

A mathematical construct is assigned a factual interpretation if it is paired off to some factual item (thing, property, state, event, or process). A mathematical theory M can be said to be *fully interpreted in factual terms* if every primitive (undefined) concept of M is assigned one member of a set F of factual items. In obvious symbols, Int(M) = F. For every member m of M, the formula Int(m) = f, where f is in F, may be called a *semantic postulate*. By changing the Int function we obtain a different interpretation of the same mathematical formalism, hence a (somewhat or very) different factual theory. Think of the alternative interpretations of the mathematical formalism of the quantum theories.

The reason for restricting the factual interpretation function to the primitive or basic concepts of a theory is that, as we saw earlier, many (actually most) of the mathematical concepts simply do not represent anything in reality. For example, in classical particle mechanics the fourth derivative of the position coordinate with respect to time, and the square root of the mass, may occur in some calculation, but they do not represent any known physical properties.

In sum, theories in pure mathematics "say" nothing about reality. It is only by enriching a mathematical formalism with an interpretation function that the former can be turned into a factual theory.

6. CONCLUSION

Whereas mathematics is exact, or at least exactifiable, the real world is messy, fuzzy, and noisy. Mathematics does not mirror reality in an exact fashion: there is a chasm between the two. For one thing, the mathematical constructs (concepts, theories, proofs, etc.) are handled—or rather brained—as if they had an autonomous existence, one above the vicissitudes of the world. Secondly, no mathematical model can account for every single empirical irregularity or impurity. Thirdly, many, perhaps most, mathematical constructs have no real counterparts. Thus, mathematics is too poor in one sense and too rich in another. The glove misses some fingers and it is too big.

Pure mathematics has neither ontological presuppositions nor ontological implications. (On the other hand, no serious modern ontology can afford not to use some mathematical tools.) However, the chasm between mathematics and the world can be spanned, nay it is being spanned all the time, by inventing and trying out alternative interpretation functions, such as "Let $Q(b)$ represent the electric charge on body b". Thus, mathematics and factual science, though distinct, are united, though not on equal terms: mathematics remains the senior partner.

Because pure mathematics is ontologically neutral, mathematical ideas are portable across research fields. This is why factual scientists and technologists have a vested interest in the advancement of pure mathematics. And this is why politicians and university administrators should abstain from demanding that mathematicians produce "useful results", i.e., that they become technologists. Such demand can only cripple or even kill pure mathematics and thus stunt the development of its main consumers, factual science and technology, which in turn feed the economy. In modern society there is no concrete output without an abstract input.

15

TWO FACES AND THREE MASKS OF PROBABILITY

(1988)

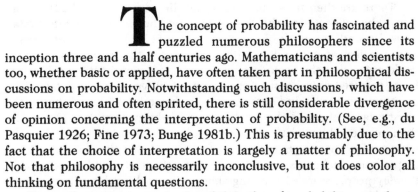

The concept of probability has fascinated and puzzled numerous philosophers since its inception three and a half centuries ago. Mathematicians and scientists too, whether basic or applied, have often taken part in philosophical discussions on probability. Notwithstanding such discussions, which have been numerous and often spirited, there is still considerable divergence of opinion concerning the interpretation of probability. (See, e.g., du Pasquier 1926; Fine 1973; Bunge 1981b.) This is presumably due to the fact that the choice of interpretation is largely a matter of philosophy. Not that philosophy is necessarily inconclusive, but it does color all thinking on fundamental questions.

Up until one century ago the philosophy of probability was dominated by subjectivism: probability was regarded as a measure credibility (or uncertainty, or weakness) of our beliefs. This interpretation had an ontological basis: since the universe was deemed to be strictly deterministic, probability had to be resorted to because of our ignorance of details. (God has no use for probability.) The paradigm cases were the games of chance and the classical kinetic theory of gases: here the basic laws were deterministic, but probability was called for because of our ignorance of the initial positions and velocities of the individual entities.

About one century ago an alternative view emerged, namely the frequency interpretation (Venn 1888). According to this view, probabilities are long-run values of relative frequencies of observed events. While this was a step in the direction of objectivity, it remained half way, because it was concerned with observations rather than with objective facts. Probability was regarded as a feature of human experience rather than as a measure of something objective. Like the subjectivistic interpretation, the frequency interpretation is still very much alive—if not de jure at least de facto.

A third interpretation of probability began to emerge at the time of World War I with reference to statistical mechanics and other stochastic theories, namely the so-called propensity interpretation. According to this view, probability-values measure the strength of a tendency or disposition of some event to happen. This objectivist interpretation, which had been adumbrated by Poisson and Cournot, can be found in Poincaré (1903), von Smoluchowski (1918), Fréchet (1946), and a few others; it was adopted independently by the writer (1951), and has been gaining ground among philosophers, particularly since it was popularized by Popper (1957b).

There are then three main views on the nature of (applied) probability: the subjectivist, the frequency, and the propensity interpretations. Up until the birth of quantum theory (1926) the first interpretation was just as compatible with realism as with antirealism, for one could argue that the basic laws of nature are deterministic, probability being required only because of our empirical ignorance of details. But quantum mechanics and quantum electrodynamics, with their basic stochastic laws, changed the relation of probability to philosophy: from then on the subjectivistic philosophy of probability is compatible only with a subjectivistic philosophy willing to hold that the stochastic laws of quantum mechanics and other scientific theories would cease to hold the day people stopped thinking about atoms, molecules, photons, and other objects with stochastic behavior.

The frequency interpretation has had a similar fate. While originally it could be espoused by realists as well as by empiricists, ever since the quantum theory was born, realists cannot accept it if only because to them the laws of atoms and the like are not supposed to depend upon our observation acts. Thus, an atom in an excited state has a definite objective probability of decaying to a lower energy state within the next second, whether or not somebody is counting the actual events of this type in a large assembly of atoms of the same kind. In other words, the propensity interpretation of probability accords well with a realistic interpretation of the quantum theory. But this argument will not persuade someone who is not a realist or who, being a realist, doubts that the quantum theory is here to stay. He will demand more general reasons, i.e., reasons that can be used with reference to all scientific theories.

The purpose of the present paper is to supply such reasons: to show that the subjectivistic and the frequency interpretations are untenable, whereas the propensity interpretation accords well with both the mathematical theory of probability and the stochastic theories of contemporary science. To this end, it will prove convenient to start by giving a brief characterization of the theory of whose interpretations are at stake, namely the probability calculus.

1. THE ABSTRACT CONCEPT

Up until half a century ago there was some confusion in the foundations of probability. The confusion consisted in a lack of distinction between the mathematical theory of probability and its various interpretations and applications. So much so that the theory was often presented as if it dealt with physical events. In particular, it was presented as the mathematics of gambling. That stage was overcome by Kolmogoroff's work (1933). This work made it clear that the probability calculus is a branch of pure mathematics—this being why it can be applied in so many different fields of research. Let us give a quick review of the gist of Kolmogoroff's axiom system in its elementary version.

The elementary calculus of probability presupposes ordinary logic (the predicate calculus with identity), elementary set theory, ring theory (a branch of abstract algebra), analysis, and measure theory. But the foundations of the elementary theory can be understood without the help of any sophisticated mathematics. Indeed, the theory has just two basic (or primitive or defining) concepts with a simple mathematical structure. These are the notions of an event (understood in a technical mathematical sense) and of probability measure, which occur in statements of the form "The probability of event x equals y". In principle, any set qualifies as an "event"; and the probability of such an "event" is the real number assigned to it by a probability function.

More precisely, a probability function Pr is defined on a family F of sets such that the union and the intersection of any two members of F be in F, and also that F be closed under complementation. In sum, F must be a σ algebra, in the sense that its members obey the laws of the algebra of classes extended to countable infinite unions. This algebraic structure is not arbitrary but is demanded by the applications of the calculus. Thus, given the probabilities of the complex events x and y, we must be able to compute the probabilities of the complex events "x and y", "x or y", and "not-x", and even the probability of an infinite disjunction of events.

Note that in the applications we have to do with events proper, not just with abstract sets. But note also that, since real events cannot be negative or disjunctive, the calculus of probability applies to possibilities, not actualities. As soon as any of the events referred to by the expression 'x or y' is actualized, the expression 'the probability of x or y' becomes pointless. More precisely, the transition from potentiality to actuality is represented by the transition from $0 < p < 1$ to $p = 0$ or to $p = 1$ (Bunge 1976b).

We are now ready for a formal definition of the probability concept, namely thus. Let F be a σ algebra on a nonempty set S (e.g., $F = 2^S$, i.e., the power set of S) and $Pr: F \rightarrow [0,1]$ a real-valued bounded function on F. Then Pr is a probability measure of F if, and only if, it satisfies the following conditions:

(i) for any countable (finite or infinite) collection of pairwise disjoint sets in F, the probability of their union equals the sum of their individual probabilities. That is, if x and y are in F, and $x \cap y = \emptyset$, then $Pr(x \cup y) = Pr(x) + Pr(y)$;

(ii) $Pr(S) = 1$.

Note that the theory based on these sole assumptions is semiabstract insofar as it does not specify the nature of the elements of the basic set S nor, a fortiori, those of the probability space F. On the other hand, the range of Pr is fully interpreted: it is not an abstract set but the unit interval of the real line. Hence, the *semi*. Were it not for the semantic indeterminacy of the domain F of the probability function, the calculus could not be applied everywhere, from physics and astronomy and chemistry to biology and psychology to sociology. As long as the probability space F is not specified, i.e., as long as no probabilistic model is constructed, probability has nothing to do with possibility, propensity, randomness, or uncertainty.

An *application* of any abstract or semiabstract theory to some domain of reality consists in enriching the theory with two different items: (a) a model or sketch of the object or domain of facts to which the theory is to be applied, and (b) an interpretation of the basic concepts of the theory in terms of the objects to which it is to be applied. Shorter: A factual scientific construct f is a mathematical construct m together with an interpretation I that assigns to m a collection P of (really) possible facts; i.e., $f = \langle m, P, I \rangle$. (For details see Bunge 1974b.)

In particular, an application of probability theory consists in joining

the above definition of probability measure (or some of its consequences) with (a) a *stochastic model*, e.g., a coin-flipping model or an
urn model or what have you, and (b) a set of interpretation (or correspondence or semantic) assumptions sketching the specific meanings to
be attached to a point x in the probability space F, as well as to its measure $Pr(x)$. As long as these additional assumptions are not introduced,
the probability theory is indistinguishable from measure theory, which
is a chapter of pure mathematics: only those specifics turn the semiabstract theory into an application of probability theory or part of it.

In other words, the general and semiabstract concept Pr of probability measure is defined (via a set of axioms) in pure mathematics. Each
factual interpretation I of the domain F of Pr, as well as of the values
$Pr(x)$ of the probability measure (for x in F), yields a *factual probability
concept* $f_i = \langle <F, Pr>, P, I \rangle$, where i is a numeral. These various factual
probability concepts belong to factual science, not to pure mathematics:
they are the probabilities of atomic collisions, of nuclear fissions, of
genetic mutations, of survival age, of learning a certain item on the nth
presentation, of moving from one social group to another, and so on and
so forth.

What the various specific (or interpreted) probability concepts have
in common is clear, namely the mathematical probability theory $\langle F, Pr \rangle$.
This explains why the attempts of the subjectivists and of the empiricists
to define the general concept of probability either in psychological terms
(degrees of belief) or in empirical terms (frequencies of observations)
were bound to fail: maximal generality requires de-interpretation, i.e.,
abstraction or semiabstraction. (For the notions of interpretation and of
numerical degree of abstraction see Bunge 1974b.)

We can now approach the problem of weighing the claims of the four
main doctrines on the nature of (applied) probability.

2. LOGICAL CONCEPT

The so-called logical concept of probability was introduced by Keynes
(1921), Carnap (1950), Jeffreys (1957), and a few others. It boils down
to the thesis that probability is a property of propositions or a relation
between propositions, in particular between hypotheses and the empirical evidence relevant to them.

This view is held almost exclusively by workers in inductive logic, or
the theory of confirmation. The centerpiece of this theory is a particular
interpretation of Bayes's theorem

$$Pr(h \mid e) = Pr(e \mid h) \cdot Pr(h)/Pr(e).$$

This formula is usually read thus: The probability of hypothesis h given datum e equals the product of the probability of e given h, multiplied by the prior probability of the hypothesis and divided by the prior probability of the evidence.

There are many problems with this reading of Bayes's mathematical correct formula. One is that the only propositions that are assigned definite values are the tautologies and the contradictions, i.e., $Pr(t) = 1$, and $Pr(\neg t) = 0$. The remaining propositions, particularly h and e, are assigned probabilities in an arbitrary fashion. For instance, nobody knows how to go about assigning a probability to scientific laws or to scientific data.

The entire enterprise seems to originate in a confusion between probability and degree of truth. The expression 'The probability that h be true' is nonsensical (du Pasquier 1926, p. 197). On the other hand, the proposition "The degree of truth of h equals v", where v is some real number comprised between 0 and 1, does make perfect sense in factual science. In any event, the logical theory of probability has found no applications in science or in technology. In these fields one assigns probabilities to states and events, e.g., to an excited state of an atom, and to the probability that the atom decays from that state to a lower energy state.

Since the "logical" interpretation has been absorbed by the more popular subjectivistic (or personalist) interpretation, we shall turn to the latter.

3. PROBABILITY AS CREDIBILITY

The subjectivistic (personalist, Bayesian) interpretation of probability construes every probability value $Pr(x)$ as a measure of the strength of someone's belief in the fact x or as the accuracy of his information about x (Savage 1954; Jeffreys 1957; de Finetti 1972). There are a number of objections to this view.

The first objection, of a logical nature, was raised towards the end of section 1, namely that one does not succeed in constructing a general concept by restricting oneself to a specific interpretation. However, a personalist might concede this point, grant that the general concept of probability belongs in pure mathematics, and claim just that the subjectivist *interpretation* is the only applicable, or useful, or clear one. However, this strategy won't save him, for he still has to face the following objections.

The second objection, of a mathematical nature, is that the expression '$Pr(x) = y$' makes no room for a subject u and the circumstances v under which u estimates her degree of belief in x, under v, as y. In other words, the elementary statements of probability theory are of the form '$Pr(x) = y$', not '$Pr(x, u, v) = y$'. And such additional variables are of course necessary to account for the fact that different subjects assign different credibilities to one and the same item, as well as for the fact that one and the same subject changes her beliefs not just in the light of fresh information but also as a result of sheer changes in mood. In sum, the subjectivist or personalist interpretation of probability is *adventitious*, i.e., incompatible with the mathematical structure of the probability concept.

Even if the former objection is waived aside as a mere technicality—which it is not—a third objection is in order, namely this. It has never been proved in the psychological laboratory that our beliefs are so rational that in fact they satisfy all of the axioms and theorems of probability theory. Rather on the contrary, there is experimental evidence pointing against this thesis. For example, most of us experience no difficulty in holding pairs of beliefs that, on closer inspection, turn out to be mutually incompatible. Of course, the subjectivist could go around this objection by claiming that the "calculus of beliefs" is a normative theory, not a descriptive one. She may indeed hold that the theory *defines* "rational belief", so that anyone whose behavior does not conform to the theory departs from rationality instead of refuting the theory. In short, she may wish to claim that the theory of probability is a theory of rationality—a philosophical theory rather than a psychological one. This move will save the theory from refutation, but it will also deprive it of confirmation.

A fourth objection is as follows. A belief may be construed either as a state of mind (or a brain state) or as a proposition (or statement). If the former, then the probability $Pr(x)$ of belief x can be interpreted as a measure of the objective strength of the propensity or tendency for x to occur in the given person's mind (or brain). But this would of course be just an instance of the objectivist interpretation and would be totally alien to the problem of the likelihood of x or even the strength of a subject's belief in x. On the alternative construal of beliefs as statements, which is the usual Bayesian strategy, we are faced with the problem of formulating objective rules for assigning them probabilities. So far as I know there are no such (unconventional) rules for allotting probabilities to propositions. In particular, nobody seems to have been able to assign probabilities to scientific hypotheses—except of course arbitrarily. Surely, the subjectivist is not worried by this objection: his whole point is that prior probabilities must be guesstimated by the subject, there being no objective tests, whether conceptual or empirical, to estimate the accuracy of his estimates. But this is just a way of saying that personalist probability is just a roundabout flight of fancy that must not be judged by the objective standards of science.

Our fifth objection is but an answer to the claim that probability values must always be assigned on purely subjective "grounds", i.e., on no grounds whatever. If probability assignments were necessarily arbitrary, then it would be impossible to account for the scientific practices of (a) setting up stochastic models of systems and processes and (b) checking the corresponding probability assignments with the help of observation, measurement, or theory. For example, genetic theory assigns definite objective probabilities to certain genetic mutations and recombinations, and experimental biology is in a position to test those theoretical values by contrasting them with observed frequencies. (On the other hand, nobody knows how to estimate the probability of either data or hypotheses. We do not even know what it means to say that such and such a statement has been assigned this or that probability.) In sum, the subjectivist interpretation of probability is at odds with the method of science: in science (a) states of things and changes of state, not propositions, are assigned probabilities, and (b) these assignments, far from being subjective, are controlled by observation, measurement, or experiment, rather than being arbitrary.

Our sixth and last objection is also perhaps the most obvious of all: if probabilities are credences, how come that all the probabilities we meet in science and technology are probabilities of states of concrete things— atoms, molecules, fields, organisms, populations, societies, or what not— or probabilities of events occurring in things of that kind, no matter what credences the personalist probabilist may assign either to the facts or to the theories about such facts? Moreover, many of the events in question, such as atomic collisions and radiative transitions, are improbable or rare, yet we cannot afford to dismiss them as being hardly credible.

The personalist might wish to rejoin that, as a matter of fact, we often do use probability as a measure of certainty or credibility, for example, when we have precious little information and when we apply the Bayes-Laplace theorem to the hypothesis/data relation. However, both cases are easily accounted for within the objectivist interpretation, as will be shown presently.

Case 1: *Incomplete information concerning equiprobable events.* Suppose you have two keys, A and B, the first for your house and the second for your office. The probability that A will open the house door is 1, and the probability that it will open the office door is 0; similarly for key B. These are objective probabilities: they are physical properties of the four key-lock couples in question. Suppose now that you are fumbling in the dark with the keys and that you have no tactual clues as to which is which. In this case the two keys are (empirically) equivalent before trying them. Whichever key you try, the probability of your choosing the right key for opening either door is 1/2. This is again an objective property, but not one of the four key-lock pairs: it is an objective property of the four you-key-lock triples. Of course, these probabilities are not the same as the previous ones: we have now taken a new domain of definition of the probability function. And surely the new probability values might be different for a different person, e.g., one capable of distinguishing the keys (always or with some probability) by some tactual clues. This relativity to the key user does not render probability subjective, any more than the relativity of motion to a reference frame renders motion subjective. Moreover, even when we assign equal probabilities to all the events of a class, for want of precise information about them, we are expected to check this hypothesis and change it if it proves empirically false. In short, incomplete information is no excuse for subjectivism.

Case 2: *Inference with the help of the Bayes-Laplace theorem.* This is of course the stronghold of the personalist school. However, it is easily stormed. First, recall that the Bayes-Laplace theorem is derivable from the mere definition of conditional probability without assuming any interpretation, whether personalist or objectivist. (Indeed, the definition is: $Pr(x \mid y) = Pr(x \cap y) / Pr(y)$. Exchanging x and y, dividing the two formulas, and rearranging, we get the theorem: $Pr(y \mid x) = Pr(x \mid y) \cdot Pr(y) / Pr(x)$.) Secondly, since there are no rules for assigning probabilities to propositions (recall our fourth objection), it is wrong to set x = evidence statement (e), and y = hypothesis (h) in the above formula, and consequently to use it as a principle of (probabilistic or statistical) inference. However, if we insist on setting $x = e$ (evidence) and $y = h$ (hypothesis), then we must adopt an *indirect*, not a literal, interpretation: $Pr(h)$ is not the credibility of *hypothesis h* but the probability that the *facts referred to by h* occur just as predicted by *h*. $Pr(e)$ is the probability of the *observable events described by e*; $Pr(h \mid e)$ is the probability of the *facts described by h, given* —i.e., it being actually the case—*that the events referred to by e occur*; and $Pr(e \mid h)$ is the probability of the *events described by e, given that the facts referred to by h happen.* This is the only legitimate interpretation of the Bayes-Laplace theorem because, as emphasized before, scientific theory and scientific experiment allow us to determine only the probabilities of (certain) facts, never the probabilities of propositions concerning facts. A by-product of this analysis is that all the systems of inductive logic that use the Bayes-Laplace theorem interpreted in terms of hypotheses and data are wrongheaded.

The upshot of our analysis is that the personalist interpretation of probability is wrong.

4. PROBABILITY AS FREQUENCY

If we cannot use the subjectivist interpretation, then we must adopt an objectivist one. Now, many objectivists believe that the only viable alternative to the personalist interpretation is the frequency interpretation.

The latter boils down to asserting that '$Pr(x) = y$' *means* that the relative long-run frequency of event x equals number y or, rather, some rational number close to y. (See Venn 1888; Wald 1950; von Mises 1972.)

A first objection that can be raised against the frequency interpretation of probability—and a fortiori against the identification of the two—is that they are *different functions* altogether. Indeed, whereas Pr is defined on a probability space F (as we saw in section 1), a frequency function f is defined, for every sampling procedure R, on the power set 2^{F^*} of a finite subset F^* of F, namely the set of actually observed events. That is, $Pr: F \to [0, 1]$, but $f: 2^{F^*} \times \Pi \to Q$, where $F^* \subseteq F$, and where Π is the set of sampling procedures (each characterized by a sample size and other statistical parameters) and Q is the set of proper fractions in $[0, 1]$.

Our second objection follows from the former: a probability statement does not *refer* to the same things as the corresponding frequency statement. Indeed, whereas a probability statement concerns usually an individual (though possibly complex) fact, the corresponding frequency statement is about a set of facts (a "collective", in von Mises's terminology) and, moreover, as chosen in agreement with certain sampling procedures. (Indeed, it follows from our previous analysis of the frequency function that its values are $f(x, \pi)$, where x is a member of the family of sets 2^{F^*} and π a member of Π.) For example, one speaks of the frequency with which one's telephone is observed (e.g., heard) to ring per unit time interval, thus referring to an entire set of events rather than to a single event—which is, on the other hand, the typical case of probability statements. Of course, probabilities can only be computed or measured for event types (or categories of event), never for unique unrepeatable events, such as my writing this page. But this does not prove that, when writing '$Pr(x) = y$', we are actually referring to a set x of events: though not unique, x is supposed to be an individual event. In other words, *whereas probability statements speak about individual events, frequency statements speak about strings of observed (or at least observable) events.* And, since they do not say the same, they cannot be regarded as identical.

To put the same objection in a slightly different way: the frequency interpretation of probability consists in mistaking percentages for probabilities. Indeed, from the fact that probabilities can *sometimes* be *estimated* by observing relative frequencies, the empiricist probabilist concludes that probabilities are identical with relative frequencies—which is like mistaking sneezes for colds and, in general, observable indicators for facts. Worse, frequencies alone do not warrant inferences to probabilities: by itself a percentage is not an unambiguous indicator of randomness. Only a (natural or artificial) selection mechanism, *if random*, authorizes the interpretation of a frequency as a measure of a probability. For example, if you are given the percentage of events of a kind, and are asked to choose blindfold any of them, then you can assign a probability to your *correctly choosing* the item of interest out of a certain reference class. In short, the inference goes like this: *Percentage & Random choice* ··········▶ *Probability*. (The line is broken to suggest that this is not a rigorous, i.e., deductive inference, but just a plausible one.)

Surely, not all frequencies are observed: sometimes they can be calculated, namely on the basis of definite stochastic models, such as the coin-flipping model or Bernoulli sequence. (As a matter of fact, all the frequencies occurring in the calculus of probability are theoretical, not empirical.) But in this case too, the expected frequency differs from the corresponding probability. So much so that the difference is precisely the concern of the laws of large numbers of the probability theory. One such theorem states that, in a sequence of Bernoulli trials such as coin flippings, the frequency f_n of successes or hits in the first n trials approaches the corresponding probability (which is constant, i.e., independent of the size n of the sample). Another theorem states that the probability that f_n deviates from the corresponding probability p by more than a preassigned number tends to zero as n goes to infinity. (Note that there are two probabilities and one frequency at stake in this theorem.) Obliterate the difference between probability and frequency, and the heart of the probability calculus vanishes. This, then, is our third objection to the frequency interpretation of probability, namely that *it cannot cope with the laws of large numbers*. For further technical objections see Ville (1939).

Our fourth argument is of an ontological nature, namely this. While a frequency is the frequency of the *actual* occurrence of facts of a certain kind, a probability may (though it need not) measure the *possibility* of a fact (or rather the strength of such a possibility). For example, as long as the coin we flipped is in the air, it has a probability 1/2 of landing "head". But once it has landed, potentiality has become actuality. On the other hand, the coin has no frequency while it is in the air, for a frequency is a property of the collection ("collective") of the final states of the coin. Consequently, identifying probabilities with frequencies (either by definition or by interpretation) implies (a) rejecting real or physical possibility, thus forsaking an understanding of all the scientific theories which, like quantum mechanics and population genetics, take real possibility seriously, and (b) confusing a theoretical (mathematical) concept with an empirical one.

The correct thing to do with regard to the probability-frequency pair is not to identify them either by way of definition or by way of interpretation, but to clarify their mutual relation as well as their relations to the categories of possibility and actuality. We submit that frequency estimates probability, which in turn measures or *quantitates* possibility of a kind, namely chance propensity (Bunge 1976b). And, while probability concerns possibles, frequency concerns actuals and, moreover, in the applications it always concerns observed actuals.

In other words: there is no valid frequency *interpretation*, let alone definition of probability; what we do have are statistical *estimates* of theoretical probability values. Moreover, frequencies are not the sole estimators or indicators of probability. For instance, in atomic and molecular physics transition probabilities are often checked by measuring spectral line intensities or else scattering cross-sections. And in statistical mechanics probabilities are estimated by calculating entropy values on the basis of either theoretical considerations (with the help of formulas such as Boltzmann's) or measurements of temperature and other thermodynamic properties. In short, probabilities are not frequencies and they are not interpretable as frequencies, although they can sometimes (by no means always) be estimated with the help of frequencies.

To be sure, frequencies, when joined to plausible random mechanisms, supply a rough indication of probability values and serve to check probability values and probability calculations. Hence, probabilities and frequencies, far from being unrelated, are in some sort of correspondence. In fact, (a) an event may be possible and may even have been assigned a nonvanishing probability without ever having been observed to happen, hence without being assigned a frequency; (b) conversely, certain events can be observed to occur with a certain frequency without, however, being assigned a nonvanishing probability.

In sum, the frequency interpretation of probability is inadmissible for a number of technical and philosophical reasons. Let us therefore look for an interpretation of probability free from the fatal flaws of the frequency interpretation.

5. PROBABILITY AS PROPENSITY

A mathematical interpretation of the probability calculus, i.e., one remaining within the context of mathematics, consists in specifying the mathematical nature of the members of the domain F of Pr, e.g., as sets of points on a plane, or as sets of integers, or in any other way compatible with the algebraic structure of F. Such an interpretation of the probability space F would yield a *full mathematical interpretation* of the probability theory. (Likewise, interpreting the elements of a group as translations, or as rotations, yields a full mathematical interpretation of the abstract theory of groups.) Obviously, such a mathematical interpretation is insufficient for the applications of probability theory to science or technology. Here we need a factual interpretation of the calculus.

A *factual interpretation* of probability theory is obtained by assigning to both F and every value $Pr(x)$ of Pr, for x in F, factual meanings. One such possible interpretation consists in taking the basic set S, out of which F is manufactured, to be the state (or phase) space of a thing. In this way every element of the probability space F is a bunch of states, and $Pr(x)$ becomes the *strength of the propensity or tendency the thing has to dwell in the state or states* x. Similarly, if x and y are states (or sets of states) of a thing, the conditional probability of y given x, i.e., $Pr(y \mid x)$, is interpreted as the strength of the propensity or tendency for the thing to go from state(s) x to state(s) y. This, then, is the *propensity interpretation of probability*.

This is not an arbitrary interpretation of the calculus of probability. Given the structure of the probability function and the interpretation of its domain F as a set of facts (or events or states of affairs), the propensity interpretation is the only possible interpretation in factual terms. Indeed, if F represents a set of facts, then $Pr(x)$, where x is in F, cannot but be a property of the individual fact(s) x. That is, contrary to the frequency view, probability is *not a collective* or ensemble property, i.e., a property of the entire set F, but a property of every *individual* member of F, namely its propensity to happen. What *are* ensemble properties are, of course, the normalization condition $Pr(S) = 1$ and such derived (defined) functions as the moments (in particular the average) of a probability distribution, its standard deviation if it has one, and so on.

This point is of both philosophical and scientific interest. Thus, some biologists hold that, because the survival probability can be measured only on entire populations, it must be a global property of a population rather than a property of each and every member of the population. (Curiously enough, they do not extend this interpretation to the mutation probability.) The truth is of course that, while each probability function Pr is a property of the ensemble F, its values $Pr(x)$ are properties of the members of F.

It is instructive to contrast the propensity to the frequency interpretations of probability values, assuming that the two agree on the nature of the probability space F. (This assumption is a pretense: not only frequentists like von Mises, but also Popper, a philosophical champion of the propensity interpretation, have stated that facts have no probabilities unless they occur in experimentally controlled situations. In fact, they emphasize that probabilities are mutual properties of a thing and a measurement setup—which of course makes it impossible to apply stochastic theories to astrophysics.) This contrast is displayed in the following table.

p = Pr(x)	Propensity	Frequency
0	x has (almost) nil propensity	x is (almost) never observed
$0 < p \ll 1$	x has a weak propensity	x is rarely observed
$0 \ll p < 1$	x has a fair propensity	x is fairly commonly observed
$p \approx 1$	x has a strong propensity	x is very commonly observed
$p = 1$	x has an overpowering propensity	x is (almost) always observed

Note the following points. First, although a probability value is meaningful, i.e., it makes sense to speak of an individual fact's propensity, it is so only in relation to a definite probability space F, e.g., with reference to a precise category of trials. Likewise, a frequency value makes sense only in relation to a definite sample-population-sampling method triple. For example, the formula "x is rare" presupposes a certain set of occurrences, to which x belongs, among which x happens to be infrequent.

Secondly, in the case of continuous distributions, zero probability is consistent with very rate (isolated) happenings. That is, even if $Pr(x) = 0$, x may happen, though rarely as compared with other events represented in the probability space. Consequently, a fact with probability 1 can fail to happen. (Recall that any set of rational numbers has zero Lebesgue measure. Entire sets of states and events are assigned zero probability in statistical mechanics for this very reason, even though the system of interest is bound to pass through them. This is what 'almost never' is taken to mean in that context, namely that the states or events in question are attained only denumerably many times.)

Thirdly, the frequency column should be retained alongside the propensity interpretation, though in a capacity other than interpretation or definition. Indeed, although the frequency column fails to tell us what "$Pr(x) = y$" *means*, it does tell us under what conditions such a formula is *true*. Long-run frequency is in short a *truth condition* for probability statements. Besides, frequency statements have a heuristic value. For example, if p means a transition probability, then the greater p, the more frequent or common the transition.

Fourthly, note again that the present propensity interpretation differs from Popper's in that the latter requires the referent to be coupled to an experimental device. No such hang-up from the frequency (or empiricist) interpretation remains in our own version of the propensity interpretation. Nor do we require that probabilities be assigned only to events proper (i.e., changes of state), as an empiricist must—since states may be unobservable. Probabilities may also be assigned to states, and in fact this is done in many a stochastic theory, such as statistical mechanics and quantum theories. (The statistical mechanical measure of entropy is a function of the thermodynamic probability of a state—or, as Planck put it, it measures the preference [*Vorliebe*] for certain states over others.) In other words, not only transition probabilities (which are conditional), but also absolute probabilities can be factually meaningful.

Fifthly, note that the propensity (or any other) *interpretation* of probability is to be distinguished from the probability *elucidation* (or exactification) of the intuitive or presystematic notion of propensity, tendency, or ability. In the former case, one "attaches" factual items to a concept, whereas in the latter one endows a factual concept with a precise mathematical structure. In science, and also in ontology, we need both factual interpretation and mathematical elucidation. The failure to draw this distinction has led some philosophers to reject the propensity interpretation for being circular (e.g., Sober 1993).

Sixthly, the propensity interpretation presupposes that possibilities can be real or physical rather than being just synonymous with our ignorance of actuality. By contrast, according to the frequency interpretation, there is no such thing as a chance propensity for an individual fact (state or event): there would be only limiting frequencies defined for entire ensembles of facts, such as a sequence of throws of a coin, or a family of radiative transitions of a kind. Indeed, the phrase '$Pr(x) = y$' is, according to the frequency school, short for something like 'The relative frequency of x in a large ensemble (or a long sequence) of similar trials is observed to approach y'. This view is refuted by the existence of microphysical theories concerning a single thing, such as a single atom, to be sharply distinguished from a theory about an aggregate of coexisting atoms of the same kind. A relative frequency is a frequency of actuals, hence it cannot be identical with a possibility (measured by a probability). Unlike frequencies, probabilities do measure real (physical) possibilities of a certain kind. Therefore, if we take real possibility seriously, i.e., if we are possibilists rather than actualists, we must favor the propensity over the frequency interpretation.

Seventhly, just as the subjectivist interpretation of probability is a necessary constituent of the subjectivist interpretation of quantum physics, so a realistic interpretation of the latter calls for an objective interpretation of probability. For example, the formula for the probability of a radiative transition of an atom or molecule, induced by an external perturbation, is always read in objective terms, not in terms of credences. So much so that it is tested by measuring the intensity of the spectral line resulting from that transition. Another example is the superposition principle. If a quantum mechanical entity is in a superposition $\psi = c_1 \cdot \phi_1 + c_2 \cdot \phi_2$ of the eigenfunctions ϕ_1 and ϕ_2 of an operator representing a certain dynamical variable (e.g., the spin), then the coefficients c_1 and c_2 are the contributions of these eigenfunctions to the state ψ, and $p_1 = |c_1|^2$ and $p_2 = |c_2|^2$ are the objective probabilities of the collapse of ψ onto ϕ_1 or ϕ_2, respectively. This collapse or projection may occur either naturally or as a result of a measurement, which in turn is an objective interaction between the entity in question with a macrophysical thing (the apparatus).

Unless the objective interpretation of the above probabilities is adopted, a number of paradoxes may result. One of them is the so-called Zeno's quantum paradox. According to it, if an unstable quantum system is observed continuously to be, say, in state ϕ_1 (excited or not-decayed), then ψ cannot raise its head: it will continue to lie on ϕ_1. But of course, this result (an interpretation of von Neumann's measurement postulate) runs counter experience, so there must be something wrong with the reasoning. What is wrong is, among other things, the adoption of the subjective interpretation of states, hence of probabilities, as being nothing but states of the observer's knowledge. The paradox dissolves upon adopting the propensity interpretation and taking the superposition principle seriously (Bunge and Kálnay 1983a).

In short, the propensity interpretation of probability is consistent with the standard theory of probability and with scientific practice, as well as with a realist epistemology and possibilist ontology. Hence, it solves the old tension between rationality and the reality of chance. None of its rivals has these virtues.

IV

PHILOSOPHY OF PHYSICS

16

PHYSICAL RELATIVITY AND PHILOSOPHY

(1979)

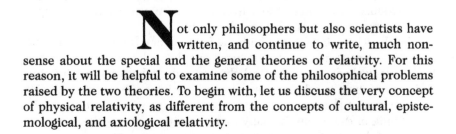

Not only philosophers but also scientists have written, and continue to write, much nonsense about the special and the general theories of relativity. For this reason, it will be helpful to examine some of the philosophical problems raised by the two theories. To begin with, let us discuss the very concept of physical relativity, as different from the concepts of cultural, epistemological, and axiological relativity.

1. RELATIVE AND APPARENT

It is commonly believed that relativistic physics supports epistemological relativism, or subjectivism, in stressing the need for explicit reference to the reference frame. (See, e.g., Latour 1988.) This would indeed be the case if reference frames were observers (or subjects), as so many authors would make us believe—but they are not. In fact, a reference frame is a physical system representable by a coordinate system and utilizable in the description of motion (of bodies, fields, or what have you). Reference frames can be natural, e.g., stars or constellations, or artificial, e.g., laboratories. In principle, any material system can serve as a reference system, but the better reference frames are those wherein a regular process occurs, since they include a (natural or artificial) clock.

Both in classical and in relativistic physics positions, hence velocities and accelerations as well, are frame-dependent. Moreover, the equations and inequalities that represent laws of nature are not true in (relative to) all reference frames, but only in (relative to) certain privileged frames called 'inertial'. Thus, the equations of mechanics, whether classical or relativistic, fail to hold exactly in terrestrial laboratories, since our planet is an accelerated system.

In classical mechanics, lengths, durations, and masses are absolute, i.e., the same in (relative to) all frames. By contrast, in relativistic mechanics lengths, durations, and masses are relative to the reference frame—hence the proliferation of the term 'relative' in relativistic writings. Therefore, every body has infinitely many sizes and masses: as many as mutually inequivalent reference frames.

It does not follow from this that everything is relative in relativistic physics. What is true is that some classical absolutes have been relativized. Others, such as electric charge and entropy, continue to be absolute. In return for the loss of certain absolutes, relativistic physics has introduced absolutes absent from classical physics, such as proper time and spatiotemporal distance. See table 1.

On the whole, if we count both the fundamental (undefined) and the derived (defined) magnitudes, it turns out that relativistic physics has just as many absolutes as classical physics. Indeed, both sets are infinite, and each of them is split into two mutually disjoint subsets: that of relative and that of absolute magnitudes, and both subsets are in turn infinite. Think of the infinitely many time-derivatives of a position coordinate, or of the equally numerous powers of the proper time.

Nor does it follow that whatever is relative is so to the observer or subject. In fact, the concept of an observer, which is a bio-psycho-social one, does not occur in any physical theory, even though it looms large in the textbooks influenced by the positivist (operationalist) philosophy, centered as it is in the observer and his laboratory operations. Physical theories deal with (refer to) physical things, their states and changes, not to biological or psychological systems. Moreover, physics refers to reality regardless of the way it may appear to some particular observer. If physical theories referred to observers and their observation acts, then they would account for their mental states and social condition—something they are obviously incapable of. The mistake originates in identifying ref-

Table 1.

Relativity has not relativized everything: it has changed the status (relative or absolute) of some magnitudes, and has added new ones unknown to classical physics.

MAGNITUDE	CLASSICAL	RELATIVISTIC
Position	Rel.	Rel.
Velocity of a body	Rel.	Rel.
Length	Abs.	Rel.
Duration	Abs.	Rel.
Mass	Abs.	Rel.
Force	Rel.	Rel.
Energy	Rel.	Rel.
Electric charge	Abs.	Abs.
Electromagnetic field intensities	Rel.	Rel.
Speed of light in a vacuum	Abs.	Abs.
Entropy	Abs.	Abs.
Temperature	Abs.	Rel.
Proper time	—	Abs.
Spatiotemporal interval	—	Abs.

erence frames with observers. This confusion is to be avoided, for although every observer can serve as a reference frame (albeit not always an inertial one), the converse is false. That is, it is not true that every reference frame is, or can be occupied by, an observer. For example, stars can serve as reference frames, but not as human dwellings.

For that reason, it is mistaken to employ in physics such expressions as 'Event *x* as seen by observer *y*', and 'Process *x* as seen from atom *y*'. Instead, we should speak just of events and processes relative to a given reference frame. For the same reason, the expressions 'apparent mass', 'apparent length', and 'apparent duration' are misleading. The correct expressions are: 'relative mass' (i.e., value of the mass relative to some reference frame in motion relative to the given body), 'relative length', 'relative duration', and the like.

The formulas of physics, whether classical or relativistic, are supposed to represent physical facts, not psychological ones: they concern reality, not appearance (which is always relative to some observer). Introducing observers into theoretical physics is smuggling. Observers are to be found in laboratories and observatories, and their actions are explained by psychological and sociological considerations that go far beyond physics.

In sum, 'relative' is short for "relative to a given reference frame", not "relative to a given observer".

2. INERTIAL REFERENCE FRAME

The concept of an inertial reference frame is central to all mechanical and electrodynamical theories. In classical mechanics, an inertial reference frame is a physical system, representable by a coordinate system, relative to which certain equations of motion (e.g., those of *Newtonian* mechanics) hold, i.e., are (sufficiently) true. We may call Newtonian such a reference frame defined in terms of the laws of classical mechanics. (In an axiomatic formulation one proceeds as follows. One first states the equations of motion, preceding them by the expression 'There exist reference frames relative to which'—and here the equations are written. And one then adds the convention: "The reference frames relative to which the above equations hold are called *inertial*". It follows that two Newtonian reference frames are equivalent if, and only if, an arbitrary free particle moves with rectilinear and uniform motion in [relative to] both reference frames.)

On the other hand, in classical electrodynamics, i.e., Maxwell's theory, a reference frame is a physical system, representable by a coordinate system, relative to which the equations of the electromagnetic field hold. We may call *Maxwellian* such a reference frame defined in terms of the laws of classical electrodynamics. It follows that two Maxwellian frames are equivalent just in case an arbitrary light ray propagates in a vacuum along a straight line with constant speed relative to both frames.

The general concept of an inertial frame, whether Newtonian or Maxwellian, enables us to state one *principle of relativity* for each of the two basic theories of classical physics. Moreover, the two principles are special cases of the following general principle: "The basic (or fundamental) laws of physics ought to be the same in (relative to) all inertial reference frames". Note that this relativity principle (a) refers to *basic* (not derived) laws, and (b) is *normative* or prescriptive rather than descriptive. Note also that it does not state that all physical events are, or ought to be, the same for all observers, i.e., that all observers ought to observe the same phenomena. This would be false, as shown, among many other examples, by the aberration of light, the Doppler effect, and the spaceship orbiting the Earth in such a manner that it is always exposed to the sunlight, so that its crew never experiences night.

Now, the inertial reference frames relative to which the equations of electrodynamics hold are not the same as those relative to which the equations of classical mechanics are expected to hold. In other words, whereas two Newtonian reference frames are related by a Galilei transformation, two Maxwellian reference frames are related by a Lorentz transformation. Also: Whereas the equations of classical mechanics do not change under a Galilei transformation, those of classical electrodynamics do change, but on the other hand they are invariant (or rather covariant) relative to a Lorentz transformation. See table 2.

Lorentz and Poincaré knew that the basic equations of classical mechanics do not hold relative to Maxwellian frames, and that those of classical electrodynamics do not hold relative to Newtonian frames. Einstein discovered this incompatibility independently but, unlike his famous predecessors, exploited it successfully. Lorentz and Poincaré tried to surmount the contradiction by keeping the equations of motion of classical mechanics but reinterpreting them in terms of contractions and dilatations caused by the aether during the body's motion through the aether.

Einstein understood that the situation called for a radical measure: that it was necessary to choose between classical mechanics and classical electrodynamics—or else reject them both. He opted for classical electrodynamics, and this both because of his philosophical and aesthetic bias for field theories and because, of the two theories, electrodynamics was the better-confirmed one. (In fact, the optical measurements are far more precise than the mechanical ones.)

That is, Einstein assumed that classical mechanics was false, and set himself the task of building a new mechanical theory compatible with classical electrodynamics, i.e., one whose basic equations would hold relative to Maxwellian reference frames. Einstein (1934, p. 205) himself stated that "The Maxwell-Lorentz theory led me inexorably to the theory of special relativity". Thus was relativistic physics born: from an examination of theories, not from an analysis of experiments. Moreover, it has been shown conclusively that Einstein got to know about the Michelson and Morley measurements after having published his 1905 paper (Shankland 1963, 1973; Holton 1973).

3. PHYSICAL COORDINATES AND COORDINATE TRANSFORMATIONS

Both the Galilei and the Lorentz transformations are often misinterpreted as transformations of geometric coordinates. Actually, they are transformations of physical coordinates, i.e., of coordinates serving to identify the spatiotemporal positions of physical things or physical events. Let me explain.

A geometric coordinate is an element of a vector space. A physical coordinate is a geometric coordinate referring to some physical entity. For example, a point in a three-dimensional space can be identified, relative to some coordinate system, by a triple of real numbers. (In other words, there is a function that maps the given space onto a Cartesian or

Table 2.

Transformation properties of the basic equations of classical physics.

BASIC LAWS	HOLD (ARE TRUE) IN	ARE INVARIANT UNDER
Newton-Euler equations	Newtonian frames	Galilei transformations
Maxwell's equations	Maxwellian frames	Lorentz transformations

numerical space.) By contrast, we identify the position of a physical entity (particle, point on a wave front, etc.), relative to some reference frame, by a triple of real numbers that happens to be a value of a time-dependent function. (In general, for each frame there exists a function, called 'position coordinate', that assigns to each pair ⟨physical entity, time instant⟩ a triple of real numbers called 'instantaneous position of the entity relative to the given frame'. Likewise, for each physical event there exists a function, called 'time coordinate', that assigns to it a real number called 'time at which the event occurs relative to the given frame'. Shorter: In Cartesian coordinates, an arbitrary point in a three-dimensional space may be identified by a triple $\langle x, y, z \rangle$, whereas a physical coordinate looks like $\langle x(t), y(t), z(t) \rangle$.)

In the physics literature a happy-go-lucky confusion between physical and geometrical coordinates is the rule. This confusion precludes a correct understanding of the Galilei and Lorentz transformations, which relate physical, not geometrical, coordinates. In other words, these transformations relate the temporal and spatial coordinates of factual items (entities or events), not those of arbitrary points in an arbitrary space. (The points in a metric space do not move, hence their coordinates cannot be time-dependent.) The Galilei and Lorentz transformations are linear transformations of the physical coordinates, and both contain a parameter alien to geometry, namely the relative speed of the (inertial) frames in question. In addition, the Lorentz transformations also contain the speed of light in a vacuum, another parameter alien to geometry (and to classical mechanics).

The choice among systems of geometric coordinates is conventional: this decision depends upon the type of problem. To handle some problems, it is convenient to adopt rectangular coordinates, whereas other problems counsel the adoption of spherical coordinates, and so on. By contrast, the choice of a physical coordinate system (or reference frame) is not merely a matter of convention, since relative to certain frames some physical equations may cease to hold. For example, a coordinate system centered in our planet is mathematically legitimate. But, since our planet is neither a Newtonian nor a Maxwellian reference frame, relative to it neither mechanics nor electrodynamics hold rigorously. For the same reason, the transformation that relates the geocentric and the heliocentric coordinates is neither a Galilei nor a Lorentz transformation, since the latter relate only inertial frames. Consequently, the conventionalist interpretation of the Lorentz equations, according to which these support the thesis that the choice of frame is conventional, is false. See table 3.

4. INVARIANCE AND COVARIANCE

A property that does not change under a transformation of physical coordinates—one that is the same in (relative to) all frames of a certain type (e.g., Maxwellian)—is said to be *invariant* with regard to the given transformation (or to be an invariant of the given group). Electric charge and entropy are both Galilei and Lorentz invariant. One can also say that they are *absolute* properties, provided one adds the qualifier 'relative to the given transformation group'. This rider is important, because a given property may be the same in all inertial frames of a certain type (e.g., Maxwellian), but may differ in (relative to) noninertial frames. This is the case with the speed of light in a vacuum, which is the same in all Maxwellian frames, but different in (relative to) noninertial frames such as our planet. (Relative to the latter, starlight does not travel in a straight line but along a helix.)

A law (or rather *law statement*) that does not change under a transformation of physical coordinates, i.e., which is the same in all reference frames of a certain kind, is said to be *covariant* with respect to the transformation concerned. The basic equations of special relativity are Lorentz-covariant. Those of general relativity (the equations for the gravitational field) are general-covariant, i.e., they do not change under an arbitrary change of physical coordinates, being relations among tensors.

Though similar, invariance and covariance are quite different, both physically and ontologically. Whereas invariance is a property of certain physical properties, covariance is a feature of certain physical regularities or patterns (laws). For example, the angle between the electric field and the magnetic induction components of an electromagnetic field is Lorentz-invariant, even though the field intensities themselves are not invariant. Such intensities are logical consequences (solutions) of the basic laws of Maxwell's theory. Covariance concerns these basic laws (postulates), not their logical consequences (theorems). In other words, Maxwell's equations are Lorentz-covariant, but every solution to such equations is frame-dependent. For example, an electrostatic field gets transformed into ("is seen as") an electromagnetic field relative to a moving reference frame.

Table 3.

The distinction between reference frame and coordinate system, hence between physical and geometrical transformations.

COORDINATES	TRANSFORMATIONS	EXAMPLES
Geometrical	Geometrical	Cartesian → Cylindrical
Physical	Physical	Galilei, Lorentz, general, canonical (of q's and p's), geocentric → heliocentric

The covariance principles, such as the relativity principle dealt with in section 1, do not concern properties, states, or events: they refer to basic patterns or their mathematical formulations. Hence, they are not physical laws, but rules about physical laws (or rather physical statements). They can therefore be called *metanomological statements* (Bunge 1959c, 1961). Moreover, in their heuristic formulation they are prescriptive (normative) statements: they do not ensure that all of the basic laws be in fact covariant, but require that they so be. Thus, they serve as guides in theory construction. See table 4.

Covariance is usually understood as form invariance, or conservation of mathematical form. However, although form-invariance implies covariance, the converse is false. In fact, form-invariance is not always attainable even while respecting covariance. Thus, when we "go" from Cartesian to spherical coordinates, certain terms may appear that lack Cartesian analogs. What we want is that the equation (or inequality) be preserved as a whole, even though every one of its components (or even their number) may change.

This matter is related to the question, which has been debated in the literature from the early days of general relativity, of the physical meaning of covariance principles. Clearly, these principles lack an immediate physical content, since they concern propositions (formulas), not facts. However, they are not mere formal requirements, because (a) they concern statements of natural law and (b) they ensure that the formulas satisfying them be universal, i.e., true in all reference frames of certain kind—hence objective, i.e., observer-free.

The last remark signals the philosophical importance of the covariance principles. Because they state that the choice of frame (in particular of observer) has no importance, or ought to have none, as far as the basic patterns are concerned, the covariance principles are objectivity principles. Besides, because they demand the covariance of the basic laws, but not of the derived statements, those principles allow us to identify what is objective and universal, and sort it from what is objective but local, i.e., bound to some reference frame. A familiar, though still puz-

Table 4.

The difference between physical laws and statements about them.

OBJECT	CHARACTERISTIC	EXAMPLE
Physical law	Regularity or Pattern	Every electromagnetic wave carries energy
Basic formula Law of law	Represents basic pattern	Maxwell's equations
Descriptive version	States property of a basic formula	Maxwell's laws are Lorentz-covariant (theorem)
Prescriptive version	Prescribes a desirable property of a basic law statement	The basic formulas of physics ought to be covariant under general (nonlinear) transformations of the physical coordinates

zling, example is that of the centrifugal force. This force is often said to be fictitious because it does not occur in the basic equations of motion when written relative to an inertial frame. But anyone who has ever ridden a merry-go-round is likely to dispute this contention. What happens is that the force, though null in the rest frame, is vividly felt in the rotating frame, which of course is noninertial. In short, the centrifugal force, like all forces, is not absolute but frame-dependent.

5. REFERENTS OF RELATIVISTIC PHYSICS

Eminent physicists have stated that "[r]elativity concerns measurements made by different observers moving relative to one another"—as Peter G. Bergmann wrote in the 1973 edition of the *Encyclopaedia Britannica*. This is the operationist or positivist interpretation of relativity, an interpretation explicitly rejected by Bergmann's erstwhile coworker, Einstein (1949, pp. 49, 81) himself.

I submit that the above interpretation is mistaken, if only because the concepts of observer and measurement do not occur among the concepts (either basic or defined) of relativistic physics. Relativistic theories do not concern measurements but physical entities in general, whether or not they are being subjected to observation. Moreover, the concepts of yardstick, clock, observer, and measurement can only be elucidated with the help of a host of scientific theories, among them the relativistic ones. Relativistic physics was not born as, or even from, a description of measurements: recall section 1. On the contrary, it has suggested measurements inconceivable in classical physics, such as the measurements of duration "contraction" and the bending of light rays by gravitational fields.

The genuine referents of a scientific theory are not uncovered by consulting authorities, but by examining typical problems soluble with the help of the theory and analyzing its basic concepts. Take, for instance, relativistic particle mechanics. The following basic or undefined concepts occur in this theory: those of spacetime, metric of the same, reference frame, speed of light in a vacuum, particle, external force, interparticle force, and mass. Let us focus our attention on the latter.

The mass of a particle is a value of a certain function that assigns a positive real number to each quadruple ⟨particle, reference frame, instant, mass unit⟩. The physical referents of the concept are obvious: they are particles and reference frames. (Since the very definition of a Maxwellian frame presupposes Maxwell's theory or, in general, the set of theories of null-mass fields, one can also assume that the latter is a tacit referent of the relativistic mass concept.) By contrast, the concept of an observer does not occur at all—unless one makes the mistake of equating it with that of a reference frame; nor does the concept of measurement occur as a referent of the mass concept. (Which is just as well, since different measurement techniques are likely to yield different measurement values of the mass of any given particle.) We may then take it that relativistic particle mechanics refers to particles, Maxwellian reference frames, and electromagnetic (or similar) signals among particles.

Relativistic particle mechanics is a very special theory. It presupposes relativistic kinematics, which is also the basis of all the other special relativistic theories. Relativistic kinematics refers to arbitrary physical objects potentially related by null-mass fields (e.g., electromagnetic) and Maxwellian frames (Bunge 1973b). Any other special relativistic (or Lorentz-covariant) theory will be defined as a theory with a specific referent (particle, body, field of a certain kind, etc.), and whose basic laws are covariant under Lorentz transformations.

The entire special relativistic physics is the union of all the Lorentz-covariant theories: classical electrodynamics, relativistic particle mechanics, relativistic thermodynamics, Dirac's theory of spin 1/2 "particles", Kemmer's of spin 1 "particles", and so on. Since, in principle, every physical theory may be relativized, i.e., be incorporated into relativistic physics, the reference class of the latter turns out to be the set of all physical entities, i.e., the physical composition of the entire universe.

If we lose sight of the referents of the various physical theories, we run the risk of misinterpreting them. We already saw one example of such misinterpretation, namely the positivist view that special relativity is a theory of measurements performed by observers moving relative to one another. Another case is that of the celebrated formula $E = mc^2$.

According to popular wisdom, it assigns a mass to every energy. This opinion is false, for the formula is a theorem of relativistic mechanics and does not occur in the theories of null-mass fields, such as the electromagnetic theory, whether classical or quantal. Hence, within the framework of relativistic physics, it is senseless to speak of the mass of a photon or of a neutrino—at least if the current photon and neutrino theories are accepted. What the formula does is to assign an energy, in particular the rest energy m_0c^2, to every particle.

Our next question is: What is general relativity about? Here too we find a diversity of views, few of them supported by the formulas of the theory. In particular, we find the operationist view that general relativity concerns observations, in particular distance and duration measurements, or perhaps the behavior of yardsticks and clocks. This opinion rests on a philosophical dogma, not on an analysis of the basic formulas of the theory. Besides, it leads to assigning the same referents to all physical theories, namely competent, diligent, well-equipped, and well-funded observers.

It is not difficult to discover that the central formulas of general relativity are those representing the basic features of a gravitational field. This should not be surprising, since the theory is used to compute the gravitational fields accompanied by things of various kinds. Those formulas can be summarized as follows: "$G = T$", where G (the Einstein tensor) represents the geometric properties of spacetime, which are partially determined by T, the matter tensor. In turn, the latter describes the sources of the gravitational field (bodies, fields of various descriptions except for gravitational ones, and whatever else the world is made of). Therefore, if we wish to pinpoint the genuine referents of general relativity, we must analyze both G and T.

If the field sources are electrically charged fields, then the corresponding general relativistic theory will concern (refer to) such bodies and the accompanying fields, in particular the gravitational ones. If, in contrast, the sources concerned are photons, the theory will concern these as well as the gravitational fields associated with them. If we now define general relativistic physics as the union of all the specific theories where the above-mentioned equations occur, i.e., all of the general-covariant theories, it turns out that the reference class of general relativistic physics is the same as that of special relativistic physics, i.e., the set of all physical entities, or the totality of physical components of the world.

The crucial differences between special and general relativistic physics are not differences in referents, but rather differences in transformation properties and in the role assigned to gravitation: whereas most special relativistic theories neglect gravitation, no general relativistic theory can neglect it. Therefore, the central or typical referent of general relativistic physics is the gravitational field.

What happens if the matter tensor T, occurring in the field equations, vanishes everywhere? The solutions of the resulting equations, namely "$G = 0$", describe in general a curved (Riemannian) spacetime. This would seem to contradict the statement that the gravitational field produces deviations of spacetime with respect to the tangent (or pseudo-Euclidean) space. And this has puzzled a number of philosophers.

In fact, there is no contradiction, for general relativity does not state that every deviation from flat spacetime is due to gravitation, but the converse, namely: If there is gravitation, then spacetime is curved. When there is no matter ($T = 0$ everywhere) there is no gravity either, hence the general theory of relativity remains unemployed. The resulting theory, i.e., the formula "$G = 0$ everywhere", together with its logical consequences, is a mathematical formalism lacking in physical content. (A theory possesses a physical content if, and only if, it refers to physical entities such as bodies or fields). General relativity *represents* gravitation as spacetime curvature, but it does not *identify* the two; and that representation holds only when there is matter. A hollow universe is unreal and therefore not a possible referent of a physical theory: it is but a mathematical fiction. In theological terms: All God has to do to annihilate the physical universe is to set T equal to zero everywhere.

6. CONVENTION AND REALITY

All scientific theories contain some conventions, such as definitions, coordinate systems, and units. But a scientific theory is more than a batch of conventions: it is a set of propositions (formulas), some of which refer to some aspect of reality. In particular, the law statements are not conventions but well-confirmed hypotheses. Hence, they are true or false to some extent, and susceptible to modification in the light of new observational or experimental data.

Conventionalism is the philosophical doctrine that asserts that every theory, whether mathematical or factual, is nothing but a set of conventions. In particular, conventionalism holds that the axioms (or postulates) are disguised definitions, hence conventions: that they would be neither true nor false and, therefore, impregnable to empirical data. We saw an example of this thesis when dealing with the Lorentz transformations, which, according to conventionalism, relate more or less convenient descriptions. Poincaré himself may have been a victim of his own conventionalism. Indeed, he was on the brink of inventing special relativity, and may have been prevented from creating it by his half-hearted conviction that the postulates of mechanics are conventional anyway, so they need not, nay cannot, be criticized in the light of experimental results. This view was so deeply entrenched in the French scientific community at the time, that rational mechanics was taught by mathematicians, not physicists. Let us now examine another example.

Einstein himself, under the influence of the conventionalism fashionable at the beginning of the twentieth century, asserted initially that simultaneity is conventional. That is, whether or not two distant events are judged as simultaneous, would be a matter of convention. Some philosophers (notably Grünbaum 1963) hold in addition that special relativity is based on such convention rather than on a set of postulates representing natural patterns, such as the constancy (or rather frame-independence) of the speed of light in a vacuum. The controversies unleashed by this question have filled many pages over the past few decades. However, the question can be decided in one minute.

In fact, in order to find out whether two distant events are simultaneous relative to a given reference frame (e.g., a laboratory), we measure or calculate distances and travel times of light signals, using one of the basic assumptions of special relativity. This is the hypothesis that light propagates in a vacuum, relative to any Maxwellian reference frame, at a constant speed c in any direction. Example: a photocell detects, at time t_2, photons emitted at time t_1, by two light flashes equidistant from the receptor and at the mutual distance d. The relation between the emission and the reception times is $t_1 = t_2 - d/2c$. This is a physical law, not a convention. So much so that it can be checked experimentally. Moreover, the formula rests on a hypothesis that may well be refuted in the future, namely that space is homogeneous and isotropic. It might well happen that the speed of light has local microfluctuations, or that in some regions the speed in one direction differs from its speed in the opposite direction. In sum, the simultaneity relation is not conventional. What happens is that (like duration and length) it is not absolute but frame-dependent. That is, two events simultaneous relative to a given frame may cease to be simultaneous relative to a different frame. And this is a feature of nature, not a human convention.

Another conventionalist thesis advanced by Grünbaum (1963) is this. Since in general relativity the choice of geometrical coordinates is indeed conventional, the metric too must be conventional. Therefore, the assignment of spatiotemporal distances would be arbitrary. The conclusion is false: if two events are separated by a certain spatiotemporal distance (or proper-time interval), this distance remains constant under a general (nonlinear) coordinate transformation, i.e., the line element ds is invariant under general coordinate transformations.

What is true is that the values of the components of the metric tensor are relative to the reference frame. But this holds for all tensors (and vectors), so it does not prove the conventionality of the metric (or line element). Besides, unlike conventions, the formulas for the value of the metric tensor at a given place in spacetime have a precise physical content: they tell us indirectly what the value of the strength of the gravitational field at that place is. In sum, the metric tensor and the spatiotemporal distance are not conventional but lawful, for in the last analysis they are determined by the distribution of matter according to the basic field equations. Which disposes with the conventionalist misinterpretation of general relativity.

7. CONCLUSIONS

As is well known, Einstein was initially influenced by conventionalism and empiricism, but had become a realist by the time he published his first papers on the theory of gravitation. However, both special and general relativity have been interpreted from the beginning in either conventionalist or positivist terms, without any regard for either the mathematical formalism or the way this was used to pose and solve particular physical problems. Whence certain myths that textbooks have preserved. Among such myths the following stand out.

1. "Special relativity concerns measurements made by observers moving relative to one another with constant velocity, and general relativity handles measurements performed by observers in arbitrary motion relative to one another." Criticism: (a) the notions of observer and of measurement do not occur among the concepts that make up the formulas of relativistic theories; (b) both theories are completely general, but general relativity concerns particularly gravitation.

2. "Special relativity is a daughter of the Michelson-Morley experiments, and general relativity was generated by imaginary experiments about freely falling elevators, as well as by Mach's hypothesis concerning the origin of mass." Criticism: (a) no experiment, particularly if it is either negative or imaginary, can generate a theory: at most it may generate the problem of building a theory to account for it; (b) general relativity does not deal particularly with elevators, and it does not contain Mach's hypothesis—among other reasons because this conjecture was not formulated clearly during Einstein's lifetime.

3. "Special relativity shows that everything is relative and, more particularly, relative to the observer: that is, that everything is subjective." Criticism: (a) special relativity has no fewer absolutes than classical physics; (b) "relative" (to some reference frame) is not the same as "subjective" (i.e., valid only for a given subject); (c) meeting a covariance requirement is a guarantee of universality and objectivity.

4. "Special relativity negates becoming: whatever has happened or

will happen is already inscribed in some place in spacetime." (This view may have been suggested by Hermann Minkovsky's calling 'event' any point in spacetime, even if nothing happens in it. It was first advanced by Hermann Weyl, and repeated by A. Grünbaum and O. Costa de Beauregard.) Criticism: (a) the possibility of extrapolating hypothetically a spacetime trajectory (or line of universe) does not disprove becoming; it only shows that we can predict it; (b) every relativistic theory centers around equations of motion (of bodies, fields, etc.) or of evolution (of quantum-mechanical, thermodynamic, and other systems).

5. "General relativity geometrizes physics." Criticism: (a) every continuum theory contains a geometrical formalism enabling us to represent certain physical properties by geometrical objects (e.g., tensors), without thereby erasing physical concepts; (b) the sources of the gravitational field are not represented in terms of the metric tensor, but instead by the matter tensor T.

6. "Special relativity rests on a definition of the simultaneity relation, whereas general relativity is based on a definition of the metric tensor, so both theories derive from conventions." Criticism: (a) the definition of simultaneity (relative to some reference frame) presupposes a physical law; (b) the foundation of every physical theory, whether or not relativistic, is a set of law statements subject to experimental tests.

7. "The two relativities are luxury articles; hence we may dispense with them." Criticism: (a) both relativities have revolutionized the physical worldview, so that to ignore them is to go backwards; (b) special relativity has marked the entirety of particle physics, where Lorentz covariance is required as a matter of course (perhaps dogmatically so); (c) both relativities have suggested a number of observations and experiments unthinkable in classical terms, such as the comparisons of lengths and durations at rest and in motion, mass defect measurement, alteration of the frequency of an electromagnetic wave in the presence of a gravitational field, the lens effect of a pair of nearby stars, etc.

In sum, the two relativities, just like classical mechanics, electrodynamics, and thermodynamics before them, and the quantum theory shortly thereafter, have given rise to a number of ontological and epistemological problems. Whereas some of these are legitimate and still worth discussing, others are pseudoproblems generated by the attempt to cast science into an inadequate philosophical mold. Moral: Any mismatch between science and philosophy is bound to harm both.

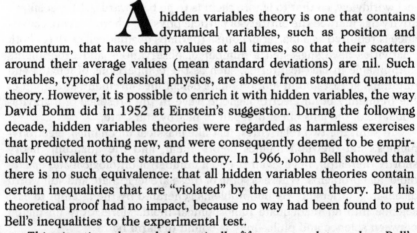

17

HIDDEN VARIABLES, SEPARABILITY, AND REALISM

(1995)

■

A hidden variables theory is one that contains dynamical variables, such as position and momentum, that have sharp values at all times, so that their scatters around their average values (mean standard deviations) are nil. Such variables, typical of classical physics, are absent from standard quantum theory. However, it is possible to enrich it with hidden variables, the way David Bohm did in 1952 at Einstein's suggestion. During the following decade, hidden variables theories were regarded as harmless exercises that predicted nothing new, and were consequently deemed to be empirically equivalent to the standard theory. In 1966, John Bell showed that there is no such equivalence: that all hidden variables theories contain certain inequalities that are "violated" by the quantum theory. But his theoretical proof had no impact, because no way had been found to put Bell's inequalities to the experimental test.

This situation changed dramatically fifteen years later, when Bell's inequalities were falsified by Alain Aspect and his coworkers (1981, 1982). These results have since been confirmed several times over. However, there still is confusion as to what exactly they prove. Some people claim that philosophical realism has been refuted and that determinism is definitely out. Still others—myself included—maintain that the loser is the conjunction of the Einstein-Podolsky-Rosen (henceforth EPR) criterion of reality and the classical principle of separability or locality. The aim of this paper is to clarify some of the key concepts involved in order to help make a decision among these alternatives.

In this paper, first the original goals of hidden variables theories are listed and examined. Along the way several philosophical concepts, particularly that of realism, are elucidated. It is shown that philosophical realism, i.e., the thesis that nature exists on its own, is different from the EPR criterion of physical reality, which is actually only a classicist tenet. Then the experiments of Aspect and his coworkers are briefly analyzed. It is concluded that, far from refuting philosophical realism, they presuppose it. It is also concluded that those experiments signal the demise of the EPR tenets that every physical property has a definite value at all times, and that the spatial separation of the components of a complex system suffices to dismantle it.

I see both the superposition principle and inseparability ("nonlocality") as the victors of this intellectual battle. By contrast, the Copenhagen or semisubjectivist interpretation of the mathematical formalism of quantum mechanics looks to me just as defective and avoidable as before the downfall of the Bell inequalities. I suggest that the realistic interpretation advanced elsewhere (Bunge 1967c) has not been touched by this event. Finally, the original EPR *Gedankenexperiment* is criticized in the appendix from a realistic viewpoint.

1. EPR: THEORY

Einstein, Podolsky, and Rosen (1935) held that the current quantum theory is incomplete, in being about ensembles of similar things rather than about individual entities. On this view, individual microphysical objects would have precise though different positions, momenta, energies, etc. Randomness would not be a basic mode of being, but it would emerge from individual differences among the components of an ensemble of mutually independent things. A complete theory, EPR claimed, should contain hidden (classical) variables only, and it should deduce the probability distributions instead of postulating them.

A hidden variable is a function possessing definite or sharp values at all times instead of being spread out: it is dispersion-free like the classical position and the classical field intensities. This type of variable was called "hidden" in contrast with the alleged "observables", for presumably referring to a subquantum level underlying that covered by quantum theory. (See Bohm 1957 for a fascinating account.) It was hoped that quantum theory could be enriched or completed with hidden variables, or even superseded by a theory containing only variables of this type. Bohm's theory (1952), which revived and expanded de Broglie's ideas on the pilot wave, was of the first kind. Stochastic quantum mechanics (de la Peña-Auerbach 1969) and stochastic quantum electrodynamics (e.g., Claverie and Diner 1977) are of the second type. Moreover, the latter theories, far from being thoroughly causal, contain random hidden variables much like those occurring in the classical theory of Brownian motion.

Hidden variables theories were expected to accomplish a handful of tasks at once. It will pay to list and examine them separately because they are often conflated, with the result that there is still no consensus on what exactly the experimental test of these theories has refuted. Those tasks are:

(i) *To restore realism in the philosophical sense of the word*, i.e., to supply an objective account of the physical world, rather than describe the information, expectations, and uncertainties of the knowing subject, in particular the observer.

(ii) *To restore realism in EPR's idiosyncratic sense of the word*. That is, to comply with the oft-quoted EPR criterion of physical reality (Einstein et al. 1935, p. 777): "If, without in any way disturbing a system, we can predict with certainty (i.e., with probability equal to unity) the value of a physical quantity, then there exists an element of physical reality corresponding to this physical quantity."

(iii) *To restore classical determinism* by deducing chance from causality, i.e., to explain probability distributions in terms of individual differences and the mutual independence of the components of statistical ensembles in the sense of Gibbs.

(iv) *To complete* the job begun by quantum theory, regarded as a statistical theory, by accounting for the behavior of individual physical entities.

(v) *To replace quantum theory*, regarded as a phenomenological (or black box) theory, with a *mechanismic* theory that would explain quantum behavior instead of providing only successful recipes for calculating it. That is, to exhibit the mechanisms underlying quantum behavior, e.g., electron diffraction and the tunnel effect, deducing quantum theory as a particular case in some limit or for special circumstances.

(vi) *To restore the separability* or independence of things that, having been components of a closely knit system in the past, have now become widely separated in space, i.e., to eliminate the distant or EPR correlations. More on this below. Suffice it for now to recall that, according to quantum theory, once a (complex) system, always a system.

The first objective (the restoration of philosophical realism) was legitimate. However, it can be attained without modifying the mathematical formalism of quantum theory and, in particular, without introducing hidden variables. Indeed, it can be shown that the Copenhagen interpretation is adventitious, and that the realistic interpretation, which focuses on things in themselves rather than on the observer, is not only possible but the one actually employed by physicists when not in a philosophical mood. For example, no assumptions about the observer and his experimental equipment are made when calculating energy levels. (See Bunge 1967c, 1973b.)

The second goal, though often mistaken for the first, has nothing to do with philosophical realism: it is simply classicism, for it involves the denial of the superposition principle, according to which the dynamical properties of a quantum object normally have distributions of values rather than sharp values. (In the nonrelativistic theory only time, mass, and electric charge have sharp values all the time, but these are not dynamical variables.) In the light of the indisputable success of quantum theory and the failure of hidden variables theories, that classicist principle must be regarded as unjustified: we shall call it the nostalgic EPR *dogma*.

The third objective (determinism) was pursued by the early hidden variables theorists, but was no longer an aim of those who worked from about 1960 on. In fact, as we recalled a while ago, there are theories containing random hidden variables, namely stochastic quantum mechanics and electrodynamics. The introduction of these theories was important to distinguish the concepts of (i) philosophical realism, (ii) classicism or EPR realism, and (iii) determinism, as we now realize with hindsight.

The fourth goal (completeness) makes sense provided quantum theory is regarded as a statistical theory dealing only with ensembles. It evaporates as soon as it is realized that the elementary theory refers to individual entities. If it did not, then statistical quantum mechanics, which deals with actual collections of microentities, would be dispensable—but of course it is not.

The fifth objective (mechanismic theory) was soon abandoned in the most advanced work on hidden variables: that of Bell (1964, 1966). In fact, Bell's hidden variables have no precise physical interpretation: they are only adjustable parameters in a theory that is far more phenomenological than quantum mechanics. Moreover, such hidden variables are not subject to any precise law statements, hence they are no part of a physical theory proper. But this is a virtue rather than a shortcoming, for it shows that Bell's famous inequalities hold for the entire family of local hidden variables theories, regardless of the precise law statements that may be assumed.

So far, only the first objective, namely the restoration of philosophical realism, seems worth being pursued. However, as noted above, it can be attained without the help of hidden variables, namely by a mere change of interpretation of the standard formalism of the theory. On the other hand, the sixth objective, namely separability, remained plausible for about half a century. Indeed, the EPR distant correlations are quite counterintuitive. How is it possible for the members of a divorced couple to behave as if they continued to be married, i.e., in such a manner that whatever happens to one of them affects the other? The quantum-theoretical answer is simple if sibylline: *there was no divorce*. However, let us not rush things: it will pay to have a look at the experiment suggested by Einstein et al. (1935). The original experiment is analyzed in the appendix. Here we shall deal with an updated version of it in the context of Bell's general hidden-variables theory, which contains the now famous Bell's inequalities.

2. EPR: EXPERIMENT

According to quantum theory, the three components of the spin of a quantum object do not commute. Hence, in our realistic interpretation, they have no definite or sharp values all the same time, as a consequence of which they cannot be measured exactly and simultaneously. Simpler: The spin components are not the components of a vector, hence there is no vector to be measured. By contrast, if the spin were an ordinary vector (hence a hidden variable), its components would commute and, consequently, it should be possible to measure all three components at the same time with the accuracy allowed by the state of the art.

Apply the preceding considerations to a pair of quantum objects, named 1 and 2, that are initially close together with spins pointing in opposite directions, which subsequently move apart without being interfered with. Compute the probability $P(x, y)$ that thing 1 has spin in the positive x-direction, and thing 2 in the positive y-direction. According to quantum mechanics, this probability exceeds the sum of the probabilities $P(x, z)$ and $P(y, z)$:

$$\text{QM} \qquad P(x, y) \geq P(x, z) + P(y, z).$$

By contrast, any hidden variables theory predicts the exact reversal of the inequality sign:

$$\text{HV} \qquad P(x, y) \leq P(x, z) + P(y, z).$$

This is one of Bell's inequalities. Bell himself, and others, derived several other inequalities that hold in all local hidden variables theories, and that involve exclusively measurable quantities such as coincidence counter rates. (See Clauser and Shimony 1978 for a masterly review.)

Thanks to this theoretical work, a number of ingenious high-precision measurements could be designed. Some of them, involving photons, have been performed. The first conclusive version was that of Aspect et al. (1981, 1982). An atom emits two photons in opposite directions at the same time. Their linear polarizations are measured by polarizers oriented in directions adjustable at will. The photons coming out of the polarizers activate photomultipliers whose coincidence rate is monitored by a counter. The result is that the polarization states of the two photons are highly correlated. That is, the state of polarization measured by one of the polarizers depends upon that measured by the other polarizer and conversely (Aspect et al. 1982). Essentially the same result obtains if the polarizers are rotated during the flight of the photons and before they strike the polarizers, to prevent any communication between them through signals propagating at a finite speed (Aspect et al. 1982).

3. WHAT HAS EXPERIMENT DISPROVED?

Although there is consensus that Bell's inequalities, and with them all of the hidden variables theories, have been experimentally falsified, it is not clear *what is at issue*: philosophical realism, determinism, the EPR dogma that all properties have sharp values at all times, or the EPR hypothesis that distant things behave independently of one another.

Most authors, e.g., Clauser and Shimony (1978), d'Espagnat (1979), Aspect et al. (1981), claim that philosophical realism is the casualty. However, they do not clarify what they mean by 'realism', and they ostensibly conflate philosophical realism with what we have called *classicism* or the *EPR dogma*. Yet, it should be clear that the very design and performance of measurements presuppose the reality (independent existence) of the entire measurement system, hence of every component of it—object, apparatus, and experimenter. So, philosophical realism is definitely not at stake: The downfall of Bell's inequalities has not refuted the principle that the physical world manages to exist without the help of those who attempt to know it. Nor is determinism at issue, since some hidden variables theories that entail Bell-type inequalities contain random variables.

So, *only the EPR dogma and the separability (locality) hypotheses are at issue*. As a matter of fact, this is the conclusion Clauser and Shimony (1978) reach after their careful analysis: *Hidden variables (sharpness) & Locality (separability)* $\Rightarrow$ *Bell's inequalities*. Hence, the experimental refutation of the latter refutes separability or the EPR dogma—or both. We shall argue in a while that the *or* is inclusive, for the EPR dogma entails separability (locality). But before doing this, we must take another glance at the Aspect experiment.

The experiment poses two problems. One is to ascertain whether the photons emitted by the source possess all their properties, in particular polarization, or acquire them when interacting with the polarizers. The second problem is to account for the correlations at a distance, or EPR correlations, between the results obtained with one of the analyzers and those obtained with the other—or, equivalently, to explain the coincidences registered by the coincidence counter.

Let us start with the first problem. According to both classical and hidden variables theories, the fragments resulting from the process at the source have sharp properties (eigenvalues) from start to finish: the analyzers only exhibit what is already there. By contrast, according to quantum theory, the entities leaving the source are in a superposition of polarization states. This superposition collapses (reduces) to eigenstates through the interaction of the system with the analyzers. In other words, the analyzers do not just measure an existing polarization, but also *produce* it. (In Pauli's terminology, they are measuring instruments of the second kind.) Clearly, this explanation will satisfy only those who believe that accurate measurement is accompanied by a projection of the state function—even though one need not believe the dogmas that this projection is subjective, instantaneous, and unrelated to the smooth processes described by the Schrödinger time-dependent equation. (See Cini 1983; Bunge and Kálnay 1983a.)

The second problem is this: If only one of the analyzers is read, we may infer the result obtained with the other analyzer without looking at it. (This cannot be done with the apparatus designed by Aspect et al., which registers only coincidences. The equivalent operation here is to alter the angle between the analyzers and note the change in coincidence rate. The experiment shows that the coincidence rate attains its maximum when the two analyzers are parallel.) This result is nicely explained by any hidden-variables theory jointly with the theorem of conservation of the total spin. Indeed, the first quantum object would arrive at the analyzer possessing a definite spin value, say +1, so that the second must have spin −1, since the total spin of the system at the source was 0. The trouble with this explanation is that it takes hidden variables for granted. Since experiment has refuted hidden variables theories, we must try to explain the same results with the help of quantum theory.

If the two quantum objects leaving the source were to become independent after a while, as they should according to classical physics and to EPR, the result obtained with one of the polarizers should be independent of the result obtained with the other. Thus, it should be possible to observe, say, the left quantum object with its spin pointing in the x-

direction, and the right one with its spin pointing in the z-direction. But this is not what the quantum theory predicts: according to this theory, there is total spin conservation. Hence, if the left polarizer projects the spin onto the positive x-axis, the right polarizer will project it onto the negative x-axis. In other words, there is a strong correlation at a distance between the polarizations of the two former components of the original system. This correlation is correctly predicted by quantum theory, which treats the "two" quantum objects as constituting a *single* entity. Indeed, only the state function for the entire system satisfies the Schrödinger equation. (In other words, the superposition principle is not optional but mandatory in this case.)

The quantum-theoretical explanation of distant correlation is some-times found unsatisfactory or even mysterious because it involves the counterintuitive ("paradoxical") hypothesis that a complex system may continue to be one, even after its components have moved far apart. In fact, nearly all known classical forces fall off quickly with increasing dis-tance. (A salient, if usually overlooked exception, is the elastic force, which is proportional to the distance.) For that reason, a host of more or less unorthodox explanations of the EPR distant correlation have been proposed. Among them we note those in terms of hidden variables and a peculiar quantum potential that does not occur in the Hamiltonian but derives from the state function of the system (Bohm and Hiley 1975); the occurrence of actions at a distance, and even psychokinesis. I reject all three explanations for the following reasons. The first occurs in a hidden-variables theory, hence one containing some of Bell's false inequalities. The second violates special relativity and electrodynamics (both classical and quantal), which involve the principle of nearby action (or locality in the field-theoretic sense). And the third is spooky: it vio-lates the principle of conservation of energy, since every message must ride on a physical process, unless, of course, one is prepared to believe in the existence of nonphysical things.

We need not look for any special *mechanism* explaining the distant or EPR correlations, any more than we need to explain the Lorentz "con-tractions" and "dilatations" in terms of mechanisms. Quantum mechanics accounts for distant correlations as follows. If two quantum objects are initially independent from one another, then the state func-tion of the aggregate they constitute is correctly represented by the product of their individual state functions. So, separability or "locality" obtains in this case just as in classical physics. But if two quantum objects have initially been part of a system, i.e., if they have interacted strongly at the beginning, then the state function of the whole cannot be so factorized even after the components have moved far apart. In other words, the state of each component is determined not only by the local conditions (i.e., the state of the immediate surroundings of the spatially separated quantum objects), but also by their *still* belonging to a system.

So, although physical separation (cessation of interaction) entails spatial separation, the converse is false. Quantum theory is then non-local in this peculiar sense or, as we prefer to say, it is *systemic* in that it does not incorporate the classical principle that widely separated things cannot belong to the same system. (Note again the ambiguity of the word 'local'. In field physics, whether classical or quantal, 'locality' means that nearby action, or nonaction-at-a-distance, obtains: All actions are assumed to proceed *de proche en proche*. Quantum theory is local in this traditional sense of the word.)

We submit that the nonseparability or (non-"locality") inherent in quantum theory is a consequence of the superposition principle together with the Schrödinger equation. In fact, once a system is in a state consisting of an inextricable merger—not just either a sum or a product but a sum of products—of the states of its components, the Schrödinger equation guarantees that the system will evolve in such a way that its state function will retain that structural property, even though the relative contribution of every individual eigenstate is likely to change in the course of time. (In the case of the EPR experiment, the total system 1 + 2 is in some such state as either $(1/\sqrt{2})(u_1v_1 + u_2v_2)$ or $(1/\sqrt{2})(u_1v_1 - u_2v_2)$. If the u's and v's are spin eigenstates, the preceding state functions are distance-independent: the components continue to be entangled no matter how far apart the two fragments move.)

The EPR paradox is such only if one presuppose the classical principle that the behavior of things that are far apart cannot be correlated. (See Einstein 1948 and Bell 1966 for statements and discussions of this classical principle of "locality" or separability.) The physicist who takes this principle for granted is bound to puzzle over how it could be possible for two things to continue interacting after they have stopped being components of a system. Quantum theory solves the "mystery" much in the same way as a detective solves certain murder cases by proving that there was no murder to begin with. Indeed, he concludes that the distant components are still part of the original system; that the "wave" function has not split into two parts, but continues to cover the entire system.

What the quantum theorist may puzzle over is a different question, namely: how, that being the case, could one effectively *dismantle* the system, and thus factorize its state function? We submit that the answer to this query is: The original system becomes dismantled only when at least one of its original components gets integrated into another system, e.g., when it is captured or absorbed by an atom.

4. FISHING IN TROUBLED WATERS

Philosophers have been quick to exploit the downfall of Bell's inequalities. Thus, Fine (1982) has interpreted it as a refutation of determinism, and van Fraassen (1982) as a refutation of philosophical realism. We have seen that it is neither. In fact, some hidden-variables theories, such as stochastic quantum mechanics, are nondeterministic in that they contain random variables. (See additional criticisms in Nordin 1979 and Eberhard 1982.)

As for philosophical realism, it would indeed be dead if, as van Fraassen contends, it involved the hypothesis that there is a *causal mechanism* underlying every correlation, since no such mechanism is known to exist in the case of the EPR distant correlations. But such a version of philosophical realism is a straw man. Philosophical realism, from Aristotle to Descartes to Russell, boils down to the thesis that nature exists even if it is neither perceived nor conceived. Moreover, philosophical realism is compatible with alternative ontologies, in particular neodeterminism, which acknowledges noncausal, in particular probabilistic, types of determination (Bunge 1959a).

If anything, the experimental refutation of Bell's inequalities, like any other experiment, has confirmed philosophical realism once more, since every well-designed experiment involves a clear distinction between knowing subject (in particular experimenter), apparatus, and object of knowledge. (See Wheeler 1978; Rohrlich 1983; Bunge 1985a.) Only "realism" in the idiosyncratic sense of EPR has been refuted along with Bell's inequalities—but that was only a classicist dogma. Actually, this dogma went down the moment quantum theory succeeded in solving the problems that classical physics had left unsolved. Since then we have learned that reality is smudged rather than neat: that every quantum object possesses some properties that, far from having sharp values all the time, have distributions of values; and that, unlike rigid bodies, but rather like fluids and fields, quantum objects are extremely sensitive to their environments, whether natural or artificial. This discovery did not alter nature and, in particular, it did not enslave nature to the supposedly omnipotent Observer. It only taught us that nature is composed not only of classical objects, but also of quantum and semiquantum objects. It altered our representation of the world, not the world itself: it was a scientific revolution, not a cosmic cataclysm.

5. CONCLUSION

Hidden-variables theorizing remained outside the mainstream of physics until Bell and others succeeded in deriving formulas capable of being subjected to rather direct crucial experimental tests. Then it came briefly to the fore, and it may soon become little more than a historical curiosity. However, the failure of hidden variables theories has taught us a valuable lesson: that, like other revolutionary theories before, such as classical electrodynamics, quantum theory has got to be understood in its own terms rather than with the help of classical analogies.

Besides, hidden variable theorists have rendered a valuable service to the scientific and philosophical communities. Firstly, they have exhibited some of the inconsistencies and obscurities of the Copenhagen orthodoxy. (Unfortunately, some of them have confused the issues,e.g., the problem of objectivity with that of determinism: see Bunge 1979b.) Secondly, they have helped others remove those inconsistencies and obscurities by reinterpreting the standard mathematical formalism of quantum theory in a strictly physical, hence realistic, fashion. Thirdly, they suggested the very experiments that were to refute classicism.

Now that the hidden-variables episode seems to be closed for good, we can look forward to theories even less intuitive than quantum theory. One wonders whether Einstein was right in believing that God (*sive natura*) is not malicious.

APPENDIX: THE EPR "PARADOX"

Let us recall and examine the original EPR "paradox" (Einstein et al. 1935). Consider a complex system, such as a molecule, that breaks up into two components that fly off in opposite directions. Call $p_1 + p_2$ the total linear momentum of the system, a quantity that is conserved, and $x = x_1 - x_2$ the relative distance between its components. Since $p_1 + p_2$ commutes with $x_1 - x_2$, both are definite (have sharp values) and can be measured simultaneously with as much accuracy as the state of the art will allow. At all times after its breakup, the system is said to be in a state represented by

$$(1) \quad \Psi(x_1, x_2) = \delta(x - a).\exp[ip(x_1 + x_2)/2h]$$

where δ is Dirac's delta, $(x_1 + x_2)/2$ is the position coordinate of the center of mass of the system, and h designates Planck's constant divided by 2π.

By measuring x_1 we can infer (without measurements) the exact position of component 2, given by $x_2 = x - a$. It would seem that component 2 "knows" what its place is, even though no information about x_1 has reached it. Alternatively, if p_1 is measured, we can infer the exact momentum of component p_2, given by $p_2 = p - p_1$. In both cases it seems clear that the property of the component that is being inferred (not measured directly) exists without measurement—which contradicts the Copenhagen doctrine. Moreover, the information on the position (or the momentum) of component 2 is not included in the state function (1). Therefore, according to EPR, quantum mechanics is incomplete.

I do not profess to understand the criticisms leveled by Bohr (1935, 1949), except that they presuppose the very operationist interpretation that Einstein and his coworkers were criticizing. It seems to me, on the other hand, that the EPR paper can be criticized as follows.

Firstly, the "wave" function (1) *does not describe correctly the system* in question, because δ is not a square-integrable function: it is not a possible eigenfunction of any operator. The delta must be replaced with a well-behaved (bounded and square-integrable) function sharply peaked at $x_1 - x_2 = a$, symmetric about a, and with a nonvanishing width (standard deviation) Δx. But this amounts to saying that the relative distance $x = x_1 - x_2$ has an *objective* spread or indeterminacy Δx. Hence, measuring x_1 exactly does *not* allow us to infer x_2 with certainty, but only within Δx. Therefore, Heisenberg's inequalities are not violated. (Dirac 1958 conjectured that the infinite length of the state vectors corresponding to sharp position eigenvalues "is connected with their unrealizability". As a matter of fact, nobody has measured a sharp position value: all one can do is to locate a quantum object within a small though nonvanishing region of space.)

Secondly, although $x_1 - x_2$ and $p_1 + p_2$ commute, in practice they cannot be measured accurately at the same time. In fact, measuring $x_1 - x_2$ involves localizing component 1 at x_1 and component 2 at x_2. But by so doing both p_1 and p_2 get smudged according to Heisenberg's inequalities. Consequently, the total momentum $p_1 + p_2$ *is not conserved* while measuring $x_1 - x_2$. The *classical* explanation (adopted by Bohr 1949) is that there are momentum exchanges p_1 and p_2 between the quantum objects and the measuring instrument. The quantum-theoretical explanation is that the quantum objects are thrown into superpositions of momentum states.

In my view, either of the above objections disposes of the EPR claim that quantum mechanics is incomplete. On the other hand, neither touches on the real crux of the matter, namely the inseparability inherent in quantum theory—which Schrödinger (1935a) regarded as "*the* characteristic trait of quantum mechanics".

18

SCHRÖDINGER'S CAT IS DEAD

(1999)

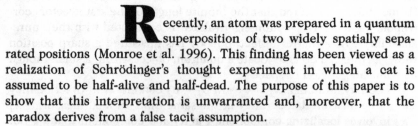

Recently, an atom was prepared in a quantum superposition of two widely spatially separated positions (Monroe et al. 1996). This finding has been viewed as a realization of Schrödinger's thought experiment in which a cat is assumed to be half-alive and half-dead. The purpose of this paper is to show that this interpretation is unwarranted and, moreover, that the paradox derives from a false tacit assumption.

Let us start by recalling Schrödinger's (1935b) popular cat paradox. A live cat is locked up for a while in a steel cage containing a phial filled with a lethal poison that may be released by the disintegration of a single atom present in a very small sample of radioactive material. Obviously, the disintegration may or may not occur during the time interval concerned. If it does, it is sure to kill the cat. But the observer won't know what happened until he or she opens the box and takes a look at the cat. So far, no paradox. The paradox leaps to life the moment the cat-box-observer supersystem is treated according to the orthodox theory of measurement proposed by von Neumann (1932), as will be sketched anon.

Assume that the cat is a quantum-mechanical system. Assume further that, while confined in the box and thus beyond the reach of the observer, the cat is in a state that is the superposition of its two possible macrostates: alive and dead. More precisely, suppose that this state is the linear combination

$$\Psi = a\Psi_{L} + b\Psi_{D}, \quad \text{with } |a|^2 + |b|^2 = 1, \tag{1}$$

where $|a|^2$ and $|b|^2$ are the probabilities that the cat be alive or dead, respectively. According to the orthodox or Copenhagen interpretation, these are the chances that the cat be *found* in either state when an observer opens the cage and looks into it. Moreover, according to this interpretation the act of observation projects Ψ onto either Ψ_{L} or Ψ_{D} with the given probabilities.

An alternative is to treat the cat-atom supersystem as satisfying quantum mechanics: this is what Monroe and coworkers (1996) do in their paper when writing their slightly more complex formula

$$\Psi = 2^{-1/2} (\Psi_{L} | \uparrow\rangle + \Psi_{D} | \downarrow\rangle), \tag{2}$$

where $|\uparrow\rangle$ and $|\downarrow\rangle$ are, respectively, the internal states of an undecayed and a decayed atom. However, this does not change the result: the new Ψ still represents a "smeared cat".

This result is rightly regarded as paradoxical, because the very idea that a cat may be neither alive nor dead—or, equivalently, that it be half-alive and half-dead—is plainly false. Some people, notably Schrödinger and Einstein, took this result for an indication that there is something seriously wrong about quantum mechanics and, in particular, the superposition principle.

But the received opinion is that, far from being paradoxical, Schrödinger's "smeared cat" illustrates the standard theory of measurement. In particular, it would exemplify von Neumann's projection (or collapse) postulate, according to which the measurement of a dynamical variable ("observable") of a quantum-mechanical object causes the instantaneous projection of the state function of the object onto one of the eigenstates of the variable in question. The reason is, of course, that an infinitely accurate experiment is supposed to yield a sharp value, which must be one of the eigenvalues of the dynamical variable concerned. However, the same result might come from a gradual, if swift, shrinking of the wave packet (Cini 1983; Cini and Lévy-Leblond 1990).

Yet, if Schrödinger's cat-in-the-box does illustrate the standard measurement theory, then Ψ_{L} and Ψ_{D} should be the eigenstates of some dynamical variable—an operator in the system's Hilbert space. But they are not. At least, no operator of this kind is known. Hence, no dynamical variable is being measured when looking inside the cage to check whether the cat is still alive. Therefore, we are not in the presence of a quantum-mechanical measurement.

In short, Schrödinger's thought experiment does not highlight the measurement process. Nor should any defender of the standard quantum theory of measurement wish the famous cat to illustrate it, for, if the cat story is indeed paradoxical, so must be the theory in question. (Whether this measurement theory is in fact true is at least doubtful, for it has never been put to the experimental test. But this is irrelevant to our argument.)

We submit that the cat paradox derives from the *tacit assumption* that the cat (or the cat-radioactive atom system) is a quantum-mechanical entity that can be adequately described in terms of macrostates (or of their combination with the states of the radioactive atom). I claim that this assumption is false. In fact, the above superposition formula is *meaningless* because the cat's alleged possible states Ψ_L (live) and Ψ_D (dead) *are not specified in quantum mechanical terms*. Indeed, they are not the eigenfunctions of any known hermitian operator(s). Nor are they solutions to the Schrödinger equation for the cat-phial system, let alone for the cat-phial-cage-observer supersystem. This remark dissolves the paradox, for showing that it originates in an unwarranted tacit assumption.

What are then the Ψ_L and Ψ_D occurring in the equations (1) and (2) besides having the outward appearance of the wave functions one encounters in a bona fide two-state quantum-mechanical problem? I submit that they are nothing but names of unspecified and therefore unknown functions—that is, just letters. (The same holds for the state functions of the brain and for the universe, found occasionally in the literature: they too are just letters.) Hence, their superposition makes no mathematical sense nor, a fortiori, physical sense either. Indeed, how much is *blah* plus *bleh*?

Even assuming that Ψ_L and Ψ_D were to stand for mathematically well-defined functions, their superposition would not yield a quantum state, whence the cat-in-the-box system could not be regarded as a quantum-mechanical object. Indeed, if a state space for the system could be built, its component Ψ_L should vanish wherever Ψ_D does not, and vice versa. Hence, the "interference" term $\Psi_L^* \Psi_D + \Psi_L \Psi_D^*$ would vanish everywhere—a clear indicator that the system is not quantum-mechanical (Bunge 1985a).

To put the same thing in mathematical terms, the *existence* of the functions designated by the symbols Ψ_L and Ψ_D has been tacitly assumed rather than proved. And, once the existence of something is taken for granted, the most outlandish results can be proved—as theologians and mainstream economists know so well. For example, using this strategy it can easily be proved that 1 is the largest integer. The proof runs like this. Let N be the largest integer. There are two possibilities: either $N = 1$ or $N > 1$. Now, the second option is false because N^2, which is an integer, is larger than N. Hence $N = 1$. The root of this paradox is of course the initial sentence 'Let N be the largest integer'. This looks like an innocent notation convention, but actually it is an existence hypothesis. The parallel with the symbols for the states of Schrödinger's cat is obvious.

Furthermore, quantum mechanics knows nothing about cats. In fact, the very notion of life is alien to that theory and, indeed, to the whole of physics. (Life happens to be an emergent property of material systems whose complexity defeats physics. The latter is necessary but not sufficient to understand living beings.) In other words, we do not know how to set up, let alone solve, Schrödinger's equation for a cat or even for a bacterium. Actually, quantum chemists do not even know how to write a classical Hamiltonian for a nucleic acid. And the whole of quantum chemistry, far from being a straightforward if laborious application of quantum mechanics, is full of subsidiary assumptions and classical models (see Polanyi and Schreiber 1973; Bunge 1982a).

One might try to dispose of this objection by saying that the cat can be replaced with a macrophysical system, such as a Geiger counter, sensitive to radioactivity. In this case, as long as the cage is closed, one would assume the superposition of the charged and discharged Geiger states. These would certainly mirror the microphysical superposition of the decayed and undecayed states of the radioactive atom. But the problem is again that those macrostates are only being named, for we do not know how to write down, let alone solve, the Schrödinger equation for a Geiger counter. This is why one explains the mechanism of a Geiger in semiclassical terms.

To take an apparently simpler case, consider coin tossing. One might think that, while moving above the ground, the coin is in a superposition of the head and tail states with equal weights, namely $2^{-1/2}$. One might further think that this superposition collapses onto either the head state or the tail state when the coin comes to rest on the ground—or, in the Copenhagen interpretation, when the coin is observed after having come to rest. But, again, this would be an exercise in hand-waving, for we have no idea how to write down, let alone solve, the Schrödinger equation for a coin-flipping process. Nor do we know which hermitian operator could have those bogus eigenfunctions.

However, back to the unfortunate cat. If the quantum theory is assumed to hold for cats, why not suppose that it also holds for people and their social and even moral properties? On this assumption it is easy and amusing, though hardly productive, to invent a number of analogous paradoxes. Consider, for instance, a person who may be either honest or corrupt, but who has not yet been examined by anyone in this respect.

A Schrödinger cat fan might wish to describe the moral state of such a person, before being subjected to a judicial inquiry, by a superposition of the honest and corrupt states:

$$\Psi = a\Psi_H + b\Psi_C.$$

But here again the "functions" remain unspecified. Still, the cat fan is likely to go on and claim that the verdict of the court causes the projection of the superposed moral state Ψ onto either Ψ_H or Ψ_C, which would be the eigenfunctions of some unknown moral operator. That is, he holds that only a court of law can make the subject either honest or corrupt. (Note, incidentally, that this view fits in with the fashionable constructivist-relativist epistemology and sociology of knowledge.) But, clearly, any upright judge would reject such an interpretation, for it challenges the objectivity and impartiality of the entire judicial process. He would say instead that the defendant was corrupt (or honest as the case may be) to begin with, and that the investigation has only probed and disclosed his previous moral standing. Quantum states can superpose provided they are well-defined; moral or social states cannot, if only because they are not specified in quantum-theoretical terms. Still, it is more realistic to hold, with Robert Louis Stevenson, that people are half-good and half-bad, than to say that cats can be half-alive and half-dead, for, at least in principle, we could count the relative frequencies of the good and bad actions of a person.

True, most textbooks on quantum theory claim that the latter is universal, i.e., that it holds for everything concrete. Regrettably, they fail to exhibit any evidence for this claim. Moreover, no such evidence is at hand. The best that can be said for this extravagant reductionist claim is that it is a programmatic conjecture on a par with the eighteenth-century belief that all natural and even social processes would eventually be accounted for in terms of classical mechanics. The worst that can be said for the universality claim is that it is an article of faith.

Ironically, whereas we know nothing about Ψ_L and Ψ_D, we do know something about their respective weights a and b in formula (1). In fact, these depend exclusively upon the radioactive atoms whose disintegration would release the poison. Indeed, to a first approximation the square of the absolute value of a should decline exponentially in the course of time, whereas that of b should increase in time according to $|b|^2 = 1 - |a|^2$. This is never mentioned in the usual discussions of the paradox, perhaps because it entails that Ψ evolves continuously, whereas, according to orthodoxy, all measurement is instantaneous. In particular, according to it, either a or b vanishes instantaneously upon observing the cat in the box. But the latter assumption overlooks two points. One is that, as a matter of fact, every real measurement takes a nonvanishing time interval (Bunge and Kálnay 1983b). The second point is the stochastic nature of the radioactive process upon which the cat's fate hangs in part—and *this* is paradoxical to say the least.

In short, the superposition formulas for the caged cat are phoney: they belong neither in quantum theory nor in any other scientific theory. They are just a bunch of signs. However, such is the power of symbolism, particularly when wielded by authority, that hundreds of able physicists and philosophers have fallen into this trap. This is why I suggest that *von Neumann's trap* is a better name than *Schrödinger's cat* for the conundrum in question. After all, John von Neumann (1932) was the first to posit, contrary to Niels Bohr, that quantum mechanics holds for *all* concrete things, as well as the first to indulge in hand-waving when attempting to describe mythical macrophysical entities such as all-purpose measuring instruments. (Bohr insisted that all measuring instruments are macrophysical and, that to detect a microphysical entity, they have to involve an amplifier of some kind, i.e., a micro-macro bridge.)

So much for the trap. What about the unfortunate cat? Both cat lore and biology will tell us that, at any given moment, the animal is either alive or dead (assuming that the poison kills instantly). As a matter of fact, this assertion can be put to the test by having a camera videotape the cat during its sojourn in the box. (Needless to say, the camera is a passive recording system: it does not interfere with the cat's life, so it cannot possibly cause the collapse of its state function even supposing it had one.) If the cat is found dead we can summon a forensic expert. She is likely to find that the kit was killed by the poison, not by the evil eye of an observer.

Finally, what about the quantum theory: Is it affected by our solution to the problem? It would be if Schrödinger's cat (or von Neumann's trap) did illustrate the superposition principle. But the above arguments have shown that it does not, for only some letters get "superposed" or, rather, juxtaposed. So, the theory remains unscathed. Not even the orthodox, albeit implausible, quantum theory of measurement is tainted by the wild story about the smeared cat, since, as we saw above, no particular dynamical variable ("observable") is being measured when observing the animal. (Still, this particular theory can be indicted independently as being both unrealistic and untested: see Bunge 1967c). Moreover, there are a few beginnings of realistic alternatives to it: see Cini and Lévy-Leblond 1990.) Consequently, Albert (1994) notwithstanding, Bohm's interpretation in terms of hidden variables and quantum potentials is unnecessary to avoid being snared by von Neumann's trap.

In conclusion, as one would say in German, *diese ganze Geschichte ist für die Katz*. And, as with every paradox, the moral is: Always examine the hidden assumptions, for that is where the cat may lie buried.

V

PHILOSOPHY
OF PSYCHOLOGY

■

19

FROM MINDLESS NEUROSCIENCE AND BRAINLESS PSYCHOLOGY TO NEUROPSYCHOLOGY

(1985)

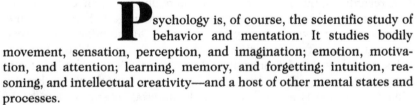

Psychology is, of course, the scientific study of behavior and mentation. It studies bodily movement, sensation, perception, and imagination; emotion, motivation, and attention; learning, memory, and forgetting; intuition, reasoning, and intellectual creativity—and a host of other mental states and processes.

There are several approaches to the study of the problems of behavior and mentation. They can be grouped into three main strategies: *behaviorism*, *mentalism*, and *psychobiology*. Behaviorism is brainless and mindless, in that it ignores the nervous system (except as a stimulus transducer) and is not interested in mental events: it restricts its attention to overt or observable behavior. By contrast, mentalism is mindful but brainless: it is concerned with the perennial problems of psychology but has no use for neuroscience, because it can explain everything very easily in terms of the immaterial (and possibly also immortal) mind or soul. Finally, neuropsychology (or biopsychology) is both brainy and mindful, for it endeavors to identify mental events with brain events, and it studies the development of the psyche as one aspect of the development of the nervous system, and the evolution of the nervous system and its abilities as one aspect of biological (and social) evolution.

We shall examine the three approaches, but first we shall clarify the rather hazy notion of approach or strategy and the more special concept of scientific approach. Then we shall settle our accounts with behaviorism and mentalism. Next we shall sketch the guiding ideas of the psychobiological approach. And finally we shall discuss the problem of whether psychobiology effects the reduction of psychology to neurophysiology, or rather a merger of biology and psychology.

1. APPROACH

An approach to problems of some kind is of course a way of conceiving and treating them: it is a problem-solving strategy. One and the same problem can often be approached in different ways, although not necessarily with the same success; and one and the same approach can be applied to several problems. In particular, the problem of the nature of mind can be approached theologically, philosophically, in the manner of traditional (mentalistic) psychology, or else neurobiologically. And the neurobiological approach, like the other strategies, can be applied to a variety of questions regarding mind, from the origin of mentality to the effects of mental events on other bodily processes. An approach suggests hypotheses and theories without sponsoring any in particular. On the other hand, every approach writes off whole bunches of theories. For example, the theological approach to mind is incompatible with the science of mind, for it endorses an unscientific and ready-made view including the thesis that the mind of a person is an immaterial and immortal substance detachable from the brain and inaccessible to experiment.

The concept of an approach can be analyzed as a *general outlook*, or conceptual framework, together with a *problematics* (problem system), a set of *aims* or goals, and a *methodics* or set of methods (not to be confused with methodology, or the study of methods). The scientific approach is a very particular kind of approach or strategy. It is the one characterized by the following quadruple:

- *General outlook*: (a) a naturalistic ontology, according to which the world is composed of concrete changing things (nothing ghostly); (b) a realistic epistemology, on which we can frame fairly true representations of things with the help of experience and reason (no supernatural or paranormal cognitive abilities); and (c) the ethos of the free search for truth (no reliance on authority, no cheating, and no withholding information).

- *Problematics*: all the cognitive problems that can be posed against the above background (nothing is irrelevant).
- *Aims*: the description, explanation, and prediction of facts with the help of laws and data (not just mere description and not just any old explanation).
- *Methodics*: the scientific method and all the scrutable, checkable, and justifiable techniques or tactics (no crystal ball and no inkblot tests).

The specialist may balk at my assertion that the scientific approach has a bulky philosophical component made of a worldview, a theory of knowledge, and a code of conduct. Yet that general outlook does guide the choice of problems, the goals of research into them, and the means or methods. We may not realize this, because we take that general outlook for granted: to absorb it is part of our scientific apprenticeship. To show that this is indeed the case, imagine cases in which the general outlook is not adopted.

Suppose someone held a supernaturalistic worldview. In this case she would count on supernatural agencies instead of restricting her theorizing and experimenting to certifiably or putatively real entities. Again, suppose our imaginary scientist were a subjectivist or a conventionalist: in either case she would hardly care for experimental tests. And if she believed in paranormal modes of cognition, such as revelation, intuition, telepathy, precognition, or what have you, she would rely on them instead of checking her hunches. Finally, suppose someone did not adopt the code of scientific ethics. In this case she might feel free to bypass problems that could embarrass the powers that be; she might feel inclined to make up her data, withhold information, or cheat in theorem proving; she might plagiarize, and she would hardly be interested in trying to prove herself wrong. So, the general or philosophical background is an indispensable component of the scientific approach. This is particularly evident in the case of psychology, so much of which is nonscientific or protoscientific, precisely because it does not partake of the general outlook of the older and harder sciences.

Presumably the other three components of the concept of scientific approach, namely the problematics, aims, and methodics, will raise no eyebrows, although admittedly every one of them can be the subject of lengthy methodological disquisitions (see, e.g., Bunge 1983a, b). Indeed, we all appear to agree that scientific research is characterized by a peculiar set of problems, aims, and methods. On the other hand, we often forget that an exclusive training in methods, to solve well-defined problems with narrowly circumscribed goals, makes technicians, not necessarily scientists. Hence our emphasis on the first component of the out-look-problematics-aims-methodics quadruple.

Let us now catch a glimpse of the approaches characterizing behaviorism, mentalism, and psychobiology, in order to assess their respective scientific merits.

2. BEHAVIORISM

To begin with, let me caution against the mistaken, yet not uncommon, confusion between behaviorism and the study of behavior. Behavior can be approached in a number of ways, among them in a behaviorist, a mentalist, or a biological fashion. From Watson (1925) to Skinner (1938) and their followers, behaviorism is not just the study of behavior. It is the approach consisting in leaving out everything else, in particular the so-called neural substrate or correlate of behavior as well as the whole of subjective experience. To put it positively, behaviorism studies the responses of organisms to various stimuli configurations, as if animals were black boxes.

More precisely, the behaviorist approach can be characterized as follows in terms of our characterization of the concept of scientific approach in section 1. First, its general outlook includes a limited naturalistic ontology. It is naturalistic because it discards disembodied entities, and it is limited because it discounts nonbehavioral facts such as emotion, imagination, and ideation. Second, behaviorism adopts an immature realistic epistemology. It is realistic because it endeavors to account for an aspect of reality (namely its skin), the existence of which it recognizes the moment it demands that research be objective. However, this epistemology is primitive because it shuns hypothetical constructs, such as desire and reasoning, and thus does not allow one to pose the problem of whether such constructs represent objective properties, states, or processes of the organism. Behaviorism can make do with a primitive epistemology because it avoids deep (nonphenomenological) hypotheses and theories, which are the ones that pose the toughest problems in the theory of knowledge. (Typically, a behaviorist learning model revolves around the concept of the probability of an organism's making a given response at the nth presentation of a stimulus of a certain kind. It deals with whole events and it makes no reference to mental states; it pretends that the latter do not matter, at least to science.) Third, behaviorism adopts a strict code of scientific conduct. Indeed, we should be grateful to behaviorists for having introduced this code into the field of psychology, where illusion and deception (unwitting or deliberate) are not uncommon. So much for the general outlook of behaviorism, which turns out to be scientific but narrow.

As for the problematics of behaviorism, it must be conceded that it is extremely narrow and that it eliminates the most interesting problems of psychology, namely all those about mental states and processes as well as their so-called neural bases or correlates. Such elimination is unsatisfactory. We all want to know not only how a woman in pain, love, or deep thought moves about. We also wish to know what pain, love, and thought are; that is, what are the neural processes we call *pain*, *love*, and *thought*. When watching two boxers exchanging blows or two psychologists exchanging views, we are not satisfied with being told that each is responding to the blows of his adversary or to the verbal behavior of his interlocutor. We want to know what motivates them to start the fight or the argument and what makes them go on or stop. When watching a neuroscientist operating on a rat, inserting electrodes into its brain, or taking instrument readings, we wish to know what makes him tick: what are his problems, hypotheses, and goals, not to speak of his doubts and hopes. Discarding motivation, affect, and ideation, the behaviorist gives a superficial and therefore unilluminating account of behavior; it is like a silent movie without titles. In his extreme ontological, epistemological, and methodological asceticism, the behaviorist denies that such problems are accessible to the scientific approach and thus leaves us in the lurch. In this manner he hands the field over to the mentalist and his wild speculations. The self-denial of the scientist becomes the self-indulgence of the pseudoscientist.

The aim of behaviorism is scientific but, again, narrow. It is scientific because it proposes to describe and predict behavior. (Actually, it describes it superficially because it ignores the internal states of the animal, and therefore it cannot predict behavior accurately.) And it is narrow because it adopts the positivistic proscription against tackling unobservable facts and trying to explain. These limitations are crippling, for we cannot attain a satisfactory description of behavior, unless we frame hypotheses about the underlying neural processes. After all, it is the brain that makes the stimulus-response-reinforcement process occur (Pribram 1971). Trying to understand behavior on the sole strength of the observation of behavior is like trying to understand motion without masses, forces, and strains, or radio without electrons and electromagnetic waves, or evolution without natural selection, or history without social forces.

Finally, the methodics of behaviorism too is scientific but narrow. It is scientific because it employs observation, measurement, and controlled experiment. But it is narrow because it makes no full use of its empirical findings; indeed, it minimizes and sometimes even denies the role of theories and in particular that of mathematical models and, consequently, has no occasion to face the problems of testing them experimentally. And because behaviorism ignores neuroscience it is spared the intriguing problem of checking its S-R hypotheses with neuroscientific data and hypotheses. As for behavior modeling, to be sure there are a few behaviorist models, most of them in the area of learning (cf. Luce, Bush, and Galanter 1963–65). However, all these models are phenomenological; they deal with appearances or externalities and are therefore shallow (for one thing, they do not bear on neural processes; for another, they ignore cognition and motivation). Worse, they are basically mistaken in assuming the Aristotelian conception of change, according to which the cause (input) alone determines the effect (response) regardless of the inner organization and state of the system. Needless to say, this view is at variance with modern physics, chemistry, and biology, all of which study not only external circumstances, but also internal structures and processes. Whether the object of study is an atom or a person, the effect of a stimulus (be it a photon or a word) depends not only upon the kind and strength of the stimulus, but also on the internal state of the object—a state that must be conjectured, since it cannot be observed directly. Moreover, what happens inside the system, e.g., the process that led to the internal state at the time the stimulus impinged, is just as important and interesting as the commerce of the system with its environment. For one thing, every concrete system, from atom to DNA molecule to cell to ecosystem, is in flux; yet according to behaviorism, such internal flux does not matter. For another, every concrete system has some spontaneous activity: it may initiate changes (e.g., spontaneous radioactive decay, spontaneous self-assembly, spontaneous formation of a melody) without environmental prodding: not every output is a response to some input. Not so according to behaviorism: all we ever do is respond to stimuli. In short, the methodological strictures of behaviorism, modern though they may appear to be, bind behaviorism to obsolete science—to say nothing of the boredom induced by observing externalities without having a clue as to their mechanism (for further criticisms see Bandura 1974).

The upshot of our examination of the behavioristic approach is that behaviorism adopts a narrow scientific approach: it is nearer to protoscience than to full-fledged or mature science. Because of its narrowness behaviorism has remained stagnant since the mid-1950s. For the same reason it has been incapable of stemming the mentalist tide—nay, it has in part provoked it. In historical perspective behaviorism may be regarded as the mother of scientific psychology. But it is an unwed mother that has refused to marry the father, namely neurobiology. Like every other mother, behaviorism deserves our love; like every mother, it must be prevented from hindering the advancement of its progeny. When criticizing the limitations of behaviorism, therefore, let us not forget that it was the beginning of a science; let our criticisms not be an excuse for reviving mentalism, which is definitely prescientific, not protoscientific.

3. MENTALISM

Mentalism is the approach that focuses on mental events, endeavors to explain them, as well as behavior, by further mental events, and relies mainly on introspection, i.e., ordinary intuition. Thus, the mentalist holds that he feels, perceives, thinks, and wills with his mind, not with his brain. He emphasizes that mind is immaterial and autonomous with respect to matter. And, of course, he regards his own view as conclusive refutation of naturalism or materialism.

Two main variants of mentalism are popular today. One is the old vulgar idea that mind is a special immaterial (and possibly immortal) substance. As such, mind is inaccessible to the scientific approach, even though it may interact in mysterious ways with the brain (cf. Popper and Eccles 1977; Eccles 1980). How such interactions between an entity and a nonentity are to be conceived is not explained. The second variant of mentalism, popular among contemporary workers in cognitive science, in particular artificial intelligence practitioners, as well as among philosophers, is that the mind is a set of programs: that it is software, structure, organization, or information, not a matter of hardware or stuff (cf. Fodor 1975, 1981; MacKay 1978; Putnam 1975; Pylyshyn 1978). We shall call these varieties of mentalism *substantialist* and *functionalist*, respectively.

Substantialist mentalism matches with ordinary (i.e., fossil) knowledge and has the blessing of theology. It is a view rather than a theory, and it contains no technical concepts, let alone mathematical ones, so anyone can understand it. In fact, substantialist mentalism has still to be formulated in precise terms, and it is doubtful whether it can ever become a testable theory. Consider, for example, the hypothesis formulated by Thomas Aquinas, and adopted by Eccles (1980, p. 240), that the mind or soul of an individual is "infused" into it by God at some time between conception and birth, so that every fruitful marriage is actually a ménage à trois; or Eccles's claim (1980, pp. 44–45) that the self-conscious immaterial mind scans and reads out the activity of the cortical modules or columns; or his postulate (1980, p. 232) of "the existence of some conscious experiences prior to the appearance of the counterparts in the specific modular patterning in the neocortex". Try and formulate these wild speculations in exact terms; try and design experiments to check them; and try to render them compatible with neurophysiology, developmental psychology, or evolutionary biology. If you fail in at least one of these attempts, confess that substantialist mentalism is anything but scientific.

The functionalist (or structuralist or information-theoretic or computational) variety of mentalism is slightly more sophisticated than substantialist mentalism. It is advertised either as being neutral between spiritualism and materialism, or as being materialist, but in fact it is good old mentalism in new garb, for it holds that form or organization is everything, whereas matter or stuff is at most the passive support of form—oh, shades of Plato! To the functionalist almost anything, from computers to persons to disembodied spirits, can have or acquire a mind: "We could be made of Swiss cheese and it wouldn't matter" (Putnam 1975, p. 291). According to this view, a psychological theory is nothing but "a program for a Turing machine" (Fodor 1981, p. 120). So why bother studying the brain? And why bother studying the peculiarities and interrelationships of perception, motivation, and cognition? An all-encompassing and stuff-free theory is already at hand: it is the theory of automata. Psychology can learn nothing from neuroscience, and it can expect no theoretical breakthrough.

Although mentalists of the functionalist or computational variety are very critical of behaviorists, their approaches are similar insofar as both are externalists and ignore the nervous system. In fact, functionalism can be regarded as the complement rather than the opposite of behaviorism. Take, for instance, Turing's criterion for telling—or rather not telling—a human from a computer, namely not to open them up but to record and analyze the net responses of the two regardless of the way they process the incoming information; that is, irrespective of the stuff they are made of (Turing 1950). This criterion is behaviorist as well as functionalist. And it will not do, because every theory of machines, in particular Turing's own theory, contains a theorem to the effect that, whereas behavior can be inferred from structure, the converse is false. (Similarity of internal structure implies similarity of behavior, but not the other way round.) This is obvious to any psychologist or ethologist. Thus, the foraging bee, the migrating swallow, and the human navigator are good at orienteering, yet each "computes" the desired path in its own peculiar fashion.

To be sure, the search for similarities, and the accompanying construction of metaphors, is useful—but it cannot replace the investigation of specifics. Trivially, any two things are similar in some respects and dissimilar in others. The question is to ascertain whether the similarities weigh more than the differences, so that both things can be grouped into the same species. Functionalists hold that this is indeed the case with regard to persons, computers, and perhaps disembodied spirits (Fodor 1981). This claim is not only offensive to parents; it is also outrageously false and misleading.

To begin with, the theory of Turing machines is far too poor to account for any real system, if only because Turing machines have a denumerable set of states, whereas the states of any real system form a nondenumerable set. Not even the neutrino and the electron, possibly the humblest things in the universe, are describable as Turing machines: they are far more complicated than that. (A Turing machine is describable by a table exhibiting the properties of its next state function. Neutrinos and electrons are describable by complicated systems of partial differential equations and other complex formulas.)

Second, the human nervous system is far more complicated than a computer, if only because it is composed of variable components capable of some degree of spontaneous activity and creativity—the last thing we want in a computer. Third, computers are artifacts, not organisms with a long evolutionary history. Fourth, computers are designed, built, and programmed to solve problems, not to find them; to process ideas, not to originate them; to supplement the brain, not to replace it; to obey, not to command. It follows that computer science can advance provided it learns from neuroscience, whereas neuroscience will stagnate if it becomes the caboose of computer science. Computers imitate brains, not the other way round.

Another feature that mentalism, whether functionalist or substantialist, shares with behaviorism is that all three regard neuroscience as being irrelevant to psychology; in the case of substantialist mentalism because psychological problems are allegedly solved by resorting to the old philosophical-theological dogma known as mind-body (or psychoneural) dualism; and in the case of functionalist mentalism because those problems are said to be solved by decreeing (not proving) that we are Turing machines or at any rate information processors and nothing else. In both cases the solutions are proposed a priori and are not checked experimentally. In neither case is the brain necessary, except perhaps to keep neuroscientists busy. And in both cases telekinesis, telepathy, reincarnation, and resurrection are possible—nay, they are sometimes the whole point of the exercise.

The scientific status of mentalism can be appreciated by checking the way it matches or mismatches the scientific approach. To begin with, the general outlook or philosophical background of mentalism involves an ontology countenancing disembodied minds (or souls, egos, superegos, etc.) or stuff-free programs, or energy-independent information—and sometimes also supernatural beings.

Accordingly, adopting this ontology turns psychology into an anomalous research field: the only one in which states are not states of concrete things, and events are not changes in the state of concrete things. The accompanying epistemology is uncritical, for it relies on intuition and metaphor, which to the scientists are just props. And the ethics of mentalism is dubious because it resorts to authority or chooses to ignore the evidence for the biological view of mind accumulated by physiological, developmental, and evolutionary psychology.

On the other hand, the problematics of mentalism is its forte and a reason for its popularity. Indeed, instead of writing off most of the classical problems of psychology, mentalism makes a point of showing that it can handle them and in this it satisfies—alas, ephemerally—our yearning to understand our subjective experience or mental life. This, then, is the one and only merit of mentalism: that it acknowledges the full problematics of psychology. A pity, that it does not approach it scientifically.

As for the aims of mentalism, they are mixed. On the one hand, it endeavors to understand the mind, but on the other it refuses to do so with the help of laws, or at any rate in terms of laws linking variables some of which are accessible to objective (not introspective) observation or measurement. In fact, the vagaries of the putative immaterial mind are untraceable with the help of scientific instruments; we can speculate about them but cannot check experimentally such speculations as long as we are restricted to introspection. As for the assertion that every psyche, whether fathered by ourselves or by IBM, is a Turing machine or some more complicated computer, it is a dogma, not a hypothesis, and it is one involving no commitment to any scientific laws. After all, every scientific law is stuff-dependent, for laws are nothing but the invariant patterns of being and becoming of concrete things, whereas functionalism claims that mind is stuff-free. In sum, mentalism does not endeavor to account for mind with the help of any laws of matter, and on this count its goals are not scientific.

Finally, the methodics of mentalism is clearly unscientific. In fact, mentalism is typically speculative, metaphorical, dogmatic, and nonexperimental. There is nothing wrong with speculation as long as it is fertile and testable in principle—or at least entertaining. But the mentalistic speculations are untestable, because they involve disembodied entities, i.e., nonentities. As for the metaphors of mentalism—"The soul is like the pilot of a boat" and "The mind is like a computer program"—they are not intended to be put to the test, for they are not scientific hypotheses. To be sure, these superficial analogies do hold in some respects, but so what? Analogies can be heuristically fertile, but in this case they are barren and misleading: the pilot metaphor, because it ushers people into the theologian's cell, and the computer metaphor, because it advises them to study machines instead of brains. (Of course, one may decide to study only that which all information systems have in common. However, this arbitrary decision does not prove that nervous systems are nothing but information processors and, therefore, accountable solely in terms of computer science. One may focus one's attention on information rather than, say, the complex neurobiological processes whereby information is transmitted— and generated and destroyed. But this does not prove that information can be transferred without energy; every single signal is carried by some physical process. To be sure, information theory is interested only in the form of signals and thus ignores matter and its properties, among them energy. But this only shows that it is an extremely general theory—so general, in fact, that it can explain no particular fact.) We conclude that the methodics of mentalism is nonscientific.

The upshot of our examination of mentalism is clear. Of the four components of the mentalistic approach, only one is acceptable, namely its problematics—and this on the charitable assumption that mentalists pose their problems in a way that is susceptible to scientific treatment, which is not always the case. The other three components of mentalism are not congenial with science. The verdict is that mentalism is nonscientific: it is just the old philosophical psychology, even if it sometimes uses fashionable terms such as *software*, *program*, and *information* (for further criticisms see Bindra 1984).

4. NEUROPSYCHOLOGY

The central thesis of neuropsychology (or biopsychology or psychobiology) is that a mind is a collection of special brain functions (Hebb 1949; Bindra 1976; Bunge 1980a). On this thesis the mind is not a separate substance (substantialist mentalism), nor is it a stuff-free program (functionalist mentalism). Instead, the mind is a collection of peculiar activities or functions of systems composed of numerous neurons (or their analogs in minding beings on other planets).

This neurobiological view of mind is not a stray philosophical opinion or a dogma entrenched in theology, but is part and parcel of every naturalistic (materialist) worldview. As such it is abhorred by all those who seek to explain the world in terms of the unworldly. The neurobiological view of the mind transforms the mystery of mind into a problem, or rather a problem system (problematics), that can be approached scientifically. The thesis denies the autonomy of mental life as well as the independence of psychology: it renders our inner life a part of our ordinary life and turns psychology into a branch of biology and, more particularly, of neuroscience. In other words, in the neurobiological perspective, psychology is the branch of neuroscience that investigates the specific functions or activities of neuron assemblies which, in ordinary parlance, we call *perceiving*, *feeling*, *learning*, *imagining*, *willing*, *evaluating*, *reasoning*, and so on.

Neuropsychology is revolutionary, but it does not ignore the contributions of alternative psychological currents. For one thing, it adopts the methodological rigor of behaviorism; for another, it accepts the genuine problems posed by mentalism. Neuropsychology thus retains whatever is valuable in those two approaches while going far beyond them. In fact, it adopts a very different approach. Let us take a look at it.

To begin with, the general outlook or philosophical background of neuropsychology is a naturalistic (materialist) ontology free from non-scientific (in particular theological) strictures. The ontology underlying neuropsychology is naturalistic, because it is concerned with organisms, and it does not hypothesize disembodied minds or stuff and energy-free information flows. The epistemology of neuropsychology is realistic and mature. It is realistic because it endeavors to account for mental and behavioral reality, not just for introspective or behavioral appearance. And it is mature because, not setting limits upon theoretical constructs, it cannot evade the problem of finding out how such theories represent reality. Finally, the ethics of neuropsychological research is supposed to be that of science in general: no argument from authority; no holding back information; no recoiling before problems, methods, or hypotheses that may shock some ideology, and so on.

The problematics of neuropsychology is the entire realm of behavior and mentation: it does not reject any problem that can in principle be investigated scientifically. In particular, it includes the problem of accounting for consciousness, perhaps in terms of the self-monitoring of the brain or, better, of the monitoring of the activity of one part of the brain by another part of it. Thus, the problematics of neuropsychology has a considerable overlap with that of mentalism. However, it drops some of the problems of mentalism, reformulates others, and adds some that mentalism cannot pose. For example, neuropsychology rejects as nonscientific the problem of finding out where the mind goes in deep sleep or coma, or at death. It reformulates the problem of finding out at what point in its development a human embryo gets its soul "infused", asking instead whether its brain can be in mental states at all and, if so, from what stage on. And it adds the entire problematics of biological evolution: At what evolutionary level did mind start? What may have been the mental abilities of our hominid ancestors? What is the origin and function of lateralization? When may language have started? and so on. In short, the problematics of neuropsychology is far richer than that of behaviorism and far more precise and challenging than that of mentalism.

The aims of neuropsychology are fully scientific: not just the description of overt behavior, nor merely the accumulation of introspective reports—let alone of old husbands' tales about the ghostly—but the explanation and, if possible, also the prediction of behavior and mentation in terms of the laws of neuroscience. The production of (relevant) data is of course indispensable to this end: no factual science without observation, measurement, and experiment. Still, data are bound to be superficial or even irrelevant in the absence of theories. Besides, the ultimate goal of scientific research is not the accumulation of facts, but their understanding. And scientific explanation is achieved only with the help of theories and models—not just descriptive theories summarizing and generalizing the data, but theories explaining them in terms of mechanisms, such as neural processes.

Finally, the basic method of neuropsychology is the scientific method, even though, of course, neuropsychologists add to it their own special techniques, from voltage clamping to surgical ablation. Unlike behaviorism, which shuns theory, and mentalism, which shuns experiment, neuropsychology makes full use of the scientific method: problem, hypothesis (or, better, model), logical processing, empirical operation, inference, evaluation of hypothesis, new problem, and so on. In particular, neuropsychology can control and occasionally measure mental variables because it identifies them with certain properties of the brain.

In sum, the neuropsychological approach to the mental is scientific and, moreover, the only one fully so. The birth of neuropsychology in recent years constitutes a scientific revolution, for it adopts a new approach to an old problematics; it elicits an explosive expansion of this problem system; it is bound to succeed where the alternative approaches failed; and it promotes the merger of previously separate disciplines, from neurophysiology, neuroendocrinology, and neurology to psychology, ethology, and psychiatry. Compare this revolution with the exhaustion of behaviorism and with the mentalist counterrevolution.

Note the following peculiarities of the research strategy in neuropsychology. First, in studying behavior it proceeds centrifugally, from the nervous system to overt behavior. For example, it attempts to explain voluntary movement in terms of specific activities of certain neuron modules presumably located in the frontal lobe and connected with the motor cortex. To be sure, some investigators still try to "read off" brain processes from behavior (in particular, speech) or from global electrophysiological data (in particular, EEG tracings). This centripetal strategy is unlikely to bear fruit, because one and the same behavior pattern can be produced by many alternative neural mechanisms. The only strategy that can succeed is to try and explain the global data in terms of neurophysiological hypotheses consistent with our current knowledge of the brain.

Second, rather than regarding the nervous system as a mere information processor restricted to transducing (encoding) external stimuli, neuropsychology is learning that the central nervous system of the higher vertebrate is constantly active, and that this activity is largely spontaneous. In other words, the nervous system activity is modulated by environmental stimuli instead of being uniquely determined by them. Hence, we must speak of information generation and destruction in addition to information transmission.

Third, neuropsychology does not restrict itself to talk of information processing, for this notion is far too generic; it attempts to understand the peculiar properties of the generation, destruction, and transmission of neural information. Nobody begrudges the information theorist the privilege of talking about information in general, but he cannot explain the functioning of the brain in information-theoretic terms any more than he can explain the generation, propagation, and detection of electromagnetic waves. Similarly, the general systems theorist is entitled to theorize about systems in general, regardless of the stuff they are made of and, consequently, regardless of the special laws they satisfy. Not so the specialist in the nervous system of the higher vertebrate: he deals with a unique system possessing properties that no other system in the world has, such as lateral inhibition, synaptic plasticity, spontaneous activity, and the possibility of knowing itself.

Fourth, neuropsychology possesses an all-encompassing theory of behavior and mentation serving as a foil for the construction of partial theories (models) each devoted to a particular kind of behavior or mentation. The general framework is used as a basis and guide for the special theories, much in the same way as general mechanics is the basis for the special theories of the spring, planetary motion, and so on. And, whether extremely general, half-way general, or extremely specific, such theories tend more and more to be couched in mathematical terms, for only mathematics gives precision, deductive power, and conceptual unity.

Fifth, in building neuropsychological theories we can choose among three styles: *holistic* (whole brain), *top-down* (reduction to cellular components), or *bottom-up* (synthesis from components). We need theories at all levels and theories relating the various levels, for the nervous system happens to be a multilevel system, composed of a number of subsystems—the more numerous, the lower the level concerned. Holism is defective because, although it insists that the whole possesses emergent properties absent from its parts, it resists the attempt to explain emergence in terms of composition and structure. Microreduction correctly emphasizes the importance of the composition of a system but is blind to emergent systemic properties, such as the ability to form an image or a concept. Consequently, this purely analytic approach is bound to fail in the attempt to explain the mental, which is presumably a collective or mass activity resulting from the activity and the interaction of myriads of neurons. We are left with the synthetic or bottom-up method, which endeavors to reconstruct a system from its components and the interactions among them as well as with the environment. This third strategy has the virtues of the alternative methods: it is concerned with emergent wholes and their composition. And it lacks the defects of its rivals: it does not reject analysis and it does not ignore the multilevel reality of the nervous system. For these reasons it is the most ambitious, the hardest, and the most promising of the three. This is also why it is seldom tried, so that so far only a few precise and reasonably realistic bottom-up models of the activity of the nervous systems are available (see e.g., Bindra 1976; Cooper 1973; Cowan and Ermentrout 1979; Pellionisz and Llinás 1979; Pérez et al. 1975; von der Malsburg 1973; Wilson 1975).

The bottom-up modeling strategy is systemic and integrative. Therefore, it is the one capable of accounting for the systemic and integrative properties of the various neural systems. By the same token, it is the strategy capable of bringing together all the studies in neuroscience and psychology. It has been well tried in physics and chemistry. Thus, the solid-state physicist builds a mathematical model of his crystal structure in order to explain the molar behavior of a lump of conducting material. And the quantum chemist tries to understand the properties of compounds and reactions with the help of hypotheses and data concerning atomic composition. (In both cases a knowledge of some global properties is needed, and in each case a set of new hypotheses must be added to atomic physics.) Similarly, the neuroscientist studying a particular neural system, such as the brainstem, is expected to make a contribution to our understanding of the specific emergent functions of that system in terms of its peculiar composition, structure, and environment—such as its being in charge of awareness and wakefulness.

So much for the salient features of neuropsychology and its scientific worth by comparison with those of behaviorism and mentalism. Let us now place neuropsychology in the scheme of science.

5. REDUCTION OR INTEGRATION?

Of the three approaches we have examined, the third is the most promising for being the most scientific. Its peculiar trait is that it reformulates most psychological problems in neurobiological terms (the exception is formed by the problems in social psychology, on which more below). In this way it succeeds in going farther and deeper than its rivals. For example, instead of restricting himself to describing the behavior of an animal engaged in solving a problem, the neuropsychologist will endeavor to identify the neural processes consisting in problem solving and will attempt to trace the evolutionary ancestry as well as the ontogeny of such behavior pattern. In short, the neuropsychologist identifies the mental with certain neural functions, and in this regard he performs a reduction. However, such reduction of the mental to the neurophysiological does not transform neuropsychology into a chapter of physiology. There are several reasons for this.

First, neuropsychology (or psychobiology, or biopsychology) includes not only neurophysiology of behavior and mentation, but also the study of the development and evolution of neural systems and their functions, and therefore the investigation of the genetic and environmental determinants of behavior and mentation.

Second, neuropsychology and even neurophysiology, when studying the mechanisms of behavior and mentation, are guided by findings in traditional psychology. Thus, the study of a perceptual system involves knowing what it does (what its specific functions are), how such activity is related to the motor centers, and in what ways it serves or disserves the whole organism.

Third, unlike neurophysiology, which studies individual animals detached from their social environment, neuropsychology cannot ignore social behavior. That is, it must mesh in with sociology to constitute social psychobiology.

In other words, neuropsychology is not a mere logical consequence of neurophysiology. But it does result from enriching neurophysiology with new hypotheses concerning behavior and mentation conceived in a biological fashion. The reduction is then partial or weak rather than total or strong, to employ terms elucidated elsewhere (Bunge 1977b, 1983b, as well as essay 11).

In sum, neuropsychology identifies mental events with neural ones, but it is not a branch of neurophysiology. Instead, it is a merger of neuroscience and psychology with a pinch of sociology. This synthesis has, like every synthesis, properties of its own, not found in its components taken in isolation. The main feature of the new synthesis is that it is capable of tackling the most formidable of all scientific-philosophical problems, namely the age-old mind-body problem, instead of ignoring it or leaving it in the hands of theology or traditional philosophy. And the main advantage of the new synthesis is that it overcomes the fragmentation of the disciplines dealing with behavior and mentation—a lack of cohesion that has often been lamented and sometimes seemed unavoidable (Koch 1978). The frontiers between the various sciences of behavior and mentation are artificial and bound to be erased.

20

EXPLAINING CREATIVITY
(1993)

■

The words 'creation' and 'creativity' are fashionable, and rightly so, for industrial civilization must renew itself continuously to subsist. We honor innovators, in particular discoverers and inventors, whereas up until two centuries ago in many countries, for instance in Spain and its colonies, being a "friend of novelties" was a crime punishable by jail.

We, particularly in North America, have become so fond of novelty that we often buy things introduced long ago, provided they come in new wrappings, or old ideas as long as they are couched in newfangled words. Moreover, our love of novelty is such, that sometimes we do not pause to find out whether the latest novelty is useful or even pernicious. In short, our enthusiasm for creativity is often blind: so much so that sometimes we become its victims.

We often boast of being creative, if not in ideas or in deeds, at least in clothes or hairdos, in attitudes or in turns of phrase. Yet nobody seems to know what exactly creativity is and, consequently, how we could stimulate it—although managers and management consultants can now buy a "creativity kit". There are even those who believe that creation, if it is for real, is mysterious and will always remain so. But of course, there is no more powerful deterrent to research than the warning that it will be fruitful. Let us therefore take a look at the problem. But first a tiny bit of history, or rather prehistory, of the concept of creation.

1. ANTECEDENT KNOWLEDGE

The concept of creation has an interesting but little-known history. For one thing, it seems to be no older than a couple of millennia. In fact, the archaic and ancient religions and cosmogonies do not seem to contain the concept of creation out of nothing. In particular, the ancient gods were unable to create things the way magicians claim to do: they could only organize the formless or chaotic—or else mess up or even destroy the organized, in particular rebellious people. For example, the Book of Genesis does not speak of a creative but rather of an organizing Yahweh, a craftsman on a cosmic scale who transformed the original chaos into a cosmos or orderly and regular universe—albeit a minute one by comparison with the actual universe discovered by modern astronomy.

The idea of a divinity so powerful that it was capable of creating the universe out of nothing seems to have been invented during the first centuries of Christianity. It may have been an Oriental graft that must have shocked the few remaining scholars steeped in classical Greek philosophy. In fact, the classical Greeks were much too rational, and often much too naturalistic as well, to believe in the possibility of creation ex nihilo. In particular, the ancient atomists denied it explicitly: recall Lucretius's principle that nothing comes out of nothingness, and nothing turns into nothingness. Even Aristotle's Zeus had not created the universe but was coeval with it.

The ancient Greek philosophers denied also that humans were able to create new ideas. The idealists held that we can only grasp ideas that preexist in an ideal world, such as Plato's ghostly Realm of Ideas. And the empiricists and materialists held that we can only refine and combine ideas originating in perception. For example, the perceptions of humans and of horses allow us to form the corresponding concepts, which we can combine into the idea of a centaur. Even Voltaire, in the midst of a century rich in discovery, invention, and hence heterodoxy, denied that humans could ever create anything. This reluctance to acknowledge the possibility of creation is even found among contemporary mathematicians and scientists: they rarely say that they have invented any ideas. Rather, they say modestly (and incorrectly) that they have discovered their theories.

But of course, even a cursory glance at what happens around us suffices to notice, particularly in our day and age, a continual flux of novelty and, particularly, of human-made novelty, mainly scientific and technological. Admittedly, most novelties are modest and short-lived, and none emerges out of nothing: every one of them is the outcome of some process or other, and every such process can be traced back years or even eons. Still, novelty does emerge more or less suddenly, frequently, and all around us. Lest there be any doubts about this matter, let us list a few novelties of familiar kinds.

Creativity is exhibited by the toddler who regularizes an irregular verb and thus comes up with a new word and, moreover, a more rational one than the accepted one, e.g., when she says 'eated' instead of 'ate', or 'speaked' instead of 'spoke'. So does the rascal who invents a new excuse or a new lie. Ditto the sportsman, acrobat, or ballerina who comes up with a new pirouette; the craftsman who finds new uses for well-known tools; the engineer who designs a new machine or a new process; and the biotechnologist who designs or manufactures a new biospecies. The poet who describes a common experience in a novel way, and the novelist or playwright who introduces a new character, are creative. So is the mathematician who conceives of a new mathematical structure or unpacks a new consequence from a set of known premises. Ditto the manager who devises a firm of a new type, or who reorganizes an old firm along new principles. The same holds for the politician or bureaucrat who drafts a new bill aiming at solving (or creating) a social problem. The pharmacologist who designs a new drug to treat a certain biological disorder, and the psychologist who conceives of a new physiological or behavioral indicator of a mental process, are likewise creative. Even the design of a new technique for destroying health, life, property, or society is creative. For example, the criminals who invented "crack" and the neutron bomb were creative. By contrast, the various "law and order" measures that are being taken to fight drug addiction and trafficking are not creative: they have all been tried before—and failed. Social problems can seldom be solved by brute force alone: they call for some creativity.

What do all of the above examples of creation have in common? Firstly, they are man-made: they are not found in nature—unlike, say, the spontaneous formation of a molecule or the self-organization of a collection of living cells. Secondly, they are all products of deliberate actions, though not always of carefully planned ones: they did not come about by pure chance, although chance does always play some role in any process, simple or complex. Thirdly, they are all original in some respect: that is, they enrich the world with something new, that did not exist before the creative act.

Now, there are degrees of originality and, therefore, of creativity. The last goal in the latest soccer game was a new fact, but it did not inaugurate a new class of facts. By contrast, the invention of a new sport, such as baseball or windsurfing, was an absolute or radical creation. The computation of some values of a well-known function by means of an existing algorithm, and the measurement of a known physical magnitude by means of a familiar technique, are original if performed for the first time, but they are not absolute creations. On the other hand, the invention of a new mathematical function or of a new scientific theory, the design of experiments or artifacts of new types, the invention of a new kind of social behavior, the introduction of a new literary or musical style, and the like, are instances of absolute creation. In short, absolute or radical creation is that which inaugurates a new type. It is a case of speciation.

2. THE CREATION MECHANISM: A NEUROCOGNITIVE HYPOTHESIS

What is the mechanism of the creative process? This is a problem for psychological research. The behaviorists did not tackle it, because they were not interested in the mind. (They declared it either nonexistent or beyond the reach of science.) Nor do the information-processing psychologists wrestle with the problem of creativity, because they conceive of the mind as a computer, and computers work to rule, whereas there are no known rules for creating anything. (There are only rules for destroying, such as those of military strategy.) We shall handle the problem of computer creativity in a while. Let us now concentrate on the psychological study of creativity.

The first psychologists to tackle the problem of creativity are the members of the now defunct Gestalt school and the physiological psychologists. Regrettably, although the Gestalt school did stress that we and other primates are capable of performing creative acts, it denied that the creation process is analyzable. In fact, they held that problem solving is, like perception, an instantaneous and unitary event. The gestaltists were eventually proved wrong with regard to perception, which has been shown to be the end result of a complex process and, moreover, one preceded by sensory analysis. Since intellectual processes are usually even more complex than perceptual processes, it is likely that the episodes of creative insight take more time than the perceptual processes (which take on the order of 100 milliseconds).

The neurophysiological approach to mind suggests an explanation of the act of mental creation and an entire research project. According to physiological psychology, every mental process is a brain process. In particular, every creative mental process is the same thing as the self-organization of a new plastic system of neurons. (A connection between two or more neurons is said to be *plastic* if it may change, in particular strengthen, in a lasting way. A connection that is strengthened only for a short while is said to be *elastic*: the stimuli acting on it leave no traces. An engram is a lasting modification of the connectivity of a neuronal system.)

We stipulate that a creation is absolute or radical if, and only if, the corresponding plastic neuronal system has emerged for the first time in the history of the universe. (Obviously, since we have no access to the whole universe, or even to the entire human history, we can never be sure whether a given creation is indeed absolute. In practice we must settle for the known part of the universe and its history.) In other words, when an animal, such as a human being, thinks up something new, it is because a new system of neurons has emerged in his brain, either spontaneously or in response to an external stimulation. If two people have independently the same idea, it is because in their brains certain very similar new neuron assemblies have been formed, perhaps as a result of having tackled the same problem on the basis of similar experiences.

This hypothesis explains simultaneous discoveries and inventions as well as single innovations. This explanation is more plausible than Plato's mythical account in terms of apprehending a piece of furniture in the unworldly Realm of Ideas. Moreover, our account has a solid foundation in the experimental study of neuronal plasticity, which is one of the most promising fields in cognitive neuroscience (see, e.g., Hebb 1980, 1982; Bunge 1980a; Bunge and Ardila 1987).

The above explanation of creativity in terms of the self-organization of neuronal systems is only sketchy, hence devoid of precise experimental confirmation. We still lack a detailed theory of neural plasticity and of the particular kind of plasticity that "mediates" (is identical with) creativity. But no such theory will be forthcoming unless psychobiologists work more intensely on creativity, and unless they overcome their fear of mathematical modeling. In other words, the above account of creativity is only a programmatic hypothesis, i.e., the nucleus of a research project. It is one that should engage the psychologists interested in brain function working side by side with neuroscientists interested in the mind.

3. THE SOCIAL MATRIX

Our explanation of mental creation as the emergence of new neuron assemblies is necessary but insufficient: some reference to the social matrix is needed as well. The reason is that here are conservative societies, where novelty and personal initiative are regarded with suspicion, hence inhibited or even repressed from birth. By contrast, a plastic society, one where novelty and personal (but not antisocial) initiative are highly regarded, is bound to favor creativity—though not necessarily for the common good. Hence, if we value creativity, we must work for a plastic social order, one where social bonds are neither too strong nor too weak, and one that encourages experimenting in (prosocial) behavior.

In order for creations to be beneficial, freedom is not enough, because there are noxious creations, such as the invention or perfection of new mass murder or mass destruction weapons, new kinds of medical or psychological quackery, new types of deceitful publicity, or new kinds of covert political manipulation. Technological and political creativity ought to be controlled democratically in the interest of society. (On the other hand, pure science and mathematics as well as the humanities and the arts can only suffer from censorship. If you ask someone to come up with a design for a thing of a given kind, he or she is unlikely to dare inventing anything of a completely different kind: technologists are not free agents, except in their spare time.)

4. CREATIVE COMPUTERS?

We have restricted creativity to well-furnished brains working in societies that tolerate innovation of some kinds. Yet, the enthusiasts of artificial intelligence and of the computational view of the mind claim that computers create, or will become creative in a few years' time. Is there any truth in this claim?

In my view this claim is false and even wrongheaded, for it involves a misconception of both creativity and computation. Indeed, by definition, computers do nothing but compute, i.e., they process symbols according to well-defined algorithms (computation rules). But some living brain must supply these machines a bunch of premises, one or more algorithms, and precise instructions to operate with them, for a machine to do its job. Unlike people, computers have no problems of their own, they have no motivation to tackle any problems, and they do not create that which they process.

Computers can only handle the computational aspects of human thought and, in particular, of mathematical thought. It is a gross error to believe that everything mathematical is computational. Suffice it to recall such mathematical processes as those of "discovering" new problems, hazarding generalizations, guessing theorems, finding the premises that entail a given proposition, coming up with the suitable auxiliary constructions in proving a geometrical theorem, recognizing or inventing the abstract theory underlying two or more "concrete" theories (models), devising new methods for adding up infinite series or integrating differential equations, or creating new concepts or even whole new hypothetico-deductive systems. None of these operations is computational, hence none of them can be entrusted to a computer. Computers assist brains but cannot replace them. Hence they are poor models of creative brains.

The claim that it is possible to design creative computers amounts to the thesis that it is possible to formulate precise rules for inventing ideas. But the very idea of an *ars inveniendi* is wrong because, by definition, an invention is something not to be had by just applying a set of known rules. Once an invention is in hand, one may invent rules for applying it as a matter of routine. And when such rules are available, not before, one may use them to program a computer or a robot. First the creation, then the routine. To attempt to reverse the process, or to conflate the two stages, betrays poor logic.

5. CONCLUSION

To sum up, creativity is intriguing but not mysterious, for it can be explained, at least in principle, as the self-organization of new systems of neurons—never as a result of anything which, like a machine, is designed to work to rule. And creativity ought to be encouraged as long as it does not result in things or processes aiming at harming people. The greatest sin a creative being can commit is to create things or ideas whose sole function is to add to human misery.

VI

PHILOSOPHY OF SOCIAL SCIENCE

■

21

ANALYTIC PHILOSOPHY OF SOCIETY AND SOCIAL SCIENCE
THE SYSTEMIC APPROACH AS AN ALTERNATIVE TO HOLISM AND INDIVIDUALISM

(1988)

■

T here are two general classical approaches to the study of society: holism and individualism. According to holism, society (*Gemeinschaft* or community) is a totality transcending its membership and having properties that cannot be traced back to properties of its members or to interactions among them. By contrast, according to individualism, a society (*Gesellschaft* or association) is nothing but a collection of individuals, and every property of it is a resultant or aggregation of properties of its members.

Most philosophers of society or of social science have opted for either holism or individualism. Generally, nonanalytic philosophers have sided with holism, whereas analytic philosophers have favored individualism. The reason for the latter preference is obvious: individualism is clearer and, at first glance, deeper than holism, in eschewing the intuitionism that ordinarily goes together with holism, and in proposing instead to analyze every social whole into its ultimate components, namely individual persons.

Regrettably individualism, though effective at holism-bashing, is no viable alternative to it, because some social groups happen to be systems possessing global properties that cannot be attributed to their members. Witness national territory and natural environment, social structure and social cohesion, gross national product and its distribution, political regime and political stability, legal and educational systems, international relations and history. All of these and many more are systems or properties thereof: they are neither persons nor properties of persons or even results from aggregating individual properties.

For example, the GNP is not the sum of the outputs of the individual members of the workforce, because the division of labor makes it generally impossible to point to the individual responsible for every good or service. Likewise, the history of a society is not just the sheaf of the biographies of its members: there are such things as the history of the economy, the culture, and the political system of a society. In short, individualism does not work in the study of society, just as it does not work in the study of any other system, be it atom or molecule, crystal or star, cell or ecosystem. Individualism does not work because it ignores the very existence of systems as complex units different from mere aggregates, as well as of their emergent or nonresultant properties. (For the notion of emergence recall essay 5.)

However, the failure of individualism does not force us to embrace holism and its rejection of analysis. Fortunately, there is an alternative to both individualism and holism. This alternative, though generally overlooked by philosophers, is *systemism*—not to be confused with the once-popular "systems philosophy", which is nothing but a rehash of holism. According to systemism, a society is a concrete system of interrelated individuals. Such interrelations or social bonds generate or maintain a society as a unit belonging to a higher level than that of its individual components. Whereas some social properties are indeed aggregations of properties of its components, others are global because they emerge from links among individuals. For example, a good school is one where diligent students are taught by competent teachers.

Far from rejecting analysis, systemism employs it to account for the formation or breakdown of social wholes, and thus to explain the emergence or submergence of social properties. In fact, it accounts for emergence and submergence in terms of social bonds and of the interactions between individuals and items in their social and natural environment. Systemism is thus a sort of synthesis of holism and individualism: it keeps and elucidates the notions of whole and emergence, as well as the rational approach inherent in the analytic tradition. Let us take a quick look at it.

1. THE SYSTEMIC APPROACH

A convenient way to introduce the systemic outlook is to consider the concept of a social relation between two persons belonging to a certain social group, such as a family, a firm, a school, or a club. Holists are not interested in interpersonal relations, but only in the alleged action of the group as a whole upon every one of its members—which, of course, dispenses them with studying interpersonal bonds. As for individualists, if consistent they must deny the very existence of that which keeps a community together or causes its breakdown, namely the set of social relations that usually goes under the name 'social structure'. Thus, Winch (1958, pp. 131ff.) and Popper (1974, p. 14) have held that social relations belong in the same category as ideas, because they are not properties of individuals. Accordingly, the social scientist who were to follow the individualist prescription should either relinquish the study of social relations or approach them in an a priori fashion. Either way she would have to give up doing empirical sociology. Just as the holist sees the forest but not the trees, so the individualist sees the trees but misses the forest. Only the systemist sees the forest as well as the trees, for she studies socially related individuals as well as the systems they constitute.

All the statements found in social science can be analyzed in terms of systems with definite, though of course changing, compositions and structures, embedded in definite, but changing, social and natural environments. Every system, in particular every social system, may thus be conceived of as a composition-environment-structure triple. (Recall essay 11.) For example, a business firm is composed of its employers and employees, it is embedded in a given market, and it is held together (though also threatened by) the various relations among its members, which derive from the nature of their jobs. Again, a school is basically composed of its students and teachers, it interacts with its neighborhood, and its structure is the learning relation and the attendant social relations. (Learning may be analyzed as a ternary relation: x learns y from z.)

I submit that social science problems, as well as practical social issues, are problems about social systems: about their composition, structure, and environment at a given time (synchronic study) or over a period of time (diachronic study). Such systems are neither mere collections of individuals nor wholes opaque to analysis, and within which the individual gets lost. Instead, social systems are concrete things with emergent properties that can be understood by analyzing them at different levels: those of the individual; the family; the economic, cultural, or political group; the supersystem composed of systems of some kind; or even the world system. This view of society and its subsystems includes the valuable insights of both holism and individualism, while avoiding their defects.

Like any other broad outlook on society, systemism is characterized by an ontology-methodology pair, that is, by a view on the nature of society together with a view on the most suitable way to study its object. Let us sketch these two components of the systemic approach.

The *ontology* of systemism boils down to the following theses of emergentist materialism (Bunge 1979a, 1981a).

O1 A society is neither an aggregate of individuals nor an entity hovering above individuals: instead, it is a concrete system of interconnected individuals.

O2 Since society is a system, it has systemic or global properties. Some of these (e.g., birthrates) are resultant or aggregate, whereas others (e.g., political regimes) are emergent, i.e., though rooted in the individual components and their interplay, they are not possessed by persons.

O3 Society cannot act upon its members any more than an organism can act upon its parts. But the members of a social group can act severally upon a single individual, so that his behavior is determined not only by his genetic equipment and experience, but also by the roles he plays in various social groups. Every interaction between two societies is an individual-individual affair, where each individual occupies a definite place, hence enacts a definite role, in his group. And every social change is a change in the structure of a society—hence a change at both the societal and the individual levels.

The *methodology* of systemism should now be clear. It is a particular case of the synthesis of rationalism and empiricism proposed elsewhere (Bunge 1983a, 1985b). Its main theses are:

M1 The proper study of society is the study of the socially relevant features of the individual, as well as the investigation of the properties and changes of entire social groups.

M2 Social facts must be explained in terms of individuals and groups as well as their interactions. Individual behavior is explainable in terms of all the characteristics—biological, psychological, and social—of the individual-in-society.

M3 The hypotheses and theories of social science are to be tested against one another as well as against the anthropological, sociological, politological, economic, culturological, and historical data. But all such data are built out of information concerning individuals and small groups, for these alone are (partially) observable.

However, systemism is not enough. In fact, one might conceive of society as a system of behavior rules, values, customs, and beliefs—the way idealist anthropologists used to do. Alternatively, one might picture society as a system of disembodied actions, as Talcott Parsons did (1951). But neither view would help, for they push everything concrete—persons, artifacts, and natural resources—to the background.

Such shift of focus would have at least two negative consequences. One would be the inability to explain those very behavior patterns, e.g., in terms of interest in getting access to wealth or power. The second undesirable consequence is that sentences such as 'Every society uses some natural resources', and 'The amount of energy (per capita and unit time) a society uses is an indicator of its degree of advancement' make no sense unless society is conceived of as a concrete or material system. This conception is no sheer physicalist reductionism, but merely the admission that whatever is composed of material entities is material itself without being necessarily physical. In sum, the most realistic and therefore fertile conception of society is that of systemic (or emergentist) materialism together with scientific realism. Let us see how it helps handle a few central philosophico-scientific problems in various social sciences.

2. ANTHROPOLOGY AND SOCIOLOGY

Anthropologists hold that theirs is the science of culture, but unfortunately they are not agreed on the meaning of 'culture'. Most of them follow the German idealist tradition in equating 'culture' with 'society', which makes it impossible to speak of the culture of a society as different from its economy and polity. The same equation suggests focusing on norms, values, and symbolic forms, at the expense of matters of reproduction, production, and social relations. (See Harris 1968, 1979, for a criticism of anthropological idealism and a formulation of cultural materialism.)

Systemism suggests adopting a narrower concept of culture, namely as the subsystem of society characterized by cognitive, artistic, and moral activities and relations. Conceived in this way, every culture is a concrete system, hence one that can be studied scientifically (Bunge 1981a). The scope of anthropology may then be widened to embrace all four major subsystems of every human society: the biological or (natural) kinship system, and the three artificial or human-made subsystems: the economy, the culture, and the polity. Simpler: Anthropology is the study of human society in all of its aspects. All of the other branches of social science can then be seen as special branches of anthropology, the most general and basic of them all.

Systemism suggests also its own model of humans, namely as highly evolved animals normally living in society, which is partly an artificial (nonbiological) system with economic, cultural, and political subsystems. I submit that this biopsychosocial model of humans includes whatever is valuable in the purely biological, psychological, or sociological models. It also involves rejecting the earlier models, in particular the religious one. (Incidentally, materialism does not deny spirituality: it just states that all so-called spiritual activities—whether religious, artistic, philosophical, scientific, or technological—are brain processes. See essays 5, 6, 7, and 19.)

The systemic model of humans helps answer an old question of interest to philosophers of language, linguists, psychologists, and anthropologists, namely 'What is language?'. According to idealism, language is an ideal object existing by itself, either as a Platonic idea (Katz 1981) or as a human creation that has acquired autonomy (Popper 1972). Needless to say, neither variety of idealism enjoys empirical support. Worse, whoever regards language as an ideal object cuts the ties of pure or general linguistics with the other branches of this science, particularly neurolinguistics, psycholinguistics, sociolinguistics, and historical linguistics. Abstracting from brains and linguistic communities makes sense only when studying the purely formal (syntactic) features of a language at a given time.

The systemic-materialist answer to our question is that what are real are not languages in themselves but linguistic communities and their members, namely individuals engaged in producing, conveying, and understanding phrases. This, the production, conveyance, and understanding of speech, should be taken to be the primary linguistic fact. Everything else about language is construct, starting with the very concept of a language.

Unlike idealism, this view encourages cultivating strong links between linguistics and other research fields—which need not prevent anyone from feigning that there is such a thing as a language detached from particular biological, psychological, and social processes. In particular, the systemic-materialist view of language includes Wittgenstein's view that language is primarily a means of communication, hence a social link—a view dismissed out of hand by the Chomsky school.

This brings us to sociology. This science may be defined as the synchronic study of human social groups or, equivalently, of their structures. Regrettably, both holism and individualism have blocked the understanding of this key notion of social structure, which we characterized earlier as the collection of bonds that join people into a social whole. Holists either overlook social structure, being obsessed with the group as a totality; or, as is the case with structuralists, they conceive of structures as separate from, and superior to, individuals—which is of course a logical blunder. And individualists cannot admit that the structure of a society is an objective property of the latter that cannot be attributed to any of its members.

One way of elucidating the notion of social structure is to start by focusing on social equivalence relations, such as those of same occupation or same income bracket, or of similar cultural background and similar political orientation. Every such equivalence relation induces its own partition of the membership or composition of a society. And every such partition is nothing but the structure of a society in the given respect, e.g., its occupational structure. And the stack or family of all such partitions may be taken to define the total social structure of society. Specifying this structure at a given time, together with the composition and the environment of the society of interest, allows us to describe the latter. And the (anonymous) history of the society is nothing but the trajectory of the system described in that way. All of this can be said in precise set-theoretic and algebraic terms (Bunge 1974c, 1979a).

The systems-analytic approach allows one not only to elucidate key concepts in social science, but also to reformulate and answer some of the oldest and toughest problems in this field. One of them is 'What keeps a social group together?' or, equivalently, 'What makes for social cohesion?' and their partners concerning social breakdown. To answer these questions, one may start by setting up a measure of the degree of social cohesion of a group. This can be done by assuming that cohesion is a function of participation. In turn, the degree of participation of a given social group in the activities of another may be defined as the measure of the overlap of the two groups, i.e., the cardinality of the set composed by the individuals who belong to both the host and the guest groups. Dually, the degree of marginality in the same respect is just the complement to unity of the degree of participation.

Interestingly enough, cohesion turns out not to be proportional to overall participation. Instead, it is maximal for middling participation—which is reasonable, since nil participation is incompatible with communality, whereas the meddling of everyone in everyone else's affairs would result in anarchy (Bunge and García-Sucre 1976). Incidentally, although cohesion is a gestalt property of a system, it emerges from the interactions among the system members. It is a good illustration of the claim that systemism is a sort of synthesis of holism and individualism.

3. POLITOLOGY AND ECONOMICS

Politology may be defined as the science of the polity. In turn, the polity may be conceived of as the social system composed of all the persons capable of making or influencing, implementing or frustrating the decisions affecting the members of any social group. Every polity is composed of persons, but persons enacting social roles, e.g., in governmental departments or political parties. The ultimate aim of the specific activity of the members of a polity is the conquest, maintenance, or usufruct of political power. Hence, politology is essentially the scientific study of power relations and their changes.

The units of analysis (or referents) of politology can be grouped into three categories or levels: individuals, organizations, and communities—from the hamlet to the world system. In modern times, the central political entities in this vast and variegated collection are nation-states: all else is either a subsystem or a supersystem of a nation-state. Nation-states, such as Austria, are concrete supraindividual systems characterized by such global properties as size, population, economic and cultural levels, and form of government.

A form of government or political regime may be characterized by the degree of voluntary participation (or cooperation) in public affairs, together with the degree or freedom of public contestation (or competition). In particular, political democracy is the form of government that maximizes both participation and competition in the management of the res publica. But humans do not live by politics alone. Hence, political democracy is only one component of democracy lato sensu, or integral democracy.

In keeping with our quadripartite analysis of society, we shall distinguish five kinds of democracy: biological, economic, cultural, political, and integral, a combination of the four preceding ones. Integral democracy boils down to the freedom to enjoy all the resources of society, as well as the right and duty to participate in social (i.e., economic, cultural, and political) activities, subject only to the limitations imposed by the rights of others. (Needless to say, integral democracy is still on the drawing board. See also essay 30.) Note that, although this form of government, or any other one, is defined by reference to individual actions, rights, and duties, it is a social property. Even though some of us are democrats, none of us is a democracy; and the form of government of a society may be undemocratic even if most of the members of the polity are democrats.

Politology is closely linked to sociology and economics. After all, each of the three disciplines deals with only one of the artificial subsystems of a single thing, namely society; and every one of those subsystems interacts strongly with the remaining two, if only because every adult member of society belongs to all three subsystems. Regrettably, most economists overlook such interactions and believe that economies are closed systems that can be described by purely economic variables. The inevitable consequence of this artificial isolation is that most economic theories and policies are unrealistic. This is not the only woe of economists. Another of their root problems is that they are not clear on what economics is all about: individuals, scarce resources, goods and services, trade, money, or economic systems such as firms and markets? (More on this in Bunge 1998.)

From a systemic viewpoint it is obvious that economics refers to economic systems, not to individuals, even though it may explain some features of the economy in terms of individual needs and wants. However, most economists, at least most neoclassical microeconomists, are ontological and methodological individualists. Most of them are also moral individualists: that is, they hold that people are actuated only by self-interest—a psychological false, socially destructive, and morally ignoble view (Sen 1977). Individualism, once the philosophy of political liberalism, has become that of political reaction (Bunge 1982b, 1986, 1996, 1998).

To puncture the individualist balloon in economics, suffice it to recall that markets are not mere aggregates of independent producers and consumers, if only because goods and services are mass-produced socially, as Adam Smith emphasized more than two centuries ago. This is reflected in the fact that the supply and demand curves, central to microeconomics, do not concern individuals but the totality of units that, in a given society, are engaged in handling a good or service of a given kind. Consequently, the so-called equilibrium price of the merchandise, said to be determined by the intersection of such curves, i.e., when demand equals supply, is a property of a commodity-in-the-market.

Thus, not even microeconomics can ignore macroentities, such as markets, characterized by global properties, e.g., that of being or not being in equilibrium, or of growing or declining, or of being sensitive to environmental or political shocks. Deprive an economic agent from a society having an institutional framework allowing him to transact his business, and no business will be transacted. Consequently, it is false that all the general statements of microeconomics "are about what they appear to be: individuals" (Rosenberg 1976, p. 45).

What holds for microeconomics holds, a fortiori, for macroeconomics. For example, the notions of economic community and national budget, prime rate of interest and inflation, of international trade and foreign debt, are not reducible to those of individual, expectation, utility, and decision making. All those macroeconomic concepts are about what they appear to be, namely national, regional, or international economic, or rather economico-political, systems. Whence the futility of utility theory and rational expectations theory in macroeconomics. (See essay 22.)

The generalizations, i.e., laws, trends, and rules, of economics are in the same boat: they all refer to systems or to individuals-in-systems. For example, the statement that the international prices of foodstuffs exported by the Third World have declined steadily over the past half century contains the nonindividualistic concepts of international price, which presupposes the existence of an international market and of Third World. Likewise, the law of diminishing returns, according to which the output of any economic system increases at first, and then grows at a decreasing pace as the input increases, concerns economic systems, not individuals.

What holds for the referents of positive economics holds also for those of normative economics: this branch of sociotechnology also deals with entire systems. However, like any other technology, normative economics is supposed to be ultimately in the service of individuals. This is where economics ties in with morals, and where ontological and methodological individualism can play either a moral or an immoral role. Regrettably, most Western normative economists have been concerned with protecting the privileges of the very rich—Keynes and Galbraith being the outstanding exceptions; and most socialist normative economics has been dominated by the holistic myth that the good of the whole is prior to, and independent of, the well-being of the individual. Thus, in either case concern for the welfare of the greatest number has been overlooked. The systemic perspective helps understand these perverse stands and it may suggest ways of reconciling individual interests with the public good. However, this is *Zukunftsmusik*.

4. HISTORY

Let us finally take a quick look at historiography, or the scientific study of the trajectory of large social groups in the course of time. Analytic philosophers, e.g., Danto (1965), have usually been suspicious of any substantive philosophy of history, or ontology of history: they have admitted only the methodological analysis of the historian's craft. While it is true that many substantive philosophies of history, notably those of Vico, Hegel, Spengler, and Toynbee, have been "intellectual monsters", there is nothing to prevent us from building analytic ontologies every bit as respectable as methodology (Bunge 1977a, 1979b, 1981a).

A first task of the part of ontology concerned with the past is to elucidate the very concept of history of a concrete system and, in particular, a social group. This can be done in an exact way with the help of the state space method, which involves constructing a state function for the system of interest, and tracing the changes of its values over a period of time. (Recall essay 5.) The implementation of this program in particular cases can be as complex as desired, for it calls for a mass of data and regularities that may not be readily available, or may even be lost forever. In the case of human history, the regularities are sociological, economic, and political laws and rules. The resulting trajectory of a human group is not a historical law, just as the trajectory of a falling leaf is not a natural law, but an outcome of laws and circumstances. This simple consideration shows why laws are among the inputs, not the outputs, of historical research.

Since humankind is composed of a number of subsystems, human history may be pictured as a sheaf of rather tangled partial histories. In turn, every such strand of the total sheaf is ultimately composed, hence analyzable into, a collection of individual biographies. But these intertwine rather than running parallel to one another, for individuals are organized into systems with dynamics of their own. Even if there were stray individuals, the historian would pay hardly any attention to them: his subjects are not isolated individuals but social systems, from the prehistoric band to the present world system. Individuals interest him only as components of such systems, much as biologists are interested in molecules only as components of living cells.

Every human community is embedded in a natural environment, and is composed of four major subsystems: the biological, economic, political, and cultural ones. Consequently, the total history of every human community is composed of five intertwining strands. Exaggerated attention to any of them results in a lopsided picture of the whole process: in geographic, biological, economic, political, or cultural determinism and their concomitant historiographic schools. Only the study of the entire system can yield an (approximately) true reconstruction of past human reality. This is precisely the way the French school of the *Annales* conceives of the historian's task (see, e.g., Braudel 1976). This conception is actually part and parcel of systemic materialism as well as of scientific realism.

I submit that systemic materialism proposes a genuine materialist and realist conception of history, to be distinguished from economic determinism, whether Marxist or of the rational choice type. Being pluralistic, systemic materialism rejects the hypothesis that there must be an absolute prime mover of history. It asserts instead that social change, i.e., history, is driven by environmental, biological, economic, political, and cultural forces. And that, whereas sometimes these take turns in initiating social changes, at other times they operate jointly, and are always closely related, though not always in step with one another.

An added advantage of this systemic view is that it fosters cooperation among social scientists. In particular, the historian must avail himself of the findings of the synchronic social sciences, every one of which describes partial time cross-sections or, rather, slices of the complex sheaf that makes up human history. If we do not know what a thing is, we cannot know how it evolved. And conversely, as Darwin said, "If we know not how a thing became we know it not". Whence the need for an intimate union between the synchronic and the diachronic social sciences, nowadays largely separated from one another. Since society is a system, it should be investigated as such.

5. CONCLUSION

Social science has come a long way since the First World War and the Russian Revolution. These and other shattering events, more than any philosophical criticism, discredited the armchair speculations and illusions of the previous period, in particular the holistic as well a the individualistic social philosophies and philosophies of history. People began to realize the woeful inadequacy of any views of society and history that underrated the environmental, biological, and economic factors, while it overrated the power of politicians and of ideas. Hundreds of social scientists have been increasingly attracted to the study of large social systems with the help of analytic techniques, rather than in the light—or rather darkness—of pre-analytic intuition. In fact, by and large the extremes of holism and atomism are nowadays avoided by social scientists, and are adopted almost exclusively by philosophers. Even rational choice theorists admit the importance of environmental constraints and, if pressed, that of interpersonal bonds and conflicts as well, although they still fail to admit that every individual action occurs in the midst of some social system or other.

Serious contemporary social science has then moved towards systemism, which states that every thing in the world is either a concrete system or a component of such; that all systems possess properties that their components lack; that the only way to understand the formation, maintenance, and breakdown of a system is by analyzing it into its composition, environment, and structure; and that the best way of putting the results of such analysis back together is by building a conceptual model or theory of the system. Thus, systemism, allied to analytic methods, is the viable alternative to the two dying traditional social philosophies, individualism and holism.

22

RATIONAL CHOICE THEORY
A CRITICAL LOOK AT ITS FOUNDATIONS

(1995)

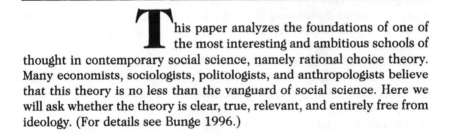

This paper analyzes the foundations of one of the most interesting and ambitious schools of thought in contemporary social science, namely rational choice theory. Many economists, sociologists, politologists, and anthropologists believe that this theory is no less than the vanguard of social science. Here we will ask whether the theory is clear, true, relevant, and entirely free from ideology. (For details see Bunge 1996.)

1. THE PROBLEM

Rational choice theory assumes that (a) choice and exchange constitute the hub of social life; and (b) all choice is motivated by self-interest and, more precisely, is guided by the principle of economic "rationality", or the maximization of expected utility. Rational choice models have been mushrooming over the past half century in all of the so-called human sciences, from psychology to history.

Rational choice theory may be regarded as the culmination of the rationalist view of humans held by such diverse thinkers as Thomas Hobbes and Baruch Spinoza, the French Enlightenment philosophers and the English utilitarians, as well as Auguste Comte, Karl Marx, and Max Weber. All of these thinkers stressed the role of interests, equated modernity with rationality, and rejected the Romantic as well as the religious models of man.

However, there are important differences between contemporary rational choice theory and its precursors. While the latter held a broad concept of rationality, rational choice theory uses the narrow instrumental concept of economic "rationality" or utility maximization formulated by the marginalist economists in the 1870s. Further, it disregards the problem of the rationality of goals (Weber's *Wertrationalität*), as well as that of the morality of both goals and means. And, whereas Comte, Marx, and Weber thought in terms of a historical rationalization process in all domains of belief and action, rational choice theory takes it for granted that preferences are constant over time, and that utility maximizing behavior is in human nature: it is an utterly ahistorical view.

Rational choice theory can be criticized either for being rational or for not being rational enough. The former kind of criticism is based on the writings of Schütz, Goffman, Geertz, Garfinkel, and Derrida, who in turn drew their inspiration from three of the main irrationalist philosophical schools of this century, namely hermeneutics, phenomenology, and existentialism. A closely related criticism of rational choice theory is the antimodernist "rhetoric of the human sciences", which claims that doing science is nothing but engaging in conversation and producing persuasive metaphors rather than maximally true data or hypotheses. (See essay 23.)

Rationalists like myself are not interested in this kind of criticism of rational choice theory, except insofar as it occasionally makes a valid point, such as that real human beings are seldom fully "rational", i.e., selfish and calculating, and that even mathematical economists use some metaphors. My own dissatisfaction with rational choice theory is that it is not rational enough, in involving an unrealistic model of humans and in skirting empirical tests. I will criticize it not for being scientistic, but for not being scientific enough.

Believers in the power of reason are expected to build, apply, teach, or at least admire rational views of human behavior, a behavior which is itself not always rational in any of the many senses of this ambiguous word. But it stands to (scientific) reason that this expectation is only justified if such views are supported by empirical evidence. In other words, rationalism is not enough: it must be combined with realism, i.e., concern for the matching of theory with facts. Moreover, realism may be regarded as a condition of integral rationality, for it is irrational, both epistemically and instrumentally, to press for untestable or untested theories about any domain of reality.

This caution, unnecessary in the case of a mature science, is called for in the case of immature research fields such as sociology, economics, and politology. Indeed, in these fields theory is often disjoined from data, while empirical research is frequently conducted in a theoretical vacuum. Worse, the philosophy underlying most social studies is either classical rationalism ("Reason is all-powerful") or classical empiricism ("Empirical data are the alpha and the omega of research"). Rational choice theory illustrates the former research strategy, whereas the trial-and-error search for statistical correlations exemplifies the latter. In the following, I intend to substantiate the first conjunct of the preceding claim.

Being a rationalist of sorts—more precisely, a ratioempiricist—my view of rational choice theories is sad rather than jaundiced. Though an enthusiast of reason, I believe that rational choice theory has failed abysmally, and it saddens me that this failure has brought discredit upon the very enterprise of serious theorizing in the field of social studies. By the same token, this discredit has brought aid and comfort to such foes of rationality as the hermeneuticists, symbolic interactionists, phenom-enologists, ethnomethodologists, rhetoricians, and deconstructionists. Therefore, the criticisms that follow are intended as a contribution to making way for a more realistic and, hence, more effective use of reason in the field of social studies.

2. BASIC PREMISES OF RATIONAL CHOICE THEORY

There is a large variety of rational choice theories, such as utility theory, decision theory, and game theory, as well as their specializations or models, such as neoclassical microeconomics, the Prisoner's Dilemma model of nuclear deterrence, Olson's model of collective action (1971), Becker's model of the marriage market (1976), and Coleman's "linear system of action" (1990). Let us begin by ferreting out the assumptions common to all such theories.

A rational choice theory or model of a domain of individual or social behavior is one that rests on the following axioms:

A1 Rationality Postulate: All agents are rational, in the sense that they act so as to maximize their (raw or expected) utilities.

A2 Ontological Individualism Postulate: All social groups are nothing but collections of individuals.

A3 Methodological Individualism Postulate: The properties and changes of any social group are understandable by studying only the behavior of the group members.

These assumptions entail the following immediate consequences:

C1 Habit, compulsion, compassion, impulse, ideology, and moral commitment play no role in determining individual behavior.

C2 Social groups have no emergent or systemic properties, i.e., they have only resultant properties.

C3 Social science has no need for irreducibly collective concepts, such as those of social structure and social cohesion.

C4 Social science is reducible to a study of the utility maximization behavior of individuals.

The first three consequences are nothing but negative restatements of the corresponding first three axioms. The fourth is a corollary of the latter.

My first complaint is that the occurrence of the word 'rational' in rational choice theory is unfortunate for two reasons. First, the word in question is notoriously polysemic: it designates at least twelve different concepts (see also Bunge 1987). In fact, we should distinguish the following concepts of rationality: (1) Semantical: clarity or minimal fuzziness; (2) logical: internal consistency; (3) dialectical: inferential validity; (4) erotetic: posing only problems that make sense in some context; (5) methodological: questioning and justifying assumptions and methods; (6) epistemological: requiring empirical support and compatibility with the bulk of the background knowledge; (7) ontological: admitting only ontological assumptions compatible with factual science; (8) valuational: choosing only attainable goals; (9) prohairetic: observing the transitivity of the preference relation; (10) moral: caring for individual and social welfare (11) practical or instrumental: choosing or devising, in the light of the best available information, the means most likely to help attain the given goals; (12) economic: self-interest. We shall group the first six rationality concepts under the rubric of *conceptual* rationality, and the last six under *substantive* rationality.

Second, it can be shown that utility maximization is often "irrational", namely, whenever it is accompanied by shortsightedness, waste of human or natural resources, exploitation, oppression, war, or destructive competition—which may end up in self-destruction.

Third and consequently, the use of the word 'rationality' in that particular sense is persuasive or ideological rather than either descriptive or normative. It may serve as a disguise for self-interest. (I am rational, you use people, he is a shameless exploiter.)

This criticism does not involve a rejection of the rational approach to social facts. It does not even involve a denial that rational decision making does play a role in human behavior, particularly in modern society, as rightly noted by Karl Marx and Max Weber among many others. I am only criticizing what I regard as a misappropriation of the ambiguous word 'rationality', as well as the thesis that self-serving calculation is the only or even the major guide of human action.

My second criticism is deeper and, I hope, more interesting. It is that rational choice theory is far too ambitious. In fact, it claims to explain everything social in terms of just three assumptions that would hold for all individuals in all social groups and in every historical period. But a Theory of Everything does not explain anything in particular. It may belong in logic or in ontology, but it is certainly out of place in factual (or empirical) science. In particular, a Theory of Everything Social cannot account for both the band and the family, the small farm and agribusiness, the small workshop and the giant industrial firm, barter and commerce, marriage and divorce, education and politics, competition and cooperation, committee decision and social welfare judgment, war and peace, progress and decline, and so on and so forth. No wonder, then, that rational choice models are either far removed from reality or, when they do capture particular traits of real systems or processes, they do so by considerably enriching, in an underhand manner, the postulate system expounded above, e.g., by assuming that the individual agents interact via the market, which in turn is not analyzed in individualistic terms.

My third criticism is that, because it is radically reductionistic (in particular individualistic), rational choice theory fails to account for the systemic constraints on individual behavior—in other words, for the effect of structure on agency. In short, the assumption that, at least in certain societies, the agents are completely free to choose, ignores the compulsions and constraints of various kinds that shape individual action. The theory overlooks the fact that one and the same individual acts differently in different social systems, just as a molecule in a liquid body behaves differently at the bottom and on the surface. True, the rational choice theorist is often apt to invoke changing "situations" or "circumstances", but, as noted before, he treats these as black boxes, and thus in a holistic manner inconsistent with his individualistic approach.

So much for general critical remarks. In what follows I shall spell these out and shall examine a number of particularly interesting, albeit highly controversial, rational choice models. I shall start with the Rationality Postulate; but since this involves the notion of utility, we must first examine the latter.

3. UTILITY

The concept of utility, or subjective value, plays the same central role in neoclassical economics and the other rational choice theories as that of objective value in classical economic theory. Although a popular dictionary of economics asserts confidently that "the concept of utility is not contentious", experts in utility theory have often expressed qualms about it, as can be seen by perusing the *Theory and Decision* journal.

I submit that the general utility concept has two fatal flaws. The first is that only its existence has been proved. Indeed, utility theory leaves it undefined—unlike, say, the sine or the logarithmic functions. While such generality is desirable in abstract mathematics, it is a severe shortcoming in any chapter of applied mathematics, where computation is of the essence. The second flaw is that, being subjective, utility is hardly measurable or even subject to interpersonal comparison. In other words, the values of the utility functions are not assigned either rationally (by calculation) or empirically (by measurement). This is why in social studies utilities are usually assigned intuitively—hardly a triumph of rationality.

In fact, the standard von Neumann-Morgenstern axioms (1953) only imply the existence of (infinitely many) utility functions, without defining any of them. And the utility functions occurring in microeconomics are only slightly more specific. In fact, from the beginning of the so-called marginalist revolution in the 1870s, the utility functions have been subject to just two qualitative conditions. These are that any utility function must be monotonically increasing and decelerated. (I.e., $du/dx > 0$ and $d^2u/dx^2 < 0$.) In other words, the "law" of decreasing marginal utility (du/dx) with increasing quantity (x) is assumed more or less tacitly.

Such lack of specificity is similar to that of the general distance function. Indeed, in the general theory of metric spaces, one only demands that the distance function d complies with three conditions: for all elements ("points") x, y, z in an arbitrary set S (the space), $d(x, x) = 0$, $d(x, y) = d(y, x)$, and $d(x, y) + d(y, z) \geq d(x, z)$. These conditions define any of infinitely many distance functions in spaces of arbitrary dimensionality. By the same token, they do not specify any particular space. In order to compute or measure the distance between two points in a particular metric space, one needs a specific distance function, such as the one defined by the Pythagorean theorem. A physicist who were to limit himself to asserting that physical space (or else spacetime) is a metric space, without committing himself to the precise metric (or distance function, or line element), would be laughed out of the laboratory. Only social scientists can get away with similar generalities concerning utility.

To put it another way, a function u about which one only knows (or demands) that it increases monotonically, though with a decreasing slope, is an unknown. Hence, to claim that one understands an action by saying that its actor is attempting to maximize his u, amounts to saying that he behaves the way he does because he has a certain occult quality. This is a case of explaining the obscure by the more obscure. It is theology, not science.

It might be rejoined that, unlike the grand theories in traditional social studies, some of the rational choice models, in particular those of the game-theoretic kind, are chock-full of numbers. True, but these values happen to be taken out of a hat. It does not help assigning the utility functions definite values (in particular maxima) if these values are posited rather than either calculated or measured. And this is precisely what most rational choice models do, namely to postulate the basic utilities instead of either computing or measuring them. The subjectivity of (both cardinal and ordinal) utilities makes it impossible to state, let alone prove, that a given person or group is objectively better or worse off than another, or that one business deal or political maneuver is objectively preferable to another. Like beauty and style, subjective utility and its changes are in the eyes of the beholder.

One consequence of this is that there would be no such thing as objective inequality or social injustice—a convenient result for the opponents of social reform. A second consequence is that, since subjective utility statements are not rigorously testable, the theories in which they play a key role, such as neoclassical economics, are in the same boat as philosophical aesthetics (as distinct from psychological or experimental aesthetics). That is, they are nonscientific. No wonder that the more methodologically sophisticated social scientists, from Vilfredo Pareto on, have treated cardinal utility in a cavalier fashion or dispense with it altogether.

What holds for subjective utility concept holds, a fortiori, for that of expected utility, which occurs in the formulation of the Rationality Principle and involves the probability concept. The probability in question is not well-defined either—except, again, in the case of games of chance. In fact, in most rational choice models probabilities are assigned in the same arbitrary manner as utilities. More precisely, they are usually taken to be *subjective* or *personal* probabilities.

The adoption of subjective or personal probability values is of course consonant with the Bayesian school of mathematical statistics as well as with Bayesian decision theory. But it is inconsistent with the way probabilities are assigned in mature disciplines, such as statistical mechanics, quantum mechanics, genetics, and even some branches of social science, such as social mobility theory. In these disciplines, probabilities (or probability distributions) are treated as objective properties, like lengths and populations, neither of which is supposed to be estimated subjectively (except provisionally and subject to experimental checking). (Recall essay 15.)

Moreover, in science probabilities are only introduced when there is reason to believe that a chance process, such as random shuffling, is at work. No probability without objective randomness. Yet Milton Friedman assures us confidently that "individuals act as if they assigned personal probabilities to all possible events". The fictionist "as if" escape clause renders the statement untestable, and thus frees the theorist from the burden of empirical testing—very convenient for the lover of wild speculation. But it leaves the statistician baffled, for in the Bayesian context there is no room for the concepts of randomness, randomization, and random sample.

Finally, a major problem with most utility theorists is that they overlook adverse empirical evidence, such as that produced by the survey conducted in 1952 by Nobel laureate Maurice Allais (1979). This survey, involving a hundred subjects well-versed in probability, showed that rational people are not bent on maximizing anything except security. Additional empirical studies have exhibited a moral dimension absent from rational choice theory. For example, Kahneman et al. (1986) found that people tend to behave fairly, even at some cost to themselves. Many other economic psychologists, by exploring the variety of human motivations, have refuted the myth that humans are above all utility maximizers. By refusing to face such adverse empirical evidence and sticking instead to nineteenth-century armchair psychology, rational choice theorists behave like true believers, not like rational agents.

4. RATIONALITY

As noted in section 2, the word 'rationality' designates at least a dozen different concepts, only one of which has been singled out by most rational choice theorists, namely that of instrumental or economic rationality, or utility-maximizing behavior. Consequently, one should be able to conceive of at least twelve "rationality postulates" in addition to the Rationality Postulate called A1 in section 2. For example, in Arrow's collective choice theory, the rationality postulate boils down to respecting the transitivity of every preference relation. And, in action theory, an action is called 'rational' just in case it is conceived of and executed in the light of the best available information.

It is instructive to place A1 alongside four alternatives found in the social studies literature. They are the following.

R1 "Agents always act in a manner appropriate to the situation in which they find themselves" (Popper).

R2 Principle of Instrumental (or Functional) Rationality: Agents always adopt the means most likely to produce the desired results.

R3 = A1 Principle of (Economic) Rationality: Agents always act so as to maximize their expected utilities.

R4 Principle of Least Effort: Agents always choose the least expensive means to attain their goals.

R5 Principle of Subjective Rationality: Agents always act on their beliefs about the situation in which they find themselves, as well as on their beliefs about the most suitable means and the possible consequences that their actions may have for themselves and others, and aim for the consequences they judge best.

Postulate R1 is so vague that it might be said to hold for an electron in an external field as well as for a human being. Indeed, both behave according to their circumstances. Moreover, R1 is hardly testable, if only because the very concept of "appropriate behavior", on which it hinges, is fuzzy. Not being rigorously testable, R1 is neither true nor false.

The principle R2 of instrumental rationality is false. Indeed, comparatively few people are intelligent and well-informed enough, as well as sufficiently resourceful and strong, to be able to employ the best available means, much less to do so always. If R2 were true, humans would seldom if ever resort to violence to settle their differences. If R2 were true, there would be no market for shoddy products such as "lemons" (bad cars); there is such a market, because the buyer ignores something that the seller knows. At best R2 could be regarded as a definition or else a norm or ideal of instrumental rationality. More on this matter at the end of this section.

The axiom R3 of economic rationality, i.e., our old acquaintance A1, is false because it implies the weaker principle R2, which is false, as we just saw. However, we shall see below that R3 may be regarded as irrefutable rather than false.

Morgenstern would have objected that R3 is neither true nor false, since it is a norm or rule of behavior rather than a descriptive statement. An obvious rejoinder is that R3 is a poor guide to action because it does not specify whether one should seek to maximize short, medium, or long-term utility. If these three possibilities are conflated, and particularly if short-term self-interest is allowed to prevail, one may end up falling into "social traps". Familiar cases of social traps are overexploitation (in particular labor and colonial exploitation, overgrazing and overcultivation), overindustrialization, overurbanization, overautomobilization, militarism, and war. In conclusion, the maximizing behavior postulate R3 is worse than either untestable or false: its observance is the ultimate cause of the global disasters humankind is facing. Some rationality!

R4, the principle of cost minimization, is false for the same reason that R2 is. Finally, postulate R5 is my own version of Boudon's principle of subjective rationality. It is so weak as to be nearly tautological, immune to refutation.

On balance, three of the principles are false, and two, to wit R1 and R5, are hardly testable. Indeed, suppose someone behaved in a manner that an expert observer, such as a social scientist, would ordinarily regard as irrational, e.g., from habit, on impulse, from compulsion to gamble, under external compulsion, out of commitment to an absurd doctrine, without sufficient information, or in an eccentric way. The agent could argue that, under the circumstances, her action was rational after all, particularly since she knows best what is in her own interest.

Of course, it would be irrational to give up rationality in social studies, adopting instead an irrationalist or Romantic approach. What we need in the social sciences and technologies are (a) a rational (and empirical) approach to human behavior of all kinds, rational and nonrational; (b) a sufficiently clear concept of instrumental rationality (Weber's *Zweckrationalität*); and (c) a normative or prescriptive, rather than descriptive, principle of instrumental rationality. The following are offered for discussion.

Definition of "instrumental rationality": An action is instrumentally rational if, and only if, it is likely to produce the desired outcome.

Note that this definition invokes neither subjective probabilities nor subjective utilities. It just equates instrumental rationality with effectiveness.

Norm of "instrumental rationality": A rational individual engages deliberately in an action M at a certain time if, and only if, (a) M is a means to his goal G, as suggested by the best available information; (b) G has precedence over other goals of his at that time; and (c) the cost of M, added to that of the undesirable effects accompanying G, is smaller than that of any other means known to him, as well as smaller than the value of G.

Obviously, this is a prescription rather than a description, because it is restricted to instrumentally rational individuals. (But it is not restricted to humans. It may well apply to subhumans as well as to humans.) It is also restricted to instrumental rationality: the norm does not touch on alternative concepts of rationality, let alone on overall rationality. In particular, no mention is made of maximizing any utilities.

An individual who makes the (practically) wrong choice of means will be said to behave in an instrumentally, objectively irrational way, not in a subjectively rational way. But the means she chooses, though instrumentally irrational, may happen to be rational in some other ways, e.g., morally. That is, she may have to trade rationality in some regard for irrationality in another. Finally, the genuinely rational agent is supposed to weigh her various options, values, and goals at any given moment, and to refrain from sacrificing everything to the overriding value or goal of the moment. Total commitment to a single value or goal, other than individual and social welfare, is self-defeating in the long run, because life is many-sided. For example, a government's choice to reduce the fiscal deficit at all costs is bound to degrade its social services.

In sum, neither of the various rationality principles found in the literature has a descriptive value. Simpler: Real people do not always act rationally, in any of the senses of 'rational'. However, the principle R2 of instrumental rationality is valuable when reworded in the normative mode, i.e., as a guide to action. Moreover, it is useful even in the original declarative mode, namely as a methodological or heuristic tool. This is in fact how Max Weber understood it. He regarded it as an idealization allowing one to explain certain human actions as outcomes of (instrumentally) rational decisions, and others as departures from rationality originating in irrationalities of all kinds, such as effects and mistakes, not to mention natural and social forces beyond the agent's control. (Mechanical analog: account for any departures from either static equilibrium or rectilinear uniform motion in terms of forces or constraints.)

In short, we uphold conceptual rationality, which is an ingredient of the scientific approach, and the Norm of Instrumental Rationality, which is a guide to efficient action. But we cannot accept any of the so-called rationality principles that occur in rational choice theory. R1 is unacceptable because it is vague and therefore untestable; R2, R3, and R4 must be rejected because they are false; and R5 is unacceptable because it is nearly tautological.

5. INDIVIDUALISM

Rational choice models are ontologically as well as methodologically individualistic. That is, they focus on individuals, and either deny the existence of social bonds and social systems, or assert these to be reducible to individuals and their actions. (Recall essay 21.) Ironically, there are few if any consistent upholders of ontological individualism or social atomism. Even such radical ontological individualists as Weber, Pareto, Simmel, Homans, and Coleman have occasionally admitted the existence of social systems, and have always resorted to such unanalyzed wholes as "the market", "the State", or "the situation"—without realizing the inconsistency.

The ontological individualist views society by analogy with a low density gas, where the individual agents are the analogs of the molecules, and the former's utility functions are the counterparts of the laws of motion. As in the elementary kinetic theory of gases, the individuals are assumed to act independently of one another, being only constrained by the container (or its analog, the market). Incidentally, neither the gas container in the elementary kinetic theory, nor its social analog in rational choice theory, is ever analyzed as a collection of individuals: both are handled as wholes—a strategy that violates methodological individualism.

But social systems, from the family and the firm to the transnational corporation and the government, are more like lumps of condensed matter than like gases. And the theories of condensed matter bear little resemblance to the theory of gases—which is as it should be, for condensation is accompanied by a qualitative change. In these theories, interaction is of the essence: it is what makes a body fluid or solid, just as face-to-face interaction turns an amorphous social group into a social system. This is why few social scientists, unlike social philosophers, practice ontological individualism in a consistent fashion.

While there are few if any consistent ontological individualists, there are plenty of methodological individualists, i.e., students of society who claim that the understanding of social facts requires only the investigation of the beliefs, intentions, and actions of all the individuals concerned. However, methodological individualism, hence rational choice theorizing, comes in at least two strengths: strong and moderate. The strong version holds that a single theory, containing concepts referring exclusively to individuals, suffices to explain all kinds of social behavior and social system. The self-styled "economic imperialists", who claim that neoclassical microeconomics suffices to explain everything social, from marriage to war, belong to this school (Bunge 1998).

On the other hand, moderate individualism holds that the universal premises concerning individuals, though necessary, are insufficient to account for social life. It asserts that different kinds of social fact call for different models sharing those universal premises, and that each model contains hypotheses concerning both the specific kind of interaction and the particular institutional framework. Moderate individualism comes close to systemism. Let us take a closer look at the two varieties.

The strong version of rational choice theorizing, particularly as championed by Gary S. Becker, holds that all social groups—small and large, formal and informal, voluntary and involuntary—must be accounted for by a single overarching theory. This assumption can be seen as a consequence of methodological individualism conjoined with the hypothesis of the uniformity and constancy of human nature. Indeed, if all individuals are only led to act by self-interest, and behave regardless of the way other people behave, then all social groups are essentially the same and, consequently, a single theory should fit them all. Moreover, since self-interest is the key concept of the paragon discipline, namely neoclassical microeconomics, all rational choice models must be just variations of the latter, i.e., all of social science is ultimately reducible to microeconomics. This may still be a minority view, but it has been steadily gaining ground during the 1970s and 1980s.

There are at least three objections to this radical view. The first was mentioned in section 2, and it is summarized in the classical formula "What is said about everything says nothing in particular". (In semitechnical semantic jargon, sense or content is inversely proportional to extension or truth domain.) Put another way, an extremely general theory can only cover the features common to all the members of its reference class, hence it will miss all peculiarities. For example, it will fail to distinguish altruism from crime, churches from business firms, schools from armies, and so on. And, being unable to account for differences among individuals and for the variety of social interactions, systems, processes, and institutions, the theory is bound to be unrealistic, i.e., false.

A second objection to radical individualism is that social facts, and even individual behavior, cannot be accounted for without the help of certain "macroconcepts" that do not seem definable in terms of "microconcepts". (Obviously, a macroconcept is a concept referring to a macroentity, such as a business firm or even the economy as a whole, and a microconcept is one concerning a microentity, such as an individual person.) For one thing, behavior does not happen in a social vacuum but in a social matrix. For another, as Granovetter (1985) has emphasized, "most behavior is closely embedded in networks of interpersonal relations".

A consequence of the embeddedness of the individual in social systems of various kinds is that a genuinely rational decision maker will take his social environment into account. For example, a businessman cannot make rational choices unless he takes into account the global entities or processes referred to by such concepts as occur in the following laundry list: "policy of the firm", "state of the market", "organization", "technology", "product quality", "scarcity", "bottleneck", "economic conjuncture", "development", "business opportunity", "social order", "value added tax", "inflation rate", "prime rate of interest", "labor movement", and "political stability".

None of the concepts we have just listed is about individual persons, and none of them is definable in terms of individual dispositions or activities. All of them transcend biology and psychology. They are irreducibly social because they concern social systems embedded in even larger systems, and every system possesses (emergent) properties that its components lack. For instance, a country may be at war without every one of its inhabitants taking up arms. Another example: capitalism, socialism, and their various combinations are features of entire economies, not of individuals. A third example: parliamentary democracy can only be a property of a political system.

Emergence sets objective limits to microreduction, or the construal of a whole as the collection of its components. It cannot be otherwise if our ideas are to match the real world, which is one of changeable systems. Interestingly enough, this is the case with atomic physics as well as with social science. Indeed, every well-posed problem in quantum theory involves a formulation of the boundary conditions, which represent the environment of the thing in question in a global manner rather than in terms of "microconcepts" representing microentities. (See also essay 11.)

The weak version of rational choice theorizing admits the need for subsidiary hypotheses concerning both the specific type of social group, and (though often in an underhand fashion) the institutional framework or social environment as well. Every rational choice model of this kind may then be regarded as the set of logical consequences of the union of the set G of general premises of "rational" behavior discussed in section 2, and a set of subsidiary or special assumptions describing the specific properties of the system or process in question. The role of G is thus similar to that of the general principles of evolutionary theory, which are necessary but insufficient to account for the differences between, say, bacteria and daffodils.

This weak version of rational choice theorizing is not open to the two objections we raised against the strong version. However, even this moderate version is unrealistic, because G itself is unrealistic. Indeed, people act from custom, loyalty, obligation, or passion, as well as on the strength of cold calculation. Moreover, not even calculated behavior can be perfect, and this for at least two reasons. Firstly, the information required by any realistic calculation is never complete. Secondly, the goal aimed at may not be within the reach of the agent, or it may be morally irrational.

Another unrealistic assumption of both the weak and the strong versions of rational choice theory is that of freedom of choice. Even a cursory examination of any of the major events in anyone's life suffices to refute this view. For example, we cannot choose whether or when to be born or to come of age: whether in an advanced or a backward country, in good times or bad. In hard times we cannot be choosy about jobs or lifestyles; we can seldom choose whether to live in a hovel or in an Alpine cottage; the education we get and the occupation we engage in depend crucially on the socioeconomic status of our family; the citizens of most nations, however democratic, have little if any say in choosing between peace and war; and so on and so forth. No doubt, we ought to promote the cause of freedom (and the concomitant duties), but the idea that we can attain total freedom is at best an illusion, at worst an ideological bait. And if our choices cannot be totally free, then they cannot be fully rational in any sense of this word.

In sum, neither ontological nor methodological individualism works. By ignoring the fact that individual action is embedded in social systems, individualism proposes an "undersocialized" view of humans and society, just as holism proposes an "oversocialized" picture. We should therefore steer clear from both extremes, as well as from the crypto-holism involved in any invocation of "the social constraints" or "the circumstances". The correct alternative to all three is systemism, for it accounts for both individual and system, or agency and structure. Indeed, systemism suggests modeling every system, social or not, as a triple ⟨Composition, Environment, Structure⟩, so that it encompasses the valid features of its rivals. (Recall essay 21.)

6. CONCLUSION

Rational choice theorists have offered rather simple, hence attractive, explanations of hundreds of social facts of all kinds. All these explanations assume individualism, free choice, and maximizing behavior—the core hypotheses of mainstream economic theory and utilitarian ethics theory from Adam Smith and Jeremy Bentham on.

Although rational choice models have become increasingly popular and sophisticated, nearly all of them share the following fatal flaws:

(a) most of them do not define the (subjective) utility functions, which play a central role in them; hence, except for the rare bird, they are pseudomathematical despite their symbolic apparatus;

(b) they contain the concept of subjective probability, which, being subjective, does not belong in science;

(c) they involve the concept of subjective rationality, or rationality in the eye of the agent—an escape clause that serves to save the Rationality Postulate from refutation;

(d) they are static: in particular, they assume the fixity of options and the constancy of preferences (or utilities);

(e) although they are supposedly strictly individualistic, they smuggle in unanalyzed holistic notions such as those of "the situation", "the market", and "the institutional framework"—whence they are "individholist" rather than consistently individualist (Bunge 1996);

(f) they overlook the power of elites, governments, bureaucracies, corporations (in particular oligopolistic and transnational firms), parties, churches, networks of all kinds, lobbies, and other groups and systems, all of which distort preferences and restrict the freedom of choice, and justify Lafontaine's famous statement, "La raison du plus fort est toujours la meilleure";

(g) they willfully ignore the planning that guides every modern social system, from the family and the retail store to the State and the transnational firm, and that constrains individual choice and action;

(h) they only cover one type of social interaction, namely exchange, and even so they often do so via some unanalyzed whole, such as the market;

(i) they omit all the sources of human action other than self-interested calculation; in particular, they overlook habit and coercion, as well as passion and compassion;

(j) focusing on choices and means, they underplay motivations, goals, and side effects;

(k) they do not solve the problem of the micro-macro relations and, in particular, they do not account for the agency-structure relations or for the emergence and breakdown of social systems;

(l) because of their enormous generality, they have almost no predictive power;

(m) for the same reason, they are unrealistic: they are mostly jeux d'esprit, even though some of them have been sold to big business or to the military establishment;

(n) because they are unrealistic, at times surrealistic, they are not suitable policy tools.

In sum, we have found rational choice theories inadequate. We have found them wanting not for being "trapped within reason"—the irrationalist reproach—but for being trapped within the individualist and utilitarian dogmas, for involving fuzzy basic notions and untestable key assumptions, for idealizing the free market, and for failing to match reality.

This is not to say that all rational choice theories are utterly useless. In fact, they can perform three services: (a) they may help hone the preliminary formulation of some problems, by forcing the student to distinguish the options, draw decision trees, and take costs and benefits into account; (b) some of them show that the pursuit of self-interest may lead to "social traps", i.e., collective disasters; (c) others may suggest experiments aiming at testing those very models, such as the experiments which have been conducted to check game-theoretic models; and (d) in failing to represent reality they may suggest the search for deeper and more complex theories—much as the failure of the flat earth hypothesis stimulated the elaboration of the round earth one. In sum, rational choice theories, if not taken too seriously, may serve heuristic functions. However, too many rational choice models serve only as intellectual gymnastics, for academic promotion, or to justify aggressive military strategies.

Why, then, does rational choice theory enjoy such enormous prestige? For the following reasons: (a) it is basically simple; (b) it claims to explain all human behavior; (c) it makes liberal use of symbols that intimidate the nonmathematical reader, although they only occasionally designate well-defined mathematical concepts; (d) it makes predictions—alas, not often correct ones; (e) it endows market worship with intellectual respectability; (f) it is often criticized for the wrong reasons; and (g) so far it has no serious rivals.

To sum up, rational choice theory is not a solid substantive theory of society. This failure can be traced back to the specific assumptions of the theory, not to the rational approach to social studies. Any serious study of society will be rational, but no realistic account of social facts will assume that people always act rationally, in particular selfishly.

The reader may feel that my criticism is excessive: that I am throwing the baby out along with the bath water. My reaction is that there is no baby. In fact, rational choice theory was born in 1871 in the field of economics, so it is anything but young. Worse, it died long ago from mathematical anemia, from deficiency in the enzymes required to digest the simplest social facts, and from lack of exposure to the sound and light and fury of the social weather.

23

REALISM AND ANTIREALISM IN SOCIAL SCIENCE

(1993)

■

1. INTRODUCTION

Until recently nearly all social scientists were realists: they took the reality of the external world for granted, and assumed tacitly that their task was to describe and perhaps also explain the social world as objectively as possible. They were innocent of subjectivism, conventionalism, fictionism, constructivism, relativism, and hermeneutics: these were either ignored or regarded as inconsequential philosophical games.

True, in their methodological writings the neo-Kantians Georg Simmel and Max Weber criticized realism. This was the respectable thing to do in the German academic establishment at the time, which was dominated by idealism, particularly neo-Kantianism. However, Simmel and Weber conducted their substantive sociological and historical research as realists. In particular, they did not invent the present, or even the past, but practiced Ranke's famous 1824 dictum, that the task of the historian is to show *wie es eigentlich gewesen ist,* i.e., what actually happened.

What is true is that Simmel and Weber were not social naturalists or, as one should say with hindsight, behaviorists. Indeed, they held that one must take the inner life of people into account, in particular their beliefs, motivations, and intentions. They and their followers did say that this committed them to subjectivism. But this was a mistake on their part, since in principle one can be objective about other people's subjective experiences, provided one employs reliable objective indicators, and as long as every imputation of belief, motivation, or intention be treated as a conjecture to be subjected to empirical tests.

Likewise, the neoclassical economists and their imitators in sociology, i.e., the self-styled "economic imperialists", often postulate arbitrary subjective utilities and probabilities. (Recall essay 22.) But they do not claim that social facts are their own constructions, or even that they are social conventions. Far from it: they claim that their theories are true, i.e., match preexisting facts. Even the arch-apriorists Ludwig von Mises and his disciple Friedrich von Hayek made this claim. Whether this claim is correct is beside the point right now. The point is that even the most far-fetched neoclassical economic models are offered as faithful representations of economic realities.

This situation started to change in the 1960s, when antirealism went on the rampage in the social studies community as well as in Anglo-American philosophy. This movement seems to have had two sources, one philosophical, the other political. The former was a reaction against positivism, which was (mistakenly but conveniently) presented as objectivist simply because it shunned mental states. This reaction was not progressive but regressive, for, instead of pointing out and overcoming the serious limitations of positivism, it denounced its concern for clarity and empirical test. It proposed to replace positivism with all manner of old enemies of science, such as conventionalism, fictionism, constructivism, intuitionism, Hegelianism, phenomenology, hermeneutics, and even existentialism, sometimes spiced with a dash of Marxism.

Mercifully not all those who profess any of these philosophies are consistent. For example, much of the field research conducted by the ethnomethodologists, who like to quote Husserl and Heidegger, falls squarely in the pedestrian tradition of raw empiricism and, in particular, behaviorism—a clear case of speaking with a forked tongue.

I submit that the political source of contemporary antirealism was the rebellion of the Vietnam War generation against the "Establishment". The latter was (wrongly) identified with the power behind science and proscientific philosophy. So, fighting science and proscientific philosophy was taken to be part of the fight against the "Establishment". But, of course, the people who took this stand were shooting themselves in the foot, or rather in the head, for any successful political action, whether from below or from above, must assume that the adversary is real and can be known. Indeed, if the world were a figment of our imagination, we would people it only with friends.

In this paper we shall examine the main kinds of realism (or objectivism) and subjectivism that are being discussed in contemporary social metascience. We shall be using the following definitions. An account of a fact (or group of facts) is *objective* if, and only if, (a) it makes no reference to the observer, and (b) it is reasonably true (or true to a sufficient approximation); otherwise it is *subjective*. Condition (a) is necessary, because one might truly describe one's own reaction (e.g., feeling) to a fact: this, though true, would be a subjective estimate of the situation. And condition (b) is necessary, because one might give a nonsubjective but totally imaginary account. Our second definition is this. An account of a fact (or group of facts) is *intersubjective* in a given community if, and only if, all the members of it agree on it. Intersubjectivity may be perceptual, conceptual, or both.

The two concepts are logically independent from one another. While the concept of objectivity is epistemological and semantical, that of intersubjectivity is psychological and sociological. However, within a scientific community intersubjectivity is an indicator of objectivity. Like most indicators, it is fallible. For example, all the members of a given group of people might perceive or conceptualize a given fact in the same wrong manner. Thus, during two decades almost everyone in the Soviet Union seem to have believed Stalin to be a genius, a saint, and a hero all in one. Consensus is not a test of truth: it is only a desirable, albeit temporary, outcome of a process of research and teaching.

Objectivism or *realism* is a philosophical doctrine. It is the view that, except in the arts, we should strive to eliminate all subjective elements from our views about reality. In other words, objectivism prescribes that we should strive not to include our own feelings and desires in our pictures of the external world. Objectivism does not entail the rejection of subjectivity: it only enjoins us to study it objectively, the way experimental psychologists do. In particular, a scientific theory may refer to subjective experiences, but it must not do so in a subjective manner.

2. SUBJECTIVISM

A statement is objective or impersonal if it purports to describe, explain, or predict one or more facts occurring in the external world—which, of course, includes the brains of others. It is subjective or personal if it is about one's own feelings or beliefs. For example, the proposition "The USSR broke up in 1990" is an objective statement, whereas "The breakup of the USSR makes me happy (or sad)" is a subjective one.

Social scientists deal with both objective facts and their "perceptions", so they have to reckon with subjective as well as with objective events and statements. But, to the extent that they abide by the canons of science, even their assertions about other people's subjective statements will be objective. For example, they may wish to check whether the statement "Most Soviet citizens were initially happy when they learned about the breakup of their country" is objectively true or false.

Subjectivism is the philosophical view that the world, far from existing on its own, is a creation of the knowing subject. Subjectivism is an easy explanation of differences of opinion and it spares one the trouble of putting one's beliefs to the test. Thus, Breit (1984, p. 20) asks why John K. Galbraith and Milton Friedman, two of the most distinguished social scientists of our time, could have arrived at conflicting views of economic reality. He answers: "There *is* no world out there which we can unambiguously compare with Friedman's and Galbraith's versions. Galbraith and Friedman did not discover the worlds they analyze; they decreed them". He then compares economists to painters: "Each offers a new way of seeing, of organizing experience", of "imposing order on sensory data". In this perspective the problems of objective truth and of the difference between science and nonscience do not arise. On the other hand, we are left wondering why on earth anyone should hire economists rather than painters to cope with economic issues.

Caution: Hayek (1955) and others have confused subjectivism with the recognition of the existence and importance of feelings, opinions, and interests. A realist (or objectivist) should be willing, nay eager, to admit the relevance of feeling, belief, and interest to social action, but she will insist that they be studied objectively. This attitude is to be contrasted to both subjectivism and the positivist (or behaviorist, or black box) approach to human behavior.

A clear modern example of subjectivism is phenomenology. In fact, according to its founder, the gist of phenomenology is that it is a "pure egology", a "science of the concrete transcendental subjectivity" (Husserl 1931, p. 68). As such, it is "the opposite of the sciences as they have been conceived up until now, i.e., as 'objective' sciences" (p. 68). The very first move of the phenomenologist is the phenomenological reduction or "bracketing out" (*époché*) of the external world. He does this because his "universal task" is the discovery of himself as transcendental (nonempirical) ego (p. 183). Once he has pretended that real things do not exist, the phenomenologist proceeds to uncover its essences, making use of a special "vision of essences" (*Wesensschau*).

The result is an a priori and intuitive science (section 34). And this science proves to be nothing but transcendental idealism (p. 118). This subjectivism is not only epistemological, but also ontological: "the world itself is an infinite idea" (p. 97).

This wild egocentric fantasy can only have either of two negative effects on social studies. One is to focus on individuals and deny the real existence of social systems and macrosocial facts: these would be the products of such mental procedures as interpretation and aggregation. The second possible negative effect is to turn students away from empirical research, thus turning the clock back to the times of armchair ("humanistic") social studies. The effect of the former move is that *social* science is impossible; that of the second is that social *science* is impossible. Either or both of these effects are apparent in the work of the phenomenological sociologists (e.g., Schütz 1967; Cicourel 1974) and the ethnomethodologists (e.g., Garfinkel 1967; Geertz 1973).

Traditional subjectivism was individualist: it regarded the knowing subject as an individual in a social vacuum. (Examples: Berkeley, Schopenhauer, and Husserl.) Consequently, it could not face the objection that there should be as many worlds as people, and thus no intersubjective agreement would be possible—unless it added the theological assumption that God takes care of the uniqueness of the world.

This objection would not worry the sociologist Luhmann (1990), according to whom there are as many realities as observers, for every one of those is "a construction of an observer for other observers". Consequently, there is no objective truth. Worse, the individual only relates to his own constructs. He cannot communicate: "only communication can communicate". (Remember Heidegger's "Language speaks", "The world worlds", and similar nonsense.) Obviously, subjectivism is not conducive to social science.

3. CONVENTIONALISM

Conventionalism is the view that scientific hypotheses and theories are useful conventions rather than more or less true (or false) representations of facts. Conventionalism rings true in pure mathematics, which is a free mental creation. But even here one distinguishes definitions on the one hand from postulates and theorems on the other. Only the definitions are strictly conventional. Everything else is either basic assumption or logical consequence from basic assumptions and definitions. Moreover, since definitions, though handy, are in principle dispensable, the conventional ingredient of mathematics is negligible. (This only holds for definitions proper, such as "1 equals the successor of 0". It does not hold for the so-called axiomatic definitions, which are creative and therefore nonconventional. See also essay 13.)

Radical conventionalism is obviously impracticable with reference to factual science, because here we need empirical data, which are anything but conventions. But one might try a moderate version of conventionalism, according to which the choice among rival theories which account equally well for the data is ultimately a matter of convention. (As we shall see in section 9, this was Cardinal Bellarmino's thesis against Galileo.) Let us look into this possibility.

Any given body of empirical data can be "covered" by an unlimited number of different hypotheses: this is known as the problem of empirical underdeterminacy. A common case is this: the points on a plane (or a higher-dimensional space) representing experimental results can be joined by any number of continuous curves (or surfaces). It would seem that the choice among such competing hypotheses is arbitrary, e.g., we may choose the simplest of them. This leads to equating truth with simplicity, or at least to regarding simplicity as a test of truth (Goodman 1958).

Is this how scientists actually proceed? Let us see. A scientist is likely to prefer the simplest of all the hypotheses compatible with a certain body of data, as long as she knows nothing else about the matter in hand. But if she pursues her research, she will want to explain the data in question. To this end she will cast for and, if need be, invent more encompassing or deeper hypotheses or even theories. She will then check not only whether any of them matches the old data, but also whether they predict any new ones. In the end she will prefer the hypothesis or theory with the largest coverage or the greatest explanatory or predictive power, even if it does not exactly fit the original data. And she will expect that further research may result in an even more powerful hypothesis or theory, likely to be more complex and to refer to even deeper mechanisms. Her goals are truth and depth, not simplicity: she is a realist, not a conventionalist. For this reason, she regards (approximate) empirical fit as only an indicator of factual truth: she also requires compatibility with a comprehensive and if possible deep theory (Bunge 1983b). In sum, conventionalism is false with reference to factual science.

Fictionism combines conventionalism with pragmatism. According to it, scientific hypotheses and theories are at best useful fictions. It is not that things are so and so, but that they look *as if* they were so and so, and some of our ideas work *as if* they were true (Vaihinger 1920). This doctrine has a small grain of truth, for, in fact, every factual theory includes fictions in the form of idealizations or simplifications (Weber's ideal types). Still, these are not always mere fictions like those of fantastic literature or surrealist art. So much so that they are only accepted when approximately true. When not, they are improved or rejected.

Fictionism has survived in two doctrines which are quite familiar to social scientists. One is the view that scientific theories are metaphors rather than literal representations of real things (Hesse 1966; Ricoeur 1975). But of course, metaphors may be heuristically fertile or barren, but they cannot be true or false, hence, unlike scientific theories, they are not subject to tests. (See essay 12.) The other fictionist view is the methodology of economics defended by Milton Friedman (1953), according to which the assumptions of a theory need not be true: all that would matter is that their consequences be realistic. But of course, one may validly infer true propositions from the wildest assumptions.

As can be seen from the following examples, fictionism is at variance with modern science. Electrons behave in certain circumstances *as if* they were particles—but they are not. "Chaotic" systems look *as if* they were random—but they are not. DNA molecules work *as if* they contained instructions for the synthesis of proteins—but they don't. Spiders and computers behave *as if* they were intelligent—but they are not. Social systems look *as if* they were alive—but they are not. Firms look *as if* they had a mind and a purpose of their own—but they don't. In all of these cases a shallow analogy was proposed, checked, and rejected. In all of these cases appearances were found to be misleading, and fictionism false.

Traditional conventionalists and fictionists were methodological individualists: they held that conventions and fictions are proposed by the individual scientist and then adopted or rejected by their fellow scientists. In recent times a kind of collectivist conventionalism has spread in the sociology of science, combined with subjectivism, fictionism, relativism, and hermeneutics. Let us take a look at it.

4. SOCIAL CONSTRUCTIVISM

Social constructivism is a blend of antirealism and collectivism (or holism). In fact, it claims that all social facts, and possibly all natural facts as well, are constructions of "thought collectives", such as a scientific community. Moreover, different "thought collectives" would hold different and possibly mutually "incommensurable" views of the world. Thus, constructivism, whether collectivist or individualist, is relativist: it denies universal truths such as "2 + 3 = 5" and "The earth is round" (Fleck 1935; Berger and Luckman 1966; Bloor 1976; Latour and Woolgar 1979; Barnes 1977, 1983; Knorr-Cetina and Mulkay 1983).

Constructivists systematically confuse reality with our representations of it: the explored with the explorer, facts with data, objective laws with law statements, assumptions with conventions. This is certainly not the way scientists proceed. Thus, while not in a philosophical mood, an anthropologist is likely to claim that the concepts of a human being occurring in the various anthropological views are theoretical, while at the same time admitting that there are real people out there, whether or not we observe them or theorize about them—and that there were people before the birth of anthropology. Likewise, a sociologist will admit that the concepts of social stratification are theoretical, while at the same time holding that modern societies are objectively stratified, and that every scientific study of social stratification attempts to represent it as truthfully as possible.

In short, all but the radical empiricists agree that constructs (concepts, hypotheses, and theories) are constructed; and only subjectivists claim that all facts are constructed as well. Thus, while epistemological constructivism is in order up to a point, ontological constructivism is not, for it flies in the face of evidence. Birth and death, health and sickness are not constructions, social or otherwise. Nor are child rearing, work, trade, war, or any other social facts. Facts are facts are facts, even when produced in the light of ideas.

The social constructivists deny that scientific knowledge is different from ordinary knowledge—they must, in order to write about it in good conscience. Consequently, they think they can study a tribe of scientists as if it were an ordinary social system, such as a tribe of hunters and gatherers, or a street-corner gang. They deny that the former has an extremely specialized function, that of producing scientific knowledge through processes that, unlike gathering, hunting, or fist fighting, are not in full view. Consequently, a layman visiting a laboratory can only observe some behavioral manifestations of the mental processes locked in the brains of the researchers and their assistants. To the layman the problems that trigger the research activity are even less intelligible than its results. Hence, he is bound to take only a superficial look, much as the behaviorist psychologist limits his task to describing overt behavior.

How would relativist-constructivists, such as Latour and Woolgar (1979), know that scientific activity is "Just one social arena", and a laboratory just "a system of literary inscriptions", if they lack the scientific background required to understand what scientists are up to? And, given their deliberate confusion between facts and propositions, how would they know when "a statement splits into an entity and a statement about an entity", or when the converse process occurs, during which reality is "deconstructed"—in ordinary parlance, a hypothesis is refuted? It is on the strength of such elementary confusions and borrowings from anti-scientific philosophies that they conclude that the world does not exist independently of the knowing subject. (For further criticisms of constructivism see Bunge 1991a.)

5. RELATIVISM

Epistemological relativism—not to be confused with physical relativity (for which see essay 16)—is the view that truth, like beauty, is in the eyes of the beholder. In other words, truth would be relative to the subject, the social group, or the historical period: there would be no objective universal (cross-cultural) truths.

Relativism is an offshoot of subjectivism. Indeed, if reality is a construction, and facts are statements of a certain kind, there can be no objective universal truths. In other words, if there is nothing "out there", the very expression 'correspondence of ideas with facts' makes no sense. And if there is no objective truth, then scientific research is not a quest for truth. Or, to put it in a somewhat milder way, "what counts as truth can vary from place to place and from time to time" (Collins 1983, p. 88). This is the kernel of epistemological relativism, which is, in turn, part and parcel of radical philosophical skepticism and anthropological relativism.

Relativists argue from the multiplicity of simultaneous or successive rival theories about one and the same domain of facts. But this multiplicity only goes to show that scientific research does not guarantee *instant, complete*, and *definitive* truth. Still, as uncounted observational and experimental tests show, we often do hit on *partially true* hypotheses. And, as the history of science shows, if a hypothesis is interesting and sufficiently true, it will stimulate further research that may result in truer or deeper hypotheses. What holds for hypotheses and theories holds also, mutatis mutandis, for experimental designs. There is scientific progress in certain periods, after all, because there is such a thing as objective (though usually only partial) truth.

As for the suspicion that, if a scientific project has been motivated or distorted by material or ideological interests, it cannot yield objectively true results, it is an instance of what philosophers have called the *genetic fallacy.* It consists in judging a piece of knowledge by its birth (or baptism) certificate. (The *argumentum ad hominem* is a special case of the genetic fallacy.) A hypothesis, datum, or method may be correct (true in the case of a proposition) regardless of the motivation of the research that produced it. Or it may be false even if produced with the purest of intentions. In short, the correctness of an idea is independent of its origin and utilization, and it must be established by strictly objective means. The same holds for the content of an idea. For instance, Durkheim held that all logical ideas, in particular that of class inclusion, have a social (in particular religious) *origin,* but he did not claim that they have a social (in particular religious) *content.*

Although the constructivist-relativists say that they have no use for the concept of truth, they cannot ignore the fact that everyone makes mistakes. Only, they do not *define* the concept of a mistake or error in terms of departure from truth, as it is done in the theory of errors of observation and in epistemology: they simply leave it undefined. Moreover, some of them would seem to value error more than truth. For example, Latour (1983, pp. 164–65) assures us that scientists "can make as many mistakes as they wish or simply more mistakes than any others 'outside' who cannot master the changes of scale. Each mistake is in turn archived, saved, recorded, and made easily readable again [. . .]. When you sum up a series of mistakes, you are stronger than anyone who has been allowed fewer mistakes than you". Thus the laboratory "is a technological device to gain strength by multiplying mistakes" (p. 165). So, instead of advancing by trial and error, scientists would hoard errors. The reader is not told whether the relativist-constructivist attempts to mimic this strange behavior.

Scientific controversy is alleged to be in the same boat. According to the relativist, all scientific controversies are conceptually unending, because there is no objective truth. Hence, *"even in the purest of sciences,* if debate is to end, it must be brought to a close by some means not usually thought of as strictly *scientific"* (Collins 1983, p. 99, emphases in the original). In other words, there would be no crucial observations or experiments, no new predictions, no logical or mathematical proofs, no decisive counterexamples, no tests of (internal or external) consistency, and so on. There is only either the arbitrary choice of the "core set" or mafia in power, or a negotiation and final compromise between the rival factions. "Politicking" would be the name of the scientific game. If the reader suspects that the relativist-constructivists mistake science for politics, he or she is right. In fact, Latour and Woolgar (1979, p. 237) wrote that "there is little to be gained by maintaining the distinction between the 'politics' of science and its 'truth' ".

Epistemological relativism must not be mistaken for methodological skepticism or fallibilism. According to the latter, all *propositions de fait* are *in principle* fallible—but also corrigible. The scientific researcher doubts only where there is some (logical or empirical) *reason* to doubt, and she never doubts *everything* at *once,* but weighs what is doubtful in the light of the bulk of his background knowledge. And she does not doubt at all some of the very philosophical principles that the new sociology of science rejects, among them those of the independent existence of the external world and its objective intelligibility, for without them scientific research would be pointless. Shorter: most of the truths about the world are likely to be only *partial,* but they are *truths* nonetheless, not just fables. (See Bunge 1991b.)

Moreover, scientific truths are supposed to be cross-cultural. If a view holds only for the members of some social group, then it is aesthetic or ideological, not scientific. Even when an idea originates within a special group, it must be *universalizable* to count as scientific. Unless this scientificity criterion is accepted (along with others), it proves impossible to distinguish science from ideology, pseudoscience, or antiscience—which is of course one of the claims of the constructivist-relativists. (For further criticisms of relativism see Archer 1987; Siegel 1987; Livingston 1988; Boudon 1990a, b; Bunge 1992, 1996, 1998.)

6. HERMENEUTICS

Philosophical hermeneutics, or textualism, consists of a pair of striking if grotesque theses, one concerning the nature of the world, and the other that of our knowledge of it. The ontological thesis is that the world, and in particular society, is a text. To paraphrase Berkeley, to *be is to be an inscriber or an inscription*. Let us examine the epistemological concomitant of this extravagant view.

If the world is the longest speech or the longest book, it follows that, if one wishes to understand it, all one has to do is to listen or read, and interpret. In particular, one must "interpret" human action (i.e., guess its purpose) and treat it as a discourse, subjecting it to hermeneutic or semiotic analysis. Dilthey, the founder of philosophical hermeneutics, restricted the range of the latter to the human sciences. Gadamer, his follower and Heidegger's, has claimed that hermeneutics is valid even for natural science. So do certain self-proclaimed sociologists of science.

The central thesis of the hermeneutic philosophy of society and social studies is that everything social is purposive as well as symbolic or textual, and moreover it must be "interpreted" in terms of intentions rather than explained in terms of causes. This thesis conveniently overlooks all the environmental, biological, and economic factors—in short, it ignores everything material, or attempts to reduce it to a bunch of conventional symbols.

No doubt, human behavior cannot be accounted for without taking beliefs, intentions, valuations, decisions, choices, and the like into account. Nor can we ignore conventions, symbols, and communication. In short, social naturalism (or behaviorism) is insufficient in social studies—which should come as no surprise, since human beings are social and largely artifactual. But this does not entail that social facts are like texts and that, as a consequence, their study is a hermeneutic activity.

Purpose, symbol, and all that can and should be studied objectively. For example, when "interpreting" a certain action as motivated by fear or bravery, greed or generosity, we form a hypothesis. And, if we proceed scientifically, we shall form an empirically testable hypothesis and shall put it to the test, or at least shall hope that someone else will do so. By proceeding in this way, i.e., by employing the scientific method, we may expect to find some truths about social facts. This is why mainstream anthropology and sociology can boast about a number of findings, whereas hermeneutic social studies have produced nothing but programmatic utterances, trivialities, and falsities.

Like Pythagoreanism, Cabbalism, and psychoanalysis, philosophical hermeneutics views all things as symbols of others. (A character in Umberto Eco's novel *Foucault's Pendulum* declares that the penis is a phallic symbol.) It is a regression to magical thinking. On the other hand, the distinction between symbol and denotatum, as well as that between fact and fiction, are basic features of rational thinking. Thus, the sentence 'The cat is on the mat' bears no resemblance to the fact it describes, or even to the sound made when uttering it. Hence, an intralinguistic, e.g., syntactic or stylistic, analysis of the sentence will not reveal what it stands for. The conflation of fact and symbol is wrongheaded and may indicate insanity.

Neither people nor social systems, any more than atoms and plants, have syntactic, semantic, or phonological properties. Not even our ideas about things can be identified with their linguistic wrappings, if only because these differ from one language to the next. In particular, theories have logical, mathematical, and semantic properties, not linguistic or literary ones. This is why scientific theories are created and studied by scientists, logicians, and scientific philosophers, not by semioticians or linguistic philosophers, let alone by literary critics. Therefore, hermeneutic philosophy has nothing to teach social scientists. (For further criticisms see Albert 1988; Ferrater Mora 1990.)

7. PROBABILITY: OBJECTIVE AND SUBJECTIVE

The mathematical concept of probability has been the subject of philosophical controversy for over two centuries (see du Pasquier 1926; Bunge 1981b; essay 15). This controversy has largely been a conflict between realism and subjectivism. The modern mathematical concept of probability, introduced by Kolmogoroff (1933), is neutral in these controversies because, in the calculus of probability, probabilities are assigned to sets of nondescript elements. Philosophical doubts arise as soon as the probability concept is used to reason about facts in the real world, i.e., as soon as the arguments of the probability functions are interpreted as representing something dwelling outside mathematics.

The main controversy concerns the question of whether probability is a measure of objective chance, or else a measure of our uncertainty or ignorance concerning the real facts. For example, if we flip a fair coin, it has equal probabilities, namely 1/2, of landing head or tail. According to the objectivist school, these probabilities are objective and, once the coin has landed, the probabilities have vanished, and this whether or not we look at the coin. Or, if preferred, when the coin lands, one of the probabilities expands to 1 while the other contracts to 0. Hence, to the objectivist it makes no sense to ask of the blindfolded person, after the coin has landed, what is the probability that she will see the coin showing its head. Beliefs cannot be assigned probabilities—except in the context of a stochastic psychological theory of belief.

By contrast, the subjectivist (or personalist, or Bayesian) believes that probability is a state of mind. Hence, he will assign a probability not only to a random event, but also to his belief in the outcome of the event. For example, he will say that the probability that the blindfolded spectator finds that the coin has landed head up is 1/2. But he has no way of testing this assertion. To take another example, suppose a woman is known to have two children, one of them a boy. Obviously, the other child is either a boy or a girl. Not knowing the sex of the second child, the subjectivist will say that the probability of it being a boy is 1/2. The objectivist, in contrast, will refuse to assign a probability to the belief in question. He will argue that such an assignment makes sense only during the very short period of egg fertilization, for this is indeed a process of random gene (or rather sperm) shuffling, where each sex has nearly the same chance. Moreover, he will accuse the subjectivist of confusing the probability of an event with the degree of certainty of his belief in the occurrence of the event—a case of mistaking embryology for psychology.

What practical difference can it make which of the two interpretations of probability is chosen? A great deal. First, since Bayesians have no use for the concept of randomness, they cannot give adequate definitions of randomization and random sample. Second, objectivism promotes the scientific investigation of the objective random mechanisms in question, while subjectivism fosters idle speculation. Indeed, subjective estimates are no serious substitutes for measurement or theoretical calculation, however approximate these may be. Third, the subjectivist interpretation is rife with paradoxes. Let us examine one of them.

Consider again the above example of the woman who admits to having two children, one of them a boy, but lets the subjectivist guess the sex of the second child. At first sight the probability of this child being a boy too is 1/2. But, if he is inquisitive, the subjectivist will go further. He may reason that, if there are two boys, and they are not twins, one is the older: call him B_1, and B_2 his kid brother. Now there are not two but three equally likely beliefs: B_1 and a girl, B_2 and a girl, and B_1 and B_2. Only one of these beliefs can be true, and the "probability" of each of them being true is 1/3, not 1/2 (Gardner 1992, p. 131). The moral I draw from this story is that, whatever credences (belief strengths) may be, they do not behave like probabilities. It is up to cognitive psychologists to find out which, if any, their metric is.

Let us next examine, in the light of the foregoing, the way probabilities are interpreted in an important school of social studies, namely rational choice theory. A central concept of this theory is the notion of expected utility, or sum of the products of the utilities of the possible outcomes of an action by their respective probabilities. In the vast majority of cases both the utilities and the probabilities are taken to be subjective, for what is at stake is the manner in which the agent "perceives" his options. Consequently, different persons are likely to assign different expected utilities to one and the same action. And neither of them can be said to be objectively more correct than the others. It is all a matter of opinion, not science. Only the use of mathematical symbols gives it a scientific appearance. (Recall essay 22.)

In the hard sciences probabilities are treated as objective properties and they are only introduced when there is reason to believe that a chance process, such as random shuffling, is at work. No probability without objective randomness and without the possibility of objective measurement. Yet Milton Friedman (1976, p. 84) assures us confidently that "individuals act as if they assigned personal probabilities to all possible events". The fictionist *as if* trick renders the statement untestable and thus frees the theorist from the burden of empirical testing—very convenient for the lover of free speculation. But it leaves the statistician baffled, for in the Bayesian context there is no room for the concepts of randomness, randomization, and random sample.

The adoption of subjective probabilities can have disastrous practical consequences, e.g., in the case of risk evaluation. A recent dramatic case was the explosion of the *Challenger* space shuttle in 1986. When asked to investigate the causes of the disaster, Nobel laureate Richard P. Feynman (1989, pp. 179–80) discussed the matter with some of the engineers and managers involved in it. A competent and forthright expert told him that five out of 127 rockets that he had looked into had failed—a rate of about 4 percent. Assuming that a manned flight would be safer than an unmanned one, he estimated that there was about a 1 percent chance of failure. But the NASA managers did not believe in estimating probabilities on the strength of relative frequencies. They insisted that the probability of failure was only one in 100,000, i.e., one-thousandth the figure estimated by the engineer. How did they come by that fantastic figure? They did not and could not say. Obviously, theirs was a subjective or personal probability of the kind occurring in most rational choice models.

A less dramatic case is that of the risks faced by insurance companies. An ordinary insurance company will calculate its insurance premiums on the basis of actuarial tables for life expectancy, fire, hail, or automobile accident. It will not insure against any risk unless it can avail itself of such tables, which give objective probabilities estimated on the strength of the corresponding relative frequencies. Only the famous Lloyd's of London was willing to issue insurance policies against such comparatively rare events as the theft of a van Gogh, the shipwreck of a jumbo tanker, or an earthquake, acting on the hope that no string of calamities of this sort would happen in a single year. But 1990 happened to be such an unlucky year, and as a consequence Lloyd's came to the brink of bankruptcy. The moral is clear: acting on subjective probabilities amounts to gambling, and gambling is a mad bad sad business.

In short, in any rigorous scientific discourse probabilities (a) are part of a model concerning some random (or stochastic) thing or process, (b) they represent objective properties of the things in question, and (c) they are supposed to be objectively measurable, though not necessarily in a direct manner. Not so in most rational choice models. In conclusion, the expected utilities occurring in the vast majority of rational choice models are neither mathematically well-defined nor objectively measurable. Subjective probabilities are in the same boat with subjective values.

Note that I am not suggesting that the outcomes of our actions are determinate rather than probable. To be sure chance is for real, not just a synonym of ignorance. Thus, in many cases, and in all choice situations, we are confronted with real (not just conceptual) possibilities, and it is often in our power to actualize or frustrate some of them. But the point is that we rarely have a clue as to the precise values of the corresponding probabilities, if only because there are few reliable (probabilistic or other) mathematical models of human action. On the other hand, in the "hard" sciences, probabilities (or probability amplitudes, or probability densities) occur in exact theories, such as quantum mechanics, where they are related to other magnitudes, some of which are measurable either directly or via indicators. (Thus, in physics one may measure probabilities indirectly through such variables as energy, temperature, or light intensity.)

Decision theory and its applications have been constructed having games of chance in mind. (This is ironic because, in the case of games of chance, utilities and probabilities are objective and knowable, hence they do not have to be guessed, much less be made up.) Now, even though life is chock-full of chance events, it is not a lottery. It is not only that our life stories are constantly being interfered with by other people's, and that we can make some events happen at will, though usually with the unforeseen side effects. Nor is it only that we usually ignore the odds and utilities of the possible outcomes of our actions.

A major point about the uncertainty of life is that in most cases we do not even know in advance the full set of possible outcomes of our actions or inactions—this being why we encounter surprises at every turn. Hence, even if every foreseeable branch in a decision tree could be assigned a probability on some reasonable grounds, the sum of the probabilities for the various known branches originating in a node could be short of unity, for the simple reason that we do not know all the branches. To put it another way, no decision tree can include all the possible outcomes of a real action. In compensation, when disaster threatens we can often react to prevent it from happening—something we are not allowed to do once the dice have been rolled. Because, in principle, we can either nip in the bud or alter in midcourse almost any course of deliberate action, and because we ignore so many factors, decision theory and its kin, modeled as they are on games of chance, are not good guides to rational action. Rational people are not gamblers: they attempt to master chance or even avoid it instead of throwing themselves at its mercy.

Lest the foregoing be construed as despondency over human ignorance, I hasten to submit that a frank recognition of our present ignorance should only (a) put us on guard against simplistic models that take the possibility of complete knowledge for granted, and (b) spur us to engage in further research. This attitude is then very different from irrationalism as well as from Hayek's skepticism over the possibility of ever building genuine sciences of society.

Furthermore I am far from suggesting that social scientists should ignore subjective phenomena, such as beliefs, uncertainties, expectations, and intentions. We must try to find them out and examine them critically. But subjectivity has got to be studied scientifically, e.g., by means of reliable objective (physiological or behavioral) indicators. The arbitrary assignment of probabilities to states of mind, or to the possible outcomes of intended actions, is not a scientific procedure, precisely for being arbitrary. 'Subjective probability' is just a fancy, scientifically sounding name for strength of belief or credence.

There can be no reasonable objection to studying "subjective probabilities" in an objective or scientific manner in the context of cognitive psychology. But it is likely that, if such study is performed, "personal probabilities" will prove not to satisfy the laws of the calculus of probability. For example, the credences (subjective "probabilities") of mutually exclusive alternatives need not add up to unity. If they don't, then the standard probability calculus won't apply to them. And, as long as an alternative (psychological, not mathematical) calculus is not introduced and justified, no precise calculation of credences is possible.

8. OBJECTIVE STUDY OF SUBJECTIVITY

The force of belief is such that, if a person believes that fact X is real, she will behave as if X were real even if X is actually nothing but a figment of her imagination. (See Merton 1957, pp. 421ff.) Since subjectivity is an important fact of human life, the realist should favor its (objective) study. In fact, social scientists are interested in studying the way subjective factors, such as beliefs, valuations, and attitudes, influence objective ones such as actions and, in turn, how the actions of other people influence our subjective experiences. In other words, scientists are interested not only in objective situations, but also in the way people "perceive" them. To see how subjective and objective factors can be combined, let us examine the problem of objective versus "perceived" justice.

Whenever benefits and burdens can be quantitated, we may define the degree of *objective* justice done an individual over a given time span as the ratio of his benefits (b) to his burdens (d) during that period, i.e., $J = b/d$. Perfect justice or equity is represented by the straight line at 45° on the d-b plane. Injustice or inequity is represented by the region below this line (underprivilege) and the one above (privilege). So much for objective justice.

Now, social scientists, from Aristotle to de Tocqueville and Marx, have known that conformity and nonconformity with regard to the distribution of benefits and burdens depend on "perceived" rather than objective fairness or inequity. (In particular, the underdog as well as the top dog tend to justify inequity in terms of desert, real or alleged.) Therefore, in addition to the concept of objective justice, we need a subjective measure of justice.

In a pioneering paper the sociologist Guillermina Jasso (1980) proposed the following formula for "perceived" or *subjective* justice:

$$J_p = k \ log(b/b_f) \ ,$$

where b_f denotes the "perceived" fair share of benefits, and k is a constant, characteristic of each person. If a person is easily satisfied, she has a large k; if picky, her k is small. Perceived justice is positive (privilege), null (equity), or negative (underprivilege), according as the actual benefit, b, is respectively greater than, equal to, or smaller than the "perceived" fair share, b_f, of benefits. The preceding formula will ring a familiar bell with psychophysicists and utility theorists.

Jasso's formula captures the "rights" side of justice but overlooks its "duties" side—which in my view is just as important as the former (Bunge 1989a; essay 28). This omission is easily remedied by dividing the argument of the logarithm by the ratio d/d_f of actual or fair (but still "perceived") burden. The resulting formula is

$$J_p = k \log [(b/b_f)/(d/d_f)] = k \log (b \cdot d_f / d \cdot b_f).$$

According to this formula, a person will feel (justifiably or not) that justice has been done her if, and only if, $b/b_f = d/d_f$, i.e., if the ratios of actual to "fair" benefits and burdens are the same. Obviously (mathematically) and interestingly (psychologically), the above condition is satisfiable in infinitely many ways. One (sufficient) condition is of course $b = b_f$ and $d = d_f$, which may be referred to as ideal subjective justice. However, $b = cb_f$ and $d = cd_f$, where c is an arbitrary real number, will do just as well. In particular, the following combinations are possible:

$$b = 2b_f \text{ and } d = 2d_f, \, b(1/2)b_f \text{ and } d = (1/2)d_f.$$

That is, doubling the "fair" share of duties can be compensated for by doubling the "fair" share of benefits. And halving d_f is balanced by halving b_f.

So far we have tacitly interpreted b_f and d_f as outcomes of self-appraisals or subjective evaluations. However, they can also be interpreted as figures arrived at by persons other than the individual in question. For example, the manager of a firm, or the chairman of an academic department, may determine what constitutes the "fair" benefits and burdens for any individual who performs a given role. And he may do so using objective criteria. Still, the individual who is being evaluated may have a different "perception".

People do not just "perceive" society: they sustain or alter it by acting on others. This points to a difference between the knowing subjects in the theory of knowledge about nature and in the theory of social knowledge. While in the former the knowing subject studies natural things, in the latter she studies people who not only know, but also act on the strength of their knowledge or, rather, belief. In particular, while the theories in natural science do not refer to observers, experimenters, or theorists, some of the theories in social science cannot help referring to people who are led (or misled) by social theories. For example, social movements differ from the movements of bodies in that their members have social goals and are inspired by ideologies.

This difference has led some scholars to challenge the belief in the possibility of social science, and others to suggest that, although social science is possible, the corresponding epistemology must be changed.

The former claim is easily disposed of by recalling that social science does exist, even though, admittedly, not always on a high level. On the other hand, the second claim is more interesting and, at first blush, correct. Indeed, it would seem that, since social facts are the work of people, they are not out there: everything social would be constructed or invented, nothing would be discovered. Consequently, realism, which might work for natural science, would be inadequate to social science: here we would need a constructivist epistemology, which shuns objectivity. But this argument holds no water. The fact that people create social facts and are influenced by their own beliefs does not render objectivity impossible. All it does, is to force us to impute to agents (conjecturally) beliefs, interests, intentions, and other mental processes. Shorter: The fact that, unlike other things, people feel and think, and act accordingly, disqualifies social naturalism, or rather behaviorism, but not realism.

9. REALISM

Philosophical realism, or objectivism, is the view that the external world exists independently of our sense experience, ideation, and volition, and that it can be known. The first conjunct is an ontological thesis while the second is an epistemological one. It is possible to assert the former while denying the latter. That is, one may hold that material (natural or social) objects exist externally to us but cannot be known, except perhaps by their appearances. Or, one can hold that the world is intelligible because we construct it ourselves, much as we construct myths and mathematical theories.

Realism is opposed to subjectivism in all of its forms. In particular, it clashes with conventionalism, fictionism, constructivism, and phenomenalism. On the other hand, realism is consistent with some moderate forms of immaterialism, such as Thomas Aquinas's. In particular, it is possible to be a realist while believing that disembodied souls and angels are flying around. By the same token, realism must not be confused with materialism, which is an ontological, not an epistemological view. (Marxists indulge in this confusion.) Nor must realism be confused with empiricism or positivism, which restricts the knowable to the experientiable, and is therefore at least partially subjectivistic.

The ontological thesis of realism can be restated thus: there are things in themselves. Its epistemological companion can then be restated as follows: we can know things in themselves (not just as they appear to us). I submit that these two theses are presupposed in any scientific research. For example, regardless of what certain popularizers of quantum physics claim, physicists model electrons, photons, and other imperceptible things as things in themselves, i.e., independently of any observers and their measuring instruments. This is not a dogmatic assertion: it can be proved by inspecting the formulas describing such entities. In fact, no variables referring to either the observer or his apparatus occur in any of the basic formulas of the quantum theory (Bunge 1967c, 1973b; see also essays 17 and 18). Nor does the psychologist or the social scientist put himself in the picture. The same holds in all the other sciences: they are impersonal even though they are, of course, the creation of persons and even though some of them refer to (third) persons.

One of the recurrent objections to mainstream mathematical economics is that it is not realistic enough, e.g., that it assumes perfect competition when actually the bulk of industry and wholesale trade are controlled by oligopolies and subject to government regulations. Still, even the most orthodox neoclassical economist will admit only so much fiction. For example, it may happen that the price (or the quantity) equations for an economy have two mathematically exact solutions, one for positive and the other for negative prices (or quantities). Since real prices and quantities are positive, the economist will declare the negative solutions economically meaningless for being unrealistic. That is, she will tacitly endorse realism even while playing with highly idealized models.

Another example of tacit realism is this. When a new hypothesis or theory fails exacting empirical tests, it is rejected for failing to fit the facts, i.e., for not being realistic. But this may not be the end of the story. Firstly, the test results may prove to be false. Secondly, if they are true one may try modifications of the original hypotheses or theories in the hope of coming up with truer, i.e., more realistic ones. Should this move fail, one may attempt to construct entirely different ideas, or even try a different approach. In either case one strives for objective truth—the trademark of realism.

Three varieties of realism are often distinguished: naive, critical, and scientific. *Naive* or *commonsensical* realism asserts that things are such as we perceive them. Either it does not distinguish between the thing in itself and the thing for us, or it demands that every concept have a real counterpart. In other words, naive realism holds that true knowledge (or language in the case of the first Wittgenstein) "mirrors" reality. The naive realist is uncritical and, consequently, prey to sensory deception and self-deception. And, since he believes in the possibility of attaining complete and definitive truths about matters of fact, he can explain neither error nor the effort to correct it by constructing ever more complete theories containing concepts increasingly far removed from perception and intuition. Naive realism is particularly unsuitable to the study of things such as electrons and social systems, and processes such as atomic collisions and stagflation, which are not directly perceptible and have counterintuitive properties.

There are two ways of reacting to the inadequacies of naive realism: to reject it altogether, or to try and refine it. The former is the antirealist reaction. The antirealist reasons that, since scientists keep changing their ideas and even their data, truth is unattainable. Shorter: reality, if it exists at all, is unknowable. This is a primitive and defeatist reaction to a naive doctrine. It overlooks the fact that scientific error is corrigible in principle: that we can frequently go from error to partial truth to a better approximation. Antirealism is blatantly unrealistic, i.e., false, because the point of scientific research is to explore the real world in order to get to know it. In particular, what would be the point of checking scientific hypotheses against facts if they did not purport to represent facts? And what would be the point of technological proposals if they did not intend to alter certain features of real things, to assemble new ones, or to dismantle existing ones? Scientists and technologists are not paid to play games but to explore or help alter reality.

The *critical* realist realizes that perception is limited and can be deceptive, and that complete truth is hard to come by. She admits that the way we perceive facts, particularly in the social realm, depends partly upon our beliefs and expectations. All this inclines her to adopting a critical or skeptical attitude: she is a fallibilist. She also realizes that perception must be corrected and supplemented with the construction of concepts, hypotheses, and theories referring to such imperceptible things as social networks, monetary systems, institutions, and nations. Moreover, the critical realist realizes that scientific theories cannot be isomorphic to their real referents because they contain (a) simplifications and idealizations, as well as (b) conventional elements, such as units and the choice of scale and coordinate system. In short, scientific theories contain constructs without real counterparts.

Scientific realism is a refined version of critical realism. In addition to the ontological and epistemological postulates of realism, it asserts (a) the methodological principle that scientific research is the best (most rewarding) mode of inquiry into any matters of fact, even though it is not infallible, and (b) the article of (justified) meliorist faith that scientific research, through fallible, can give us increasingly true representations of the world. These two additional principles can jointly be called *scientism*. (This signification of 'scientism' is the traditional one: see, e.g., Lalande [1938]. On the other hand, Hayek [1955] has proposed the Pickwickian definition of 'scientism' as a "slavish imitation of the method and language of science". To be scientistic, in the traditionally accepted sense of the word, is to practice the scientific approach, not just to ape it. The imitation of science is properly called 'pseudoscience': see, e.g., Bunge 1991b, c.)

Scientific realism is not a recent philosophical fad. It was explicitly defended by Galileo and it was at the center of his infamous trial. As we all know, Galileo held that the heliocentric planetary astronomy was true. This view contradicted commonsensical realism, phenomenalism, and the Book of Genesis. Galileo's inquisitor, Cardinal Bellarmino, took the phenomenalist view—which had been defended earlier by Ptolemy—that the astronomer's task is to account for appearances, not to find out how things really are. However, the Inquisition did not press Galileo to adopt the old geocentric view: it just wanted him to declare that the two rival views were equivalent, being both compatible with the data, so that the new astronomy did not refute the Scriptures.

In short, the Inquisition fought the newly born scientific realism and defended phenomenalism and conventionalism—thus betraying Aquinas's realistic teaching. (Ironically, three centuries later the logical positivists, in particular Philipp Frank, Hans Reichenbach, and Herbert Dingle, repeated the cardinal's contention that the two "systems of the world" are equivalent.) The Church terminated the theological controversy, but scientific realism was vindicated a few years later. In fact, Newtonian celestial mechanics justified the heliocentric hypothesis arguing that, because the Sun is at least a thousand times more massive than any of the planets, the latter really revolve around the former. (When Kant revived phenomenalism, science had superseded it one century earlier.)

To conclude. Scientific realism is tacitly embraced by all practicing scientists. If a researcher were not a realist, she would remain content with recording appearances and constructing egocentric (or at most lococentric) views. She would make no effort to explain appearances in terms of hidden, though presumably real, entities and processes. She would not bother with empirical tests for truth. She might even turn to nonscience or even antiscience for illumination, instead of doggedly pursuing objective truth. (She may certainly hold unscientific beliefs while not engaged in scientific research. But this only shows that weekend philosophy may be inconsistent with workday philosophy.) And if she had no faith in the possibility of correcting error and converging gradually to the truth, the researcher would not attempt to improve theories or experimental designs. In short, she would not be a scientist.

10. CONCLUSION

By denying the autonomous existence of the external world, or at least the possibility of knowing it objectively, antirealism discourages its scientific exploration. Shorter: antirealism is inimical to science. On the other hand, realism is not just one more philosophical fantasy: it is inherent in factual science and technology. It inheres in the former because the declared aim of scientific research is to represent reality, in the latter because the technologist's job is to design or redesign realizable artifacts (things or processes) or feasible policies capable of altering the natural or social environment.

Moreover, the philosophy we adopt tacitly in daily life is realist—albeit, of the commonsensical variety. There is a powerful biological motivation for this: know thy world or perish. No complex animal can survive unless it is capable of modeling its environment correctly (truthfully), at least in some respects. It would starve to death unless it could identify what it can eat, and it would get lost unless it were able to map it surroundings and locate in them those who might eat it. Realism is thus necessary for animal survival as well as for understanding and altering the world in a rational manner. If ever there were subjectivist animals, they either died very young from exposure to the world they denied, or they were appointed professors of philosophy.

VII

PHILOSOPHY OF
TECHNOLOGY

■

24

THE NATURE OF APPLIED SCIENCE AND TECHNOLOGY

(1988)

■

The purpose of this paper is to describe applied science and technology and to locate them in the map of contemporary culture and, indeed, in the wider system of production and circulation of knowledge, artifacts, and services. (For details see Bunge1976b, 1980b, 1983b, 1985b.)

We start by distinguishing applied science from both pure science and technology, characterizing the former as the investigation of cognitive problems with possible practical relevance. Here is a *pêle-mêle* list of fields of applied scientific research: materials science, in particular metallurgy; the analysis of natural products with a view to isolating some useful chemicals; the experimental synthesis of promising polymers, e.g., synthetic fibers; the investigation of the biology of plants with a possible industrial interest, such as the *guayule* (that produces rubber); the whole of pharmacology and food science; the entire field of medical research, from internal medicine to neurology; psychiatry and clinical psychology; the entire science of education; and the applications of

345

biology, psychology, and social science to investigating social problems, such as unemployment and marginality, drug addiction and criminality, cultural deprivation and political apathy, with a view to designing social programs aiming at their eradication.

Applied science does not follow automatically from basic science. Doing applied science, like doing basic science, is conducting research aiming at acquiring new knowledge. The differences between applied research and basic research are differences in intellectual debt to basic science, in scope, and in aim. The first difference consists in that applied research employs knowledge acquired in basic research. This does not entail that applied research is necessarily a matter of routine: if it did not yield new knowledge, it would not be research proper, and to deliver new knowledge, it must investigate problems of its own. However, the task of the applied scientist is to enrich and exploit a fund of knowledge that has been produced by basic research. For example, the space scientists who developed new materials and studied human physiology and psychology under unfamiliar conditions (zero gravity, isolation, etc.) tackled new problems that could not possibly be solved with the sole help of existing knowledge—and they came up with new knowledge. The applied scientist is supposed to make discoveries, but is not expected to discover any deep properties or general laws. He or she does not intend to.

Secondly, the domain or scope of applied science is narrower than that of basic science. For example, instead of studying learning in general, the applied psychologist may study the learning of a given language by a given human group in particular social circumstances, e.g., the way Mexican Americans learn English in a Los Angeles slum. And instead of studying social cohesion in general, the applied sociologist may study the social cohesion of, say, a marginal community in a Latin American shantytown, with an eye on possible ways of improving its lot.

Thirdly, all applied research has some practical target, even if it is a long-term one. For example, the forester is not only interested in forests in general, but also, and particularly, in forests with a possible industrial utility. And the pharmacologist, unlike the biochemist, focuses her research on chemicals that may be beneficial or noxious to certain species, particularly ours. In every case we expect from the applied scientist to end up every one of her reports by asserting not just that she has discovered X, but that she has discovered X, which seems to be useful to produce a useful Y or prevent a noxious Z. But we will not ask her to design an artifact or a procedure for actually producing Y or preventing Z: this is a task for the technologist. Table 1 illustrates what has been said so far, and prepares the ground for what comes next.

Applied science lies between science and technology, but there are no borderlines between the three domains: each shades into the other. The

Table 1. Some applied, technological, and economic partners of some basic investigations.

BASIC SCIENCE	APPLIED SCIENCE	TECHNOLOGIES	INDUSTRY, COMMERCE, SERVICES
Mathematics	All	All	Consulting
Astronomy	Optics of light, radio, and X-ray telescopes; photometry, bolometry, etc.	Design of astronomical instruments; invention and development of photographic plates; observatory architecture, etc.	Optical industry; photographic industry; maintenance and repair of astronomic instruments, etc.
Atomic physics	Physics of semi-conductors, electronics	Design of radios, TV sets, computers, calculators, etc.	Manufacture and maintenance of electronic apparata
Chemistry	Hydrocarbon chemistry	Petroleum engineering (prospection, drilling, refining, etc.)	Building and maintenance of petroleum machinery, installation and maintenance of oil wells and refining plants
Biology	Biology of edible plants	Phytotechnology of edible plants: creation of new varieties, study of new cultivation methods	Agriculture, food industry, food marketing
Sociology	Development sociology	Development plans	Implementation of development plans

outcome of a piece of basic research may suggest an applied research project, which in turn may point to some technological project. Once in a blue moon a single investigator spans all three. More often than not, each task demands people with peculiar backgrounds, interests, and goals. Whereas the original scientist, whether basic or applied, is essentially a discoverer, the original technologist is essentially an inventor of artificial things or processes. Indeed, the invention of artifacts or of social organizations is the very hub of technological innovation. No matter how modest, an invention is something new that did not exist before, or that existed but was beyond human control. (See also essay 20.)

Most of the inventions proposed until the beginning of the Modern period owe hardly anything to science: recall the domestication of most plants, animals, fungi, and bacteria; the plow and metallurgy, architecture and coastal navigation. Things started to change in the seventeenth century and, particularly, since about 1800. The pendulum clock and the Watt governor are based on mechanics; the electric generator and the electric motor, on electrodynamics; the synthetic products used in industry and medicine, on chemistry; the electronic gadgets, on atomic physics; the supercultivars, on genetics, and so on. In short, since about 1800 and on the whole, technological breakthroughs have followed scientific discoveries. The most common pattern is nowadays: *scientific paper—applied science report—technological blueprint*. And, as is well known, the time lag between these stages is decreasing rapidly.

Saying that an invention "is based on" a bit of scientific knowledge does not signify that science suffices to produce technology, but that it is employed to some extent in designing the artifacts or the plan of action. For example, Joseph Henry designed the first electric motor on the basis of his knowledge of electrodynamics (to which he himself had contributed); and Marconi invented the radio by exploiting Maxwell's theory as well as Hertz's experiments. The modern inventor need not know much science, but he cannot ignore it altogether, for what is called "the principle" of a modern invention is some property or law usually discovered by some scientific research. Thus, the "principle" of jet propulsion is Newton's third law of motion, and the antihistaminic "principle" is the antigen-antibody reaction discovered by the immunologists. What characterizes the inventor is not so much the breadth and depth of his knowledge as his knack for exploiting what he knows, his wondrous imagination, and his great practical sense.

Invention is only the first stage of the technological process. Next comes the so-called development stage, which is where most inventions sink. The blueprint must be translated into a prototype, or a handful of seeds of a new cultivar, or a few milligrams of a new drug, or a plan for a new social organization. Once these artifacts have been produced, it is necessary to put them to the test to see whether they are effective. In the case of a new drug the tests may take some years, and millions of dollars.

The third stage is the design of the production in the case of artifacts, and of the implementation in the case of social programs. This may demand the building of an entire pilot plant, which poses new problems, which demands new inventions. (In technology, as in basic science and in life in general, one thing leads to another.) Even if built, the pilot plant may not work satisfactorily for some reason or another; and even if it is technically satisfactory it may prove to be too expensive, or to produce socially undesirable effects, such as unemployment or high pollution. No wonder that most of the R&D budget goes into the development stages.

If the third and last stage of the technological process has been accomplished successfully, production may begin and, eventually, marketing may be undertaken. These two stages may require new inventions

Table 2. From lab to market.

ACTIVITY	CARRIED OUT BY	MAINLY AT	OUTPUT
Basic research	basic scientists	universities	data, formulas, graphs, experimental designs, blueprints of scientific instruments, calculations, methods, etc.
Applied research	applied scientists	universities, industrial labs, state labs, etc.	
technological invention	technologists	industrial labs	designs and models of artifacts or procedures
development	technologists and managers	industrial labs, workshops, and experimental stations	prototype, pilot plant, production plant, social programs
production	workers, foremen, technicians, managers, clerks, workers	industrial plants, farms, mines, forests, seas	goods or services
marketing	marketing experts, managers, salesmen, ad men, etc.	offices, stores, warehouses, etc.	

and new tests concerning the production as well as its organization and that of the distribution of the product. Such new inputs are likely to be the more novel and frequent, the more original the invention and the more massive the production line and the marketing network. However, these problems usually lie beyond the horizon of the first inventor, unless he happens to be, like Bell, Edison, or Land, a businessman himself. Again, this is largely due to differences in personality traits: the inventor is primarily motivated by curiosity and love of tinkering, not by hopes of fortune.

Table 2 describes schematically the process that ends up in the market. In the case of an artifact, such as an electronic calculator, or a procedure such as a medical treatment, all stages are gone over. On the other hand, in the case of a more modest product or service, such as a canned food or the organization schema of a cooperative, the scientific stages are usually skipped: not that scientific knowledge is totally absent, but it is borrowed rather than freshly produced. In other words, whereas some inventions call for new investigations, others can be produced with the help of extant knowledge.

So far we have been concerned with the basic science–applied science–technology–economy flux, typical of the contemporary period. But there is also a constant reflux in the opposite direction. The laboratory worker uses instruments, materials, drugs, and even experimental animals produced in mass and uniformly by industry. Applied science and technology supply basic science new materials, new instruments, and especially new problems. In short, every one of the components in table 2 acts on all the others, not counting the remaining branches of culture and politics. All of these items compose a system that is characteristic of the modem times, namely the *system of production and circulation of knowledge, artifacts, and services*. See the following flow diagram.

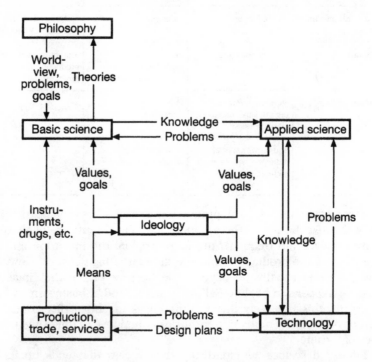

Finally, note that we have included in technology far more than the traditional branches of engineering. In fact, we conceive of technology as the *design of things or processes of possible practical value to some individuals or groups with the help of knowledge gained in basic or applied research*. The things and processes in question need not be physical or chemical: they can also be biological or social. And the condition that the design make use of scientific knowledge distinguishes contemporary technology from *technics*, or the traditional arts and crafts.

Our definition of "technology" makes room for all the action-oriented fields of knowledge as long as they employ some scientific knowledge. (See also essay 26.) Hence, in our view technology subsumes the following fields.

Physiotechnology: civil, mechanical, electrical, nuclear, and space engineering.

Chemotechnology: industrial chemistry, chemical engineering.

Biotechnology: pharmacology, medicine, dentistry, agronomy, veterinary medicine, bioengineering, genetic engineering, etc.

Psychotechnology: psychiatry; clinical, industrial, commercial, and war psychology; education.

Sociotechnology: management science (or operations research), law, city planning, human engineering, military science.

General technology: linear systems and control theory, information sciences, computer science, artificial intelligence.

In conclusion, basic science, applied science, and technology have commonalities as well as differences. All three share essentially the same worldview, mathematics, and the scientific method. They differ mainly in their *aims*: that of basic science is to understand the world in terms of patterns; that of applied research is to use this understanding to make further inquiries that may prove practically useful; and that of technology is to control and change reality through the design of artificial systems and plans of action based on scientific knowledge.

25

THE TECHNOLOGY-SCIENCE-PHILOSOPHY TRIANGLE IN ITS SOCIAL CONTEXT

(1999)

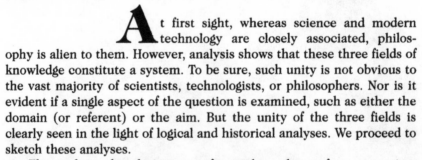

At first sight, whereas science and modern technology are closely associated, philosophy is alien to them. However, analysis shows that these three fields of knowledge constitute a system. To be sure, such unity is not obvious to the vast majority of scientists, technologists, or philosophers. Nor is it evident if a single aspect of the question is examined, such as either the domain (or referent) or the aim. But the unity of the three fields is clearly seen in the light of logical and historical analyses. We proceed to sketch these analyses.

The traditional technics or crafts, such as those of stone carving, primitive agriculture, and the art of killing by hand, owe nothing to science, and their relation to philosophy is weak. They only presuppose the philosophical principles that the world exists on its own, that it can be modified by human action, and that it can be known at least in part. By contrast, the modern technologies are science-based and they involve further philosophical principles, as will be seen shortly.

For instance, modern civil and mechanical engineering are based on mathematics and theoretical mechanics, as well as on materials science. Without these sciences, engineers would be unable to compute anything, and they would not know what or how to measure any variable. Moreover, they would be unable to design the complex artifacts and processes peculiar to modern industry. True, the Roman engineers designed roads, bridges, and public buildings that are still standing and in good use. However, these works were more remarkable for size and durability than for ingenuity and variety. They all involved enormous waste in materials and manpower. It is not just that, as a Marxist would say, the Romans had no economic incentive to perform more exact calculations, because slave work was cheap. The main cause of their limitation was the absence of theories (of, e.g., elasticity) and experiments (in, e.g., the resistance of materials). They proceeded exclusively by trial and error.

How are civil and mechanical engineering related to philosophy? In various ways. For one thing, all of the modern technologies have philosophical presuppositions. For example, they take it for granted that the natural resources exist independently of the mind, and that nature is lawful and can be known, at least in part and gradually. In short, modern technology is tacitly naturalist and realist. It does not assume that there are ideas detached from brains, or that they can act directly upon things. Nor does it admit the possibility of magical or miraculous tricks to avoid work altogether. Nor, finally, does it accept the existence of unfathomable mysteries: it only admits the existence of more or less difficult problems. This epistemic optimism was expressed in the motto of the American corps of engineers in World War II: "The possible is done right away. The impossible will take a little longer."

Since technologists accept tacitly a realist theory of knowledge, they believe in the possibility of attaining the truth, at least approximately. Hence, in their professional practice they ignore the postmodern attacks on objectivity and truth. An engineer who did not believe in the approximate truth of his equations, tables, or experimental results could not practice his profession. This is why constructivism-relativism is not popular among engineers. In fact, it is popular in faculties of arts.

To be sure, every engineer knows, or ought to know, that scientific theories and experimental data are at most approximately true. That is, realistic technologists are fallibilists. But at the same time, they trust the corrigibility of theories and data, and they often make use of approximation methods; for instance, when using computers to graph continuous functions or to solve differential equations. In this regard, they adopt the same epistemological realism as the scientific investigators, even though neither may have heard of it.

In sum, civil and mechanical engineering are philosophically committed rather than neutral. Furthermore, the successes of mechanics and its applications suggested the mechanistic worldview, or philosophical mechanism (or mechanicism), which in turn stimulated research in theoretical mechanics—a positive feedback loop. It stands to reason: if the world is a gigantic clock, then to understand it we must craft mechanical theories of everything. Incidentally, the first two mechanistic thinkers of the modern period, namely Galileo and Descartes, were at once scientists and philosophers.

Mechanism was a whole worldview that replaced the organicist and hierarchical worldview inspired by Aristotle and Christian theodicy—and now resurrected by New Age and "feminist philosophy". Mechanism holds that the universe is a system composed exclusively of bodies and particles that interact and move in accordance with the laws of mechanics found by Archimedes, Stevinus, Galileo, Kepler, Huyghens, Newton, Euler, the Bernouillis, and their successors. Even organisms and societies would be mechanical systems. This worldview, together with the realist theory of knowledge, suggests that classical mechanics suffices to understand and control the world. It is a philosophy and a research project tailored to the mechanical or civil engineer.

The mechanical project dominated scientific and philosophical thinking, as well as a sizable part of philosophy, during the three centuries following the seventeenth-century Scientific Revolution. This monopoly ended with the birth of electromagnetism in the mid-nineteenth century. The experiments of Alessandro Volta and Michael Faraday, and the theories of André-Marie Ampère and James Clerk Maxwell, signaled the beginning of the decline of the mechanistic worldview. Indeed, though material in the philosophical sense of the word, the electromagnetic fields are not mechanical objects: they lack mass and, therefore, fail to satisfy the laws of mechanics. From the moment such fields were investigated, the mechanical paradigm coexisted alongside with the field-theoretic paradigm—a fact that, incidentally, refutes Thomas Kuhn's claim that every mature science is monoparadigmatic. The universe was pictured as a gigantic electromagnetic field (and later as a gravitational field, too) sown with particles and bodies.

No sooner was it born than electromagnetic theory was used to design electric motors and generators. Half a century later, the technology of electromagnetic waves was born, and electronics sprang one generation later. Incidentally, every one of these momentous technological innovations came in the wake of disinterested experimental and theoretical investigations. None of them was a result of tinkering. Furthermore, none of them was market-driven. On the contrary, every one of them generated a new market. Much the same is true of nuclear engineering, genetics-based agriculture, computer technology, and nearly all the other advanced technologies.

All of these advances were accompanied by spirited philosophical discussions. For example, many wondered whether the electromagnetic waves described by Maxwell's equations exist really or are just fictions useful to summarize experimental data. (This is the realism-conventionalism alternative.) The same question arose with reference to electrons. Thus, as late as 1948, when electron physics celebrated its fiftieth anniversary, *Nature* published a paper that asked whether the electron had been discovered or invented. The positivists denied the independent existence of electrons, because they could only accept directly observable entities. (After all, electrons are just as imperceptible as gods: they manifest themselves only indirectly, for instance, through their impact on a TV screen.) The obvious distinction between the electron, a permanent fixture of the world, and our changing theories of the electron, helps solve the conundrum.

This debate between realists and antirealists continues nowadays among philosophers and sociologists of science (see essay 23), even though it must look grotesque to the engineers who design such devices as antennas and TV sets, which use electrons and electromagnetic waves. When locked up in a study surrounded by nonscientific works, it is easy to doubt the reality of the external world, and think instead of imaginary worlds. But, when working in a laboratory or a workshop, one takes the reality of the external world for granted, since this is what one attempts to understand or alter.

Another famous philosophical controversy over the nature of the world and science was the one posed by energetics, an alleged generalization of thermodynamics. The energeticists of around 1900 held that the failure of mechanism had forced scientists to give up explanations, and propose instead only descriptions employing exclusively observable entities and properties. In other words, the energeticists favored black boxes over translucent boxes. Not surprisingly, they attacked atomism.

However, the most advanced applied sciences and technologies, such as pharmacology, industrial chemistry, electronics, communications engineering, computer engineering, and genetic engineering, cannot benefit from descriptivism. Sooner or later, the technologist will have to open the lid of his black box, to find out how to repair or improve it. For instance, whereas the computer programmer need not know anything about the circuitry of his machine, the computer engineer needs to know a lot about electronics and solid-state physics. He cannot afford to adopt functionalism.

The search for hidden mechanisms, whether mechanical or nonmechanical, is peculiar to modern science, just as the design of equally occult artificial mechanisms is characteristic of modern technology. (Examples: the cathode-ray tube in TV sets, and vaccination.) The philosophical principle behind such endeavors is that everything hidden can be unveiled and altered. To put it negatively: there is nothing inscrutable in principle. To put it paradoxically: in contradistinction to the "occult sciences" of yore, modern science is knowledge of the occult. On the other hand, phenomenalism, according to which we should stick to appearances and abstain from hypothesizing unobservables, is an obstacle to the exploration of the most interesting things. These are the imperceptible entities, which, as the ancient atomists suspected, are the constituents of the perceptible ones.

Another fertile triangle that emerged in the nineteenth century was the one made up of basic chemistry, chemical technology, and the matching philosophy. The latter contains, first of all, the philosophical assumptions common to all the sciences, starting with naturalism (or materialism) and epistemological realism. But of course, its specific component is atomism, which has been part of the materialist worldview since about 500 B.C.E.

Certainly, modern atomic physics is quite different from the ancient atomic speculations. However, the key idea is the same: that every concrete thing is composed of particles belonging to a few species. (To update this thesis, we must replace "particles" with "particles and fields"—or just "quanta".) It is likely that the nineteenth-century chemists would have invented atomism even without knowing about its ancient roots, since the basic laws of chemical composition and electrolysis point clearly to atomism. Still, it is a historical fact that ancient atomism—rejected by Aristotle and later on by Descartes, but kept warm by a few philosophers, such as Gassendi and Hobbes—was the midwife of modern chemistry.

Another interesting fact was the anti-atomist resistance opposed by the positivists, from Comte to Mach, Duhem, and Ostwald, all of whom regarded atoms as useless fictions beyond experimental control. This conservative attitude slowed down the progress of atomic physics and gave rise to a spirited philosophical controversy around 1900. This controversy involved not only a number of distinguished scientists, but also Lenin, who this time backed the winning horse. By contrast, the technologists kept silent: they were busy applying some of the new ideas, particularly to telecommunications and industrial chemistry. However, their achievements constituted tacit endorsements of the new physical and chemical heterodoxies. Even the humble homemade galena radio strongly suggested the objective reality of radio waves and electrons.

In short, the birth of modern chemistry, modern atomic theory, and electronics, was accompanied by vigorous discussions concerning the nature of things as well as that of our knowledge about them. Since hardly anyone nowadays doubts the real existence, materiality and knowability of the building blocks of the physical universe, it is only fair to admit that naturalism and realism were ultimately the winners in those disputes.

Another philosophico-scientific idea that played an important role in the construction of modern chemistry and the birth of modern biochemistry was the hypothesis that "organic" compounds are not necessarily produced by living beings. The synthesis of urea in the laboratory, in 1828, weakened the vitalist school in biophilosophy and, by the same token, reinforced its materialist rivals. This discovery also seemed to confirm the thesis that being able to make a thing is proof that we know it. This is not the place to discuss this controversial thesis, first advanced by Vico and later on adopted by Engels. We must confine ourselves to posing the problem: Is knowledge necessary, sufficient, both, or neither for action? This is an important, if neglected, problem in praxiology (action theory).

Biochemistry, genetics, molecular biology, and molecular biotechnology are usually associated with reductionism. In those fields, this is the thesis according to which a cell is nothing but a bag of atoms and molecules, as a consequence of which biology would be just the physics and chemistry of complex systems. This view was proclaimed in 1865 by Claude Bernard, the great physiologist and biochemist, who went as far as to propose excising the word 'life' from the biological literature. And, of course, nowadays many scientists believe that molecular biology suffices to understand the whole of biology. For instance, the Genome Project was launched on the premise that the sequencing of the entire human genome would suffice to unveil once and for all the mystery of human nature. That is, the idea is that knowledge of the constituents of a whole is both necessary and sufficient to know the whole, regardless of the way those constituents are organized and interact.

This is not the suitable place to criticize radical reductionism, a strategy as fertile as it is mistaken. (See essay 11.) I only mention it as a reminder that chemistry, biochemistry, and biology, as well as their technological applications, including pharmacology and therapeutics, are far from being philosophically neutral. Hence, their foundations cannot be adequately examined without the help of some philosophy.

Biotechnology, in the broad sense, is the branch of technology that deals with the deliberate control and utilization of life processes. It was born with the domestication of animals, agriculture, and medicine. However, these ancient techniques did not become modern technologies until the mid-nineteenth century. Nowadays, biotechnology includes also bioengineering and genetic engineering: just think of pacemakers and the manufacture of new biological species through genetic manipulation. Evidently, modern biotechnology could not have been born without modern biology and biochemistry, both of which have an important Darwinian component. In particular, the design of new biospecies makes explicit use of the two key ideas of evolutionary biology, namely genic change and selection.

Evolutionary biology has a huge range: it concerns all the modern organisms and their ancestors, so that it spans about three billion years. It is thus one of the historical sciences. Moreover, it is committed to naturalism as well as to Heraclitus's dynamicist ontology. It has deeply influenced several natural and social sciences, and it cannot be ignored by any modern philosopher. Yet, it can easily be misunderstood. Witness vulgar evolutionary biology, psychology, and medicine, according to which every single biological and psychological trait, even if apparently nonadaptive, must be good since it has survived.

Genetic engineering presupposes not only evolutionary biology, but also an ontology closer to mechanism than to vitalism, since it is associated to a reductionist methodology. It also involves the thesis that human beings are not only products of evolution, but also engines of it, since they can steer selection processes, and even create new biospecies. The truth of this bold view should be obvious to anyone reflecting on the achievements of genetic engineering, or on the failure to contain environmental degradation.

The thesis in question agrees with philosophical materialism as well as with pragmatism, but it is inconsistent with supernaturalism as well as with the antiscientific and technophobic philosophies. Therefore, it should not be surprising that most of the enemies of modern technology are theologians (such as Jacques Ellul) or obscurantist pseudophilosophers (like Martin Heidegger). Incidentally, since some technologies serve evil goals, the rejection of technophobia does not amount to a blanket acceptance of all new technologies. (As the management expert Laurence Peter once said, there are two kinds of fools. The first say, "It must be good since it is old", and the second, "Since it is new, it must be good".)

Another young technology, and one rich in philosophical ideas, is knowledge-engineering, which includes information technology and artificial intelligence. This technology is based on logic, abstract algebra, electronics, and solid-state physics. Since logic is in the intersection of mathematics and philosophy, the latter must be admitted into the knowledge-engineering family. Nor is solid-state physics utterly alien to philosophy. Indeed, it is based on quantum physics, the modern inheritor of ancient atomism. And the birth of quantum physics gave rise to spirited philosophical controversies. One of them was, and still is, the famous discussion between Einstein and Bohr over the question whether quanta exist independently, or only come into being as outcomes of laboratory measurements.

As if this were not enough, knowledge-engineering has restated in a new light the old ontological problem of the relation between mind and matter. Indeed, ever since the seminal papers by Alan Turing and John von Neumann in theoretical computer science, knowledge engineers and philosophers have been discussing a handful of key philosophical questions. Among these the following stand out: Can computers think other than by proxy? Is it in principle possible to design imaginative machines, capable of creating original ideas? Could a computer be endowed with free will? Could it be that the mind is nothing but a bunch of computer programs, which can be materialized now in a brain, now in a machine? This is not the place to tackle these questions, which are as intriguing as they are practically important. Suffice it to note that they are at once scientific, technological, and philosophical. Which proves that we are dealing with another fertile triangle.

Up to now we have been examining technologies based on mathematics and natural science. Our last example will be drawn from sociotechnology, or the technology that aims at altering human behavior with the help of scientific knowledge. This is the technology involved in the design, assembly, maintenance, or reform of social organizations of all kinds, such as factories and government departments, schools and courts of law, hospitals and armies, and even entire societies in the cases of legislation and normative macroeconomics.

Management science is one of the most interesting and useful of all sociotechnologies. The most advanced branch of this field is operations research, which crafts, tests, and applies mathematical models of formal organizations with a view of optimizing their performance. Thus, there are models of queues, inventories, biddings, quality control, logistics, and much more. Like other technologies, this one is not limited to using knowledge: it supplies new knowledge, such as organization blueprints, strategic plans, marketing plans, budgets, and sales predictions. And, like the other technologies, management science presupposes that its objects of study, once set up, exist really, satisfy laws or norms, can be known up to a point, and can be redesigned and rebuilt if need be. It also assumes tacitly that human action can be planned and predicted within limits, and that planning can be better than improvisation. Now, the very notion of a plan belongs to praxiology (or action theory), a philosophical technology closely related to sociology. Moreover, since planned actions can benefit or harm, planning theory cannot ignore another philosophical technology, namely ethics. (See essay 26.) In short, management science also has philosophical components.

This is not all. The tacit central thesis of management science is that the scientific method can be used advantageously to study and design human action. This is a philosophical thesis and, moreover, one contrary to the so-called humanistic school in social studies. This school, also called 'interpretivist' and 'hermeneutic', claims that everything human is spiritual and, moreover, in the nature of a text, and thus beyond the ken of science. Obviously, this school cannot inspire successful management practices, if only because management involves dealing with concrete things, computing, and checking for efficiency, neither of which is reducible to the interpretation of symbols.

Technologists are likely to feel offended if told that they are steeped in philosophy. They feel even more offended if told that philosophy, in particular metaphysics, is livelier in schools of advanced technology than in many a department of philosophy. They look down upon philosophy, and with some reason, since most philosophers despise technology and ignore its conceptual richness. But the technologists who work on general theories of systems, control theory, optimization theory, the design of algorithms, or simulation are applied philosophers of sorts, since they handle broad cross-disciplinary concepts. Even if they avoid encounters with philosophers, technologists cannot avoid philosophical contagion, since they use philosophical concepts, such as those of event and system, and philosophical principles, such as those of the existence and lawfulness of the external world. They are philosophers in spite of themselves. And, since they philosophize, it would be better if they did it well. To do so, they must learn some philosophy. Their problem, then, is not how best to avoid philosophy, but how to find or craft a philosophy that could help them in their professional work. The recipe is obvious: technologists should shop around for the philosophy most likely to advance their work, and they should put it to the test before adopting it.

Heretofore we have noted some of the ontological and epistemological presuppositions and implications of technology. Let us now deal briefly with the moral aspect. The morality of basic research is simple: it boils down to intellectual honesty and willingness to share knowledge. The authentic scientific investigator neither makes up her findings nor steals them; she does not hide difficulties and doubts; and she does not use an obscure jargon to make believe that she is profound rather than ignorant. True, there is the occasional scientific fraud; but, if caught, the delinquent is exposed and ostracized: good research involves good conduct. In sum, the morality of basic science is internal to the scientific community, and may therefore be called an *endomorality*. Its maximal principle is: search for the truth and share it.

By contrast, the technologist faces much tougher moral problems, as became obvious at the birth of nuclear engineering in 1945. The moral and social responsibility of the technologist is far heavier than that of the basic researcher, because the former designs or controls the manufacture or maintenance of artifacts or organizations that may harm some people, either directly or through their impact upon the natural and social environment: a new technology may kill jobs, it may help deceive people, or it may be physically or culturally polluting. For example, whereas the nuclear physicist only attempts to discover the composition of atomic nuclei and the forces that hold their constituents together, as well as the mechanisms of nuclear reactions, the nuclear engineer may design or direct the manufacture of nuclear bombs, or of nuclear power stations. These tasks pose a number of interesting theoretical problems of little interest to the nuclear physicist, such as the behavior of large bodies of fissile material, and the containment of hot plasmas (for fusion). And, of course, it also poses some still unsolved practical problems, such as that of the disposal of nuclear waste.

Basic researchers do not know in advance what they may discover: if they knew, they would not undertake any research. They only know that, if they succeed, they will have produced a piece of new knowledge, and one that is unlikely to have any practical use. By contrast, the technologists know beforehand the kind of thing or process that they will design or perfect: tool or machine, building or bridge, electric network or chemical reaction, new variety of seed or bacterium, firm or nongovernmental organization, parliament bill or social program, and so on. Most of the time they know this because it is what their employers have told them to do: design an artifact (or process or organization) complying with such and such specifications.

It is true that, on occasion, scientific findings turn out to be utilizable in technology. But in most cases it is impossible to predict the practical usefulness of a scientific result. Who would have predicted that Apollonius's study of the conic sections would be used nearly two millennia later to calculate the trajectories of cannon balls? Lord Rutherford, the father of nuclear physics, firmly denied the possibility of ever tapping the energy that binds the atomic nucleus. Turing, the founder of theoretical computer science, could not possibly have predicted the invention and diffusion of Internet. Crick and Watson, the discoverers of the so-called genetic code, did not imagine that their work would soon become the basis of molecular biotechnology.

The technologist will not undertake the design of an artifact, process, or organization unlikely to be useful to someone—ordinarily her employer. Hence, she has the moral obligation to ask who will benefit or suffer from her work: firm, government, humankind, or nobody? The basic researcher does not face this problem: if she makes a contribution to knowledge, however modest, it will benefit anyone who cares to use it. If competent, she cannot be a malefactor, whereas a competent technologist can. For instance, Oppenheimer and Teller were benefactors as long as they produced good physics. But, as soon as they became sorcerer's apprentices under the military, they turned into malefactors. Something similar can be said of the economists who advised antisocial policies that make the rich richer and the poor poorer.

Basic scientists cannot harm, but technologists can devise means to harm, from weapons to programs to cut social services. In contradistinction to science, which has only an internal moral code, technology is ruled by two codes, one internal and the other external. The former, or endomorality of technology, is similar to that of science. The only difference between the two endomoralities is that the technologist is tacitly or explicitly authorized to steal ideas. He does it every time he copies an original design or introduces an insignificant change in it to skirt the patent regulations. Such thefts, small or large, are universally tolerated even though they harm the owners of the original patents. Indeed, industrial espionage is a recognized and well-remunerated profession without a counterpart in basic science.

It might be argued that industrial spies are benefactors of humankind because they contribute to the socialization of technical novelty. By contrast, the scientific communities punish plagiarists and rightly so. The reasons for these differences are obvious: (a) whereas in science truth is both end and means, in technology it is only a means; (b) whereas scientific findings have only a cultural value and belong to everybody, technological patents are merchandises. Technological designs can be sold, whereas theorems, theories, experimental designs, and observational data are priceless. In short, contrary to science, which is a public cultural good, technology is also a commercial good.

What I have called the *exomorality* of technology is the set of the moral norms that ought to regulate the technological professions in order for these to serve the public interest rather than special interests. In recent years, some engineering associations have added a few clauses concerning the protection of the public interest, in particular the protection of the environment. However, these norms are largely pious wishes, because the technologist enjoys neither freedom nor power. As the saying goes, he is always on tap, never on top. That is, far from disposing, he is always at the disposal of his employer. Indeed, the technologist is either employed by a private firm or a government, or he has an independent practice, in which case he is at the mercy of the market, unless he is so original that he can carve his own niche in the market. In particular, if he refuses to work on antisocial projects, he puts his job on the line, particularly if he blows the whistle.

What can the technologist do to save his moral conscience? The isolated individual has no choice: he must obey or be fired. The professional societies that ought to protect his moral integrity are too weak to face the economic and political powers that be. In the long run, the only practical solution is the democratic control of technology. This is the participation of the citizenry in the process of deciding what kind of projects technologists should embark on.

It is not a question of subjecting every new project to popular ballot: this procedure would be too slow and costly, and it might put technology at the mercy of know-nothings or demagogues, who could persuade the majority that certain worthwhile projects should be vetoed, whereas others, which are potentially hazardous, should be encouraged. Nor is it a question of placing a moral policeman behind every technologist, since this would suffocate the technologist. To be efficient, the democratic control of technology should be rational: it should involve an enlightened public, as well as technologists, statesmen, bureaucrats, and managers. Technology without democracy can be dangerous, whereas democracy without technology is inefficient.

The task is then to combine technology with democracy: to build a technodemocratic social order—not to be confused with a technocratic one. (See essay 30.) Evidently, such change cannot be brought about overnight, not only because it has not yet been included in the political agenda, but also because it cannot be undertaken without extensive studies and plans involving ample public consultation. It may be objected that such a project is utopian. True, but unless we do something soon to avert the social and natural catastrophes to which unbridled technological advancement is leading us, there will be little left for technologists to tinker and tamper with. Suffice it to think of such megasocial issues as environmental degradation, the decline of biodiversity, the exhaustion of nonrenewable resources, overpopulation, massive unemployment, grinding poverty, growing income inequality, poor health, illiteracy, insecurity, militarism, the national debts of Third-World countries, and the overproduction of junk of all kinds.

Basic scientists are likely to continue studying these issues, but they can do nothing to solve them: all they can do is to provide data and ideas to tackle them. And specialized technologists, such as engineers and finance experts, cannot solve them either, because most of them adopt a sectoral perspective, whereas those issues are multisectoral: every one of them has economic, political, and cultural features. Global problems can have only global solutions. Their solution can only fall to statesmen and bureaucrats in consultation with technologists of all kinds, and in cooperation with the people most affected by them.

We started stating that technology forms a triangle together with basic science and philosophy. However, a quick discussion of the global problems that threaten the survival of humankind has suggested that technology is also part of another triangle: the technology-industry-government one. So much so that modern technology can only flourish in industrial countries, which in turn call for suitable political institutions.

There is more: technology also interacts with art, particularly in the cases of architecture, industrial design, and advertising. It also interacts with ideology, which in a few cases illuminates, but in most cases obscures the goals of technological activity. But, in turn, ideology is

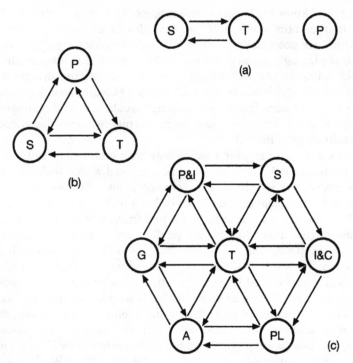

Fig. 1. (a) The traditional view: philosophy (P) is alien to the system composed by science (S) and technology (T). (b) A first approximation to the truth: technology (T), science (S), and philosophy (P) constitute a system. (c) A second approximation: philosophy-*cum*-ideology (P & I), science (S), industry and commerce (I & C), the public (PL), art (A), and government (G) interact with technology (T), thus constituting a system.

related to philosophy: indeed, every ideology has a philosophical nucleus, particularly a value system. Therefore, for practical purposes we can simplify and consider the ideology-philosophy pair as a single unit, provided we do not condone the subordination of philosophy to ideology.

To conclude. Technology, far from being autonomous, is at the center of a hexagon whose vertices are the state, the public, industry, art, the philosophy-ideology pair, and science. Thus, technology is a component of a complex and dynamical system, whose constituents interact, and therefore alter one another: see figure 1. The technologist who fails to see the hexagon runs the risk of getting bored and, worse, of making bad mistakes costly for herself, her employer, society, or all three. By contrast, she who perceives the entire hexagon may be able to better enjoy life and help live.

26

THE TECHNOLOGIES
IN PHILOSOPHY

(1999)

■

The central thesis of this paper is that philosophy comprises five disciplines that it shares with technology. These are axiology or value theory, praxiology or action theory, methodology or normative epistemology, moral philosophy or ethics, and political philosophy. The reason is this. Technology is about designing procedures, plans, processes, or things in order to get certain things either done or avoided. So are the above-mentioned philosophical disciplines. Q.E.D.

To put it into other words, all of the disciplines in question are normative: they identify values and seek to realize them. Or, if preferred, they spot gaps between the way things are and the way they ought to be, or (inclusive "or") devise means to bridge the said gap. Again, whatever means is devised in the light of reason and practice to bring an *is* in line with the corresponding *ought* deserves being called 'technological'. The methodological rules, as well as the value-theoretical, praxiological, ethical, and general political principles qualify as technological if, like the principles and rules of engineering, they can be justified both theoretically and practically.

1. THE GENERAL CONCEPT OF TECHNOLOGY

Technology is usually equated with engineering. This concept is too narrow, because it excludes or at least makes no clear room for such branches of technology as biotechnology, operations research, knowledge engineering, and the various sociotechnologies, from law and management science to urban planning and pedagogy. Given the proliferation of technologies over the past half century, we need a broader concept of technology. (See essay 24.)

I propose the following definition: Technology is the sector of human knowledge concerned with the design and redesign, maintenance, and repair of artificial systems and processes. In particular, scientific technology, unlike craftsmanship, makes use of basic science and mathematics.

The systems and processes in question may be physical, chemical, biological, social, or semiotic. Formal organizations such as schools, hospitals, and corporations qualify as artifacts along with machines, domestic animals, and high-yield grains. Likewise, managing and litigating, as well as healing and teaching, qualify as artificial processes along with steel lamination, construction, artificial hybridization, design, planning, accounting, information processing, and programming. Accordingly, modern technology includes the following branches: hardware technology, e.g., civil and nuclear engineering; biotechnology, e.g., agronomy and medicine; psychotechnology, e.g., psychiatry and marketing psychology; sociotechnology, e.g., management science and military strategy; and knowledge technology, e.g., information science and robotics.

It should be clear, then, that modern technology, though scientific, is not the same as basic science or even applied science; and that, though practical, it is not the same as manufacture or management. To clarify this point consider, for instance, the relation between industrial production P and gas emission E. The latter is proportional to the former; in fact, E equals a constant c times P, i.e., $E = cP$. This law statement belongs in applied science. To move from science to technology, or from *is* to *ought*, we replace the proportionality constant c with $c(1 - b)$. The new parameter b is a strategic or "knob" variable, i.e., one that can be regulated. More precisely, b can be varied between 0 (no control) and 1 (total prohibition). Indeed, if $b = 0$, then $E = cP$, whereas if $b = 1$, then $E = 0$. It is up to the environmental and health experts, jointly with the industrial economists, to recommend the optimal value of b, i.e., one that will protect the environment without throttling industry. In other words, the technologist turns the law statement "$E = cP$" into the rule "$E^* = c(1 - b)P^*$", to adjust the level of industrial production to a value compatible with certain social values, such as public health. In sum, the is-ought gap is being bridged in the light of scientific knowledge and social values.

The difference between technology and science is particularly important when it comes to identifying the culprits of war, exploitation, oppression, or environmental degradation. Do not blame the technologists, let alone the scientists that taught them. Blame instead their employers. Technologists do not make things: they only design them, mostly to order.

Technologists are intellectuals no less than scholars. Only the nature of their problems and their relation with their employers distinguishes them from scholars: they cannot indulge in unpromising speculation and, as a consequence, they seldom enjoy academic freedom—unless of course they happen to be academics. Moreover, their products—designs and plans—, unlike those of basic scientists and humanists, are priced and they belong to their employers, not to the public domain.

2. THE PHILOSOPHICAL TECHNOLOGIES

The philosophical technologies are all branches of philosophy that address general practical concerns in an armchair fashion, though not necessarily in ignorance of the relevant scientific and technological findings. Let us recall briefly their salient features.

To begin with, let us distinguish morals from ethics. *Morals*, or *morality*, is any system of social values and rules of conduct prevailing in a given social group—or at least the body of such norms that the group members pay lip service to. Examples: the loyalty and reciprocity values, and the ethical codes of the various professions.

Since moral codes contribute to shaping social behavior, their study behooves several biosocial and social sciences, from social psychology and anthropology to politology and history. The motley collection of disciplines about morals may be called *scientific ethics*. It is a strictly descriptive and explanatory discipline: its findings are testable and thus more or less true. Examples: the empirical study of the development of the moral conscience in the child, of the moral code of basic scientists, and of the emergence of moral norms along history.

On the other hand, *philosophical ethics*, or moral philosophy, is the branch of philosophy concerned with examining, proposing, interrelating, systematizing, and evaluating moral rules, whether actually enforced in some social group or desirable. Examples: the deontological, utilitarian, and agathonist moral philosophies. The union of scientific and philosophical ethics may be called *ethics*.

There are two major differences between scientific and philosophical ethics. One is that the former is predominantly empirical, whereas the latter is predominantly theoretical. The other difference is that scientific ethics is concerned with local morals, e.g., those of a given tribe or a given occupational group, whereas philosophical ethics reaches for a single morality for everyone.

Metaethics, or analytical ethics, is the branch of philosophy devoted to (a) analyzing such key moral and ethical concepts as those of goodness, rightness, fairness, and moral code; (b) examining the logical, semantical, epistemological, and ontological underpinnings and status of the moral discourse; and (c) unveiling its relations to value theory, science, technology, and ideology. Examples: the problem of the existence of moral facts and the corresponding truths, and the subject of this paper.

As for *praxiology,* or action theory, it is supposed to investigate the general concepts of individual and collective action, as well as the conditions of efficient action regardless of its moral value (see Kotarbinski 1965). In this regard, praxiology is nothing but the philosophical counterpart of management technology (usually called 'management science'). Examples: the investigation, in general terms, of the means-goal (or input-output) relation, and the search for general principles of efficient action, such as that of "satisficing" (instead of maximizing).

Now, an action can be efficient and satisficing to its agent, yet morally flawed for being selfish, just as it can be morally well-motivated but inefficient or even counterproductive. This shows that ethics and praxiology should not be conducted in isolation from each other, as they usually are. Only the union of the two fields can tackle the problems around the full legitimacy—both practical and moral—of action. One such problem is the design of the new behavior norms called for by the introduction of new practices or products bound to alter the everyday lives of many people, such as the downsizing of the workforce, the dismantling of the welfare state, info-addiction, and the globalization of junk culture.

Methodology may be regarded as the chapter of praxiology that studies the most efficient strategies for gaining knowledge, whether scientific, technological, or humanistic. Methodology, the normative branch of epistemology or theory of knowledge, is thus the intersection of epistemology and praxiology. Example: the R&D process can be analyzed as the sequence of the following steps: choice of field → formulation of a practical problem → acquisition of the requisite background knowledge → invention of technical rules → invention of artifact in outline → detailed blueprint or plan → test (at the workbench or in the field or on the computer) → test evaluation → eventual correction of design or plan. This sequence may be called the *technological method.* It is similar to the scientific method, except that technological tests are tests for efficiency rather than truth.

Finally, *political philosophy* is the research field whose practitioners tackle conceptual problems raised by social issues, in the light of all the other disciplines that touch on the same issues. It is thus a hybrid discipline: it lies in the intersection of political science, sociotechnology, ethics, praxiology, and ideology. It is not value-neutral like political science, but it is not supposed to be bound to special interests and dogmatic as an ideology. It must justify its value judgments and general policy recommendations, but it is motivated and guided by social ideals or "visions". Ideally, such justification should invoke not pious wishes but findings of hard-nosed social science and sociotechnology.

Consequently, technology may be split into two parts: scientific and philosophical. And the latter may be grouped together with information technology to form what may be called *software technology* in the broad sense.

The idea that ethics is a technology is less novel than it may seem at first sight. In fact, it has traditionally been regarded as the practical branch of philosophy. As for the inclusion of the other three philosophical disciplines among the technologies, the only surprise is that it does not seem to have been performed much earlier. This may be due to the traditional identification of technology with engineering.

Facing a practical problem, taking responsibility for it, and reflecting on the best means to solve it under the known constraints and in the light of the available knowledge and resources, may be regarded as a technological problem. Much the same holds for conceptual problems: these too call for the design of strategies and plans in the light of the best available knowledge. Likewise, facing a technological problem in any depth necessitates invoking general praxiological and methodological concepts and principles. And tackling any large-scale technological project with social responsibility requires some value-theoretical concepts and ethical principles, as well as principles of political philosophy. When such deliberations are skipped, environmental or social disasters are bound to happen.

These commonalities between technology, ethics, praxiology, methodology, and political philosophy coexist along with salient differences. The most obvious divergence between the philosophical and the strictly technical approaches to a practical issue is that the philosophical technologist is more interested in the universal than in the particular, and in morality rather than efficiency. Moreover, she tends to view the problems of scientific technology as mere examples that raise philosophical problems and check philosophical hypotheses.

However, in recent times public opinion has started to exert some pressure on the technological community, exhorting it not to skirt the moral aspect of human action. Indeed, this is the point of the enlightened branch of the Green movement, namely the one that does not reject science and technology. The classing of moral philosophy, praxiology, methodology, and political philosophy as technologies can only help drive this tendency forward. Or, at least, it can help deflate the popular perception that technology is necessarily at odds with the humanities.

Still, there is an additional difference between the philosophical technologies and the scientific ones: the former do not rely on any known laws, whereas the scientific ones do. Let me explain. Every technological rule, unlike the rules of thumb characteristic of the arts and crafts, is based on some scientific law. More precisely, any scientific law with possible practical application is the basis for two technological rules: one that tells us what to do to attain a given goal, and another that tells us what not to do in order to avoid a certain effect (Bunge 1967a, b). This is the root of the moral ambivalence of technology, in contrast with the moral univalence of basic science. Take, for example, the sociological law that the crime rate is a linear function of the unemployment rate. This scientific law is the basis of two rules of social policy. One of them states: To decrease criminality, create jobs. (This rule is ignored by the politicians who take credit for the decrease of criminality when unemployment is on the decline.) The dual rule is: To increase criminality, disregard the unemployment issue. (It might be thought that nobody implements this rule, but this impression is wrong. In fact, the legal crime industry, in particular the booming industries of law and order, in particular of the construction and management of jails both public and private, rely on the second rule, for it entails that politically profitable "Wars on Crime" should always take precedence over effective job creation programs and other crime-preventing programs.)

The peculiarity of technological rules is then that, far from being either conventional or sanctioned by practice alone, they are based on scientific laws. By contrast, the ethical and praxiological norms, as well as the policy recommendations are, at least so far, seldom justified in the same manner. It is arguable that they are only justified by their consequences and by such high-level principles as the Golden Rule, the utilitarian norm of the greatest happiness of the greatest number, or the agathonist maxim "Enjoy life and help live an enjoyable life" (Bunge 1989a). Likewise, the methodological rules are vindicated by their success in scientific research and technological R&D.

However, no realistic moral philosophy can afford to ignore the known biological and sociological laws and quasilaws, since moral norms are fashioned to cope with biological and social needs. Thus, although the moral norms are not explicitly based on known scientific laws, they are not incompatible with them either. Something similar holds for the praxiological, methodological, and political rules: to be realistic, these must match the rules that work successfully in real life as studied by the various technologies.

3. EXAMPLES

To qualify as a modern technology, a discipline must be able to tackle contemporary problems with the help of up-to-date scientific and technological knowledge. I claim that none of the classical moral philosophies satisfy these conditions, if only because none of them was built to cope with the moral side of such mega-issues as overpopulation, nuclear armament, mass unemployment, North-South inequities, or the manipulation of public opinion with the help of social psychology.

Something similar applies to the extant embryos of praxiology and political philosophy: they were not devised to tackle the complex problems of policy design, decision, and planning faced by corporate managers, cabinet ministers, or leaders of the opposition. Even the theories of decision and games are inadequate, despite their scientific appearance, in being simplistic and because the vast majority of them involve undefined concepts, such as those of utility and subjective probability. (Recall essay 22.) Hence, a fresh start is required: ethics, praxiology, and political philosophy must be reconstructed to become relevant to contemporary life and match, at the very least, the exactness of science and technology. The following examples are intended to sketch such a new start.

Let us begin with ethics. The classical moral philosophies are either of the duties-only or of the rights-only type. Nor surprisingly, they fail to match real-life situations, where fairness is attained only through balancing duties with rights. For example, the right to vote implies the obligation to cast the ballot in an informed way; and the duty to support one's children implies the right to earn the corresponding wherewithal. A realistic ethical theory will then include the maxim that rights imply duties and conversely. (This maxim can be rigorously derived from more basic ethical premises: see Bunge 1989a, as well as essay 28.) A system of moral philosophy both exact and relevant to modern life can thus be crafted.

Let me now exhibit a fragment of a systemic and science-oriented praxiology, namely the elucidation of the concepts of instrumental and moral rationality, which are supposed to be involved in the design and implementation of any large-scale action. (For details see Bunge 1998.) An action may be represented summarily by an ordered quadruple ⟨goals, means, outcome, side-effect⟩, or $A = \langle G, M, O, S \rangle$ for short, where S is included in O. Let us stipulate next that a means M to a goal G is *instrumentally rational* if, and only if, M is both necessary and sufficient for G. However, since the outcome O may not coincide with the goal G, the preceding formula must be revised to read: M is necessary and sufficient for O jointly with S. This forces us to supplement the previous definition of instrumental rationality with the following convention: M proves (a posteriori) to be an *instrumentally rational* means for G if, and only if, (a) M is necessary and sufficient for G, and (b) O is far more valuable than S (which may be disvaluable). For example, having an appendix removed meets both conditions. By contrast, since smoking satisfies the first condition but not the second, it is not instrumentally rational.

In some cases it is possible to quantitate the values of means and outcomes: this is normal with commodities. When this is the case, we may postulate that the degree of rationality of an action equals its efficiency, i.e., $\rho(\mathcal{A}) = V(O)/V(M)$.

The moral component is easily introduced by adding the assumption that a goal is *morally rational* just in case it contributes to meeting either a basic need or a legitimate want; that is, one whose satisfaction does not jeopardize someone else's chance of meeting her basic needs. When the needs satisfied and the side effects of the action can be quantitated, the degree of moral rationality can be defined as $\mu(\mathcal{A}) = V(N) + V(S)$.

This allows one to characterize an action as being *rational* if it is both instrumentally and morally rational. The degree of rationality of an action can then be set equal to the products of its degrees of practical and moral rationality: $\alpha(\mathcal{A}) = \rho(\mathcal{A}) \cdot \mu(\mathcal{A})$. This formula assigns to morality the same weight as instrumental rationality. Finally, we stipulate that an action is *fully rational* if, and only if, $\alpha > 0$, *nonrational* if $\alpha = 0$, and *irrational* if $\alpha < 0$.

So much for a sample of the philosophical technologies. Though tiny, it may suffice to show that ethics is logically prior to praxiology and political philosophy, because a morally justifiable action is only a special kind of deliberate action. By contrast, methodology is morally neutral. Yet it may be argued that, since efficient action relies on reasonably firm knowledge, the neglect of methodology is morally objectionable when human welfare is at stake. Shorter: methodology precedes praxiology, political philosophy, and scientific technology.

4. CONCLUSIONS

We have seen that, although there are clear differences between the philosophical technologies and the others, there are also important commonalities. This could not be otherwise, since all of them are normative disciplines concerned with getting things done in optimal ways. The realization of such commonalities has at least two consequences, one for the classing of technologies, and the other for academic activities, in particular the training of well-rounded technologists and philosophers.

The upshot of the preceding for the classing of technologies is this. We should add explicitly the branches of philosophical technology to the others. The new list looks like this:

Physical: e.g., mechanical, electrical, and mining engineering.
Chemical: industrial chemistry and chemical engineering.
Biological: e.g., agronomy and genetic engineering.
Biosocial: e.g., normative epidemiology and resource management.
Social: e.g., management science and law.
Epistemic: computer science and artificial intelligence.
Philosophical: moral philosophy, praxiology, methodology, and political philosophy.

If accepted, our thesis should have some impact on certain academic activities. First, philosophers should attempt to bridge the philosophical technologies to the standard ones. In particular, they should realize that doing ethics or praxiology should not be idle speculation in an epistemic and social vacuum: that doing the good and the right thing takes both knowledge and adequate resources.

Second, it should be realized that moral and political philosophies, as well as action theories and methodologies, should be validated in the same way as any other technologies, namely by their compatibility with the relevant sciences and by their consequences.

Third, philosophers should realize that the fashionable technophobic and irrationalist philosophies—in particular existentialism and hermeneutics—render students incapable of tackling the conceptual and moral problems posed by technological advancement, and consequently incapable of taking part in rational debates over the right way to adjust those advancements to social needs, and to adjust society to those innovations.

Another practical consequence of our thesis is that those responsible for the design of the curricula of schools of engineering, medicine, management, normative economics, law, city planning, social work, and other "hard" technologies, should realize that their good students are expected not just to apply recipes, but to find new knowledge and to tackle new issues armed with general principles of action and morals. Hence, their courses of study should include some ethics, praxiology, methodology, and political philosophy.

Such inclusion would benefit both parties. First, it would sharpen the moral conscience and the social responsibility of students, and it would stimulate philosophers to climb down their ivory towers, to become better acquainted with the day-to-day philosophical perplexities of the people who, perhaps more than anyone else, are designing the future. Second, awareness on the part of technologists that philosophers can make a contribution to technology would stimulate them to seek their cooperation instead of shunning them. After all, we know that useful epistemic novelty is often hatched in the interstices between disciplines, and that all technological megaprojects call for interdisciplinarity and generalism. No sectoral vision, however competent, can capture an essential feature of the world: that every thing is either a system or a component of some system. Philosophy can provide a unified view.

VIII

MORAL PHILOSOPHY

■

27

A NEW LOOK AT MORAL REALISM

(1993)

■

Moral realism is the view according to which there are moral facts and, correspondingly, moral truths. Though at variance with mainstream ethics, in particular with Kantianism and standard (subjective) utilitarianism, moral realism has been gaining currency in recent years. (See particularly Wiggins 1976; Platts 1979; Lovinbond 1983; Brink 1989.) Regrettably, so far this view has been presented mainly as a by-product of the philosophy of language, or rather linguistic philosophy. Hence, this variety of moral realism is unlikely to appeal to any thinkers who take the extralinguistic world as primary. Moreover, some moral realists (e.g., Platts) are intuitionists, hence distant from rationalism as well as from empiricism; and others (e.g., Wiggins) write about "the meaning of life". But no one who takes science and technology seriously, and regards them as inputs to philosophy, can condone intuitionism; and genuinely analytic philosophers distrust musings about "the meaning of life". Other moral realists, particularly Brink, profess ethical naturalism and objective utilitarianism. But the former is untenable, since moral rules are social, hence unnatural, as well as biological (or psychological); and utilitarianism does not account for the well-known fact that we often make sacrifices without expecting any rewards. For these reasons, in the following I will sketch an altogether different variant of moral (or rather ethical) realism, which is backed up by a comprehensive philosophical system including an ontology, an epistemology, and a semantics (Bunge 1974–89).

1. MORAL TRUTHS

Most philosophers do not even consider the possibility that there may be moral truths corresponding to moral facts. Perhaps their indifference to this possibility is ultimately due to their thinking of moral principles or norms as imperatives, of the type of "Thou shalt not kill". And of course an imperative may or may not be pertinent or efficient, good or evil, but it can be neither true nor false, because it neither affirms nor denies any matters of fact.

This objection overlooks the fact that the linguistic wrapping of an idea is conventional. Indeed, any ordinary knowledge idea may be expressed in different ways in any language and, a fortiori, in different languages. (The rider 'ordinary' excludes mathematical and scientific ideas, which require symbols that, though conventional, are more or less universal.) Since there are about five thousand living languages, it is possible to give at least ten thousand translations of any nontechnical statement.

In particular, any moral imperative may be expressed by a declarative sentence, with hardly any loss of content or practical value, in particular psychological or legal. For example, "Thou shall not kill" can be translated into "To kill is evil", "Murder is the worst sin", "It is forbidden to kill", and so on. The first two sentences designate propositions that hold (as true) in any moral code that affirms the right of all persons to life, and false in any code that does not admit such right. (Note that the principle in question holds only for persons, not for human embryos or for nonhuman animals.) As for the legal counterpart of the moral principle in question, i.e., "It is forbidden to kill", it is true in the vast majority of criminal codes—except of course those that include the death penalty.

In short, it is possible to propositionalize any given imperative. Such propositionalization is necessary to find the truth value (or the truth conditions) of moral maxims. Once a translation of an imperative into a declarative sentence has been performed, one should proceed to a second translation: one in terms of rights and duties, or else of virtues and vices, or of good or bad consequences of an action. For example, "Help thy neighbor" is first translated into "It is good to help your neighbor", or

"Helping one's neighbor is good". The truth of this maxim becomes obvious if expanded into "If you take into account your neighbor's needs, or wish to count on her in an emergency, or want to be at peace with yourself, you'll help her". Likewise, the precept "Saving is virtuous while waste is sinful" is true in a world suffering from increasing scarcity. (The fact that almost all economists recommend increasing spending in consumption during recessions, while decreasing it during periods of expansion, only suggests that they are just as amoral as they are short-sighted.)

The methodological rule that holds for moral truths also holds, of course, for moral falsities. For instance, the thesis that "What is good for the economy is good for the individual" is false, because "the economy" can be stimulated in the short term by spending on superfluous or even noxious commodities, such as sports cars and weapons. (However, the best economic journals, including the usually realistic *The Economist*, continue to praise managerial and economic policies that cause unemployment, as long as they benefit the stock market.) Likewise, the maxim "What is good for the government (or for General Motors, or for the Church, or for the Party) is good for all" is false every time the government (or the corporation, the church, or the party) forget about the welfare of the majority.

Now, if there are moral falsities, there must also be moral lies, i.e., sentences that express what the speaker knows are moral falsities. For example, the claim that there are inferior human races is not only scientifically false, it is also a moral falsity utilized to justify the exploitation or oppression of certain ethnic groups. And, whereas the statement of a moral falsity may be a mere error, that of a moral lie is almost always sinful. (The exception is of course the white or pious lie uttered to avoid unnecessary suffering.)

2. MORAL FACTS

On the correspondence theory of factual truth, the thesis that there are moral truths and falsities presupposes that there are moral facts. (The correspondence theory only holds for statements or denials of fact: logical and mathematical truths are an entirely different matter. See Bunge 1974b, as well as essay 13.) In turn, this presupposes that moral truths are just as factual as the truths of physics, biology, or history. Now, the vast majority of philosophers deny that there are moral facts. (Harman 1977 and Nino 1985 are exceptions.) They base this denial on the fact-value dichotomy usually attributed to Hume. Let us examine this matter.

Hume was perfectly right in asserting that what must be is not the same as what is: that norms are not of the same kind as factual propositions. However, I submit that he was wrong in holding that the value-fact gap is unbridgeable. He was wrong because we move daily from the one to the other, namely when taking action. For instance, if I tell myself that I must pay my debt to someone, and proceed to pay it, I cross the gap between ought and is. Likewise, when I take note of an unjust situation and make the resolve of remedying it, I go the inverse way.

What separates is from ought is a conceptual or logical abyss, but in practice it is not an unbridgeable chasm: it is a mere ditch that we can often jump over. It is a ditch that not only conscious beings can cross, but also any system endowed with a control device capable of leading the system from its current state to the final state judged valuable by the operator or user.

Furthermore, I suggest that every fact involving rights or duties, whether met or not, is a moral fact. For instance, poverty is a moral fact, not only a social one, because poverty involves avoidable pain and degradation. Involuntary unemployment is a moral fact because it violates the right to work, the source of all legitimate income necessary to meet needs and legitimate wants. Analogously, job creation is a moral fact, not only an economic one, because it satisfies the right to work. Another case of moral (or rather immoral) fact is the dumping of toxic waste in a lake, and this because it violates the right we all have to enjoy a clean environment. For the same reason the dragging of a lake to remove the toxic waste is a moral fact.

3. MORAL FACTS ARE SOCIAL

Moral facts are for real, but they are not of the same kind as physical facts: the former are social, while the latter are invariant with respect to social changes—notwithstanding the constructivist-relativist sociology and philosophy of science (see essay 23). How should moral facts be identified or individuated? One possibility is this: A fact is moral if, and only if, it poses a moral problem to some person in some culture. In turn, a moral problem is a problem whose examination and possible solution calls for, among other things, moral norms. And of course the latter pose ethical and metaethical problems, which may have solutions leading to altering some moral precept. And, in turn, such change may have an impact on individual behavior and therefore on the original moral problem, e.g., contributing to solving or worsening it.

Moral facts and our moral, ethical, and metaethical reflections on them are then components of a feedback circuit:

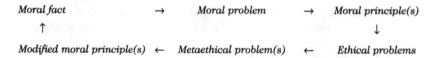

Moral fact → *Moral problem* → *Moral principle(s)*

↑ ↓

Modified moral principle(s) ← *Metaethical problem(s)* ← *Ethical problems*

In the end, then, moral truths resemble the truths of science and technology. But they differ in one important respect: unlike the latter, the former are contextual or relative. Indeed, since moral truths concern in the last analysis rights and duties, and since both are relative to the culture and its moral code, moral truths are contextual. In this regard, and only in this, moral truths resemble mathematical truths, all of which hold in certain theories but not in others. However, this dependence with respect to context is relative, because all viable moral codes share certain principles, such as that of reciprocity. In other words, some rights and duties are basic or unrenounceable, while others are dispensable or negotiable. The right to life and the duty to help the needy are absolute, whereas property rights and religious duties are not.

(Natural law theorists deny of course this partial relativity. In fact, they hold that all moral rights and duties are absolute, i.e., context-free. But, since as a matter of fact some moral codes do not admit certain rights and duties, the natural law theorist cannot hold that the corresponding propositions are true. And if, this notwithstanding, he insists on the claim, then he cannot hold that such moral truths correspond to any facts. Instead, he must either state that they are just as formal as the mathematical theorems, or he must adopt a nonrealist theory of truth. Given these difficulties, the natural law theorist will perhaps take refuge in the traditional opinion that the moral norms are neither true nor false, but only effective or ineffective, just like the cooking recipes or the medical prescriptions. But if he adopts this instrumentalist stand, he cuts the very roots that he claims his theory has in nature, so that he ceases to have the right to call himself a *natural* law theorist.)

In short, moral facts are social, not natural: they belong to the fabric of society, not to the furniture of the natural world. Moreover, they are artificial, i.e., made, not found.

4. TESTABILITY AND TRUTH VALUE OF MORAL NORMS

If all the moral norms, even the relative or contextual ones, have to do with moral facts, then it must be possible to subject them to empirical tests. In fact, I submit that a moral norm is *justifiable* if, and only if, (a) it is a moral truth for fitting a moral fact, and (b) it is practically effective for promoting the right conduct. There is only one exception to this rule, namely the maximal norm of the moral code in question, since it is the standard against which all the other components of the code are gauged. Let us see how a moral norm can be put to the test.

I suggest that the moral norms can be tested in three different, though mutually complementary, ways. The first test is that of compatibility with the higher-level principles. The second is that of compatibility with the best available relevant knowledge (ordinary, scientific, or technological). The third is that of the contribution to individual or social welfare. Let me explain.

Moral norms, just like scientific hypotheses and techniques, ought to be *compatible with the highest-level principles*, in this case the moral and metaethical maxims of the system in question. In the case of my own system, *agathonism*, the maximal principle is "Enjoy life and help live" (Bunge 1989a). Any rule that facilitates the implementation of this principle will be declared morally correct, otherwise incorrect. Example of the former: "Take care of yourself and your neighbor". An example of the latter kind is any rule leading to discrimination on the ground of sex, race, class, or religion.

The second of our tests is that of *compatibility with the best available knowledge*. For example, we should discard any rule that were to ignore moral feelings, such as compassion and goodwill. The same holds for any rule that were to overlook the fact that moral rules are human constructions and that they affect human relations. For instance, a rule that orders to make human sacrifices to the divinity is not only cruel: it is also inconsistent with modern knowledge about myths and conventions.

The third and last of our tests is that of *efficiency*. Evidently, it consists in finding out whether the given norm facilitates or hampers the realization of the basic underlying values, such as those of survival and self-government, and the freedoms to love and learn. In this regard moral rules resemble the rules of modern technology: both are based on laws, and both are tested by their efficiency in reaching certain goals. Moreover, the efficiency of moral rules, like the technical ones, can be estimated by the scientists and sociotechnologists who study or channel human conduct. A good moral rule is good only if it is efficient. But in order to be efficient, it must start by being viable, which in turn presupposes scientific realism. For example, "If struck, offer your other cheek" is not a viable rule. (Clearly, natural law theorists and Kantians disagree.)

However, even if a moral rule were to pass all three tests, it should be regarded as imperfect. Nearly five centuries of scientific and technological experience, not to mention social experience and philosophical critique, ought to prevent us from chasing infallibilist mirages, in particular that of a perennial ethics fashioned for perfect humans in the perfect society. We ought to know that no moral code and no ethical theory can guarantee right conduct. We may only hope to improve on the existing codes and theories through conceptual analysis and comparisons with empirical data. (For example, we must reject moral positivism, as well as legal positivism, for being totally relativist and conformist.) Such ethical fallibilism is far from both moral intuitionism and the ethical theories attached to religious dogmas.

But, unless accompanied by meliorism, fallibilism is destructive. I suggest that we can and must seek moral and ethical progress through theoretical work and social action, voluntary as well as institutional. Ethical meliorism may be justified as follows. Firstly, morals coevolve with society, and ethics with the rest of philosophy as well as with the social sciences and technologies. Secondly, if we try, we may discover and correct moral deviations and ethical mistakes, although sometimes either calls for some intellectual or even civic courage.

Take for instance the slogan *Freedom or death*, shouted since ancient times by uncounted ideologists and political leaders, particularly nationalist ones. At first sight all friends of liberty ought to embrace it. But, on closer inspection, the precept ought to be optional. To begin with, even a slave's life may be worth living, particularly if it makes room for some hope. Second, no one, not even the state, has the (moral) right to force anyone to fight till death for any cause. Third, the price exacted by any war without quarter can be staggering, not only because of the number of deaths, but also for making more difficult the life of the survivors. The above slogan is justified only in two cases: when there is evidence that the enemy won't give quarter, and when the price of personal liberty is treason.

5. STATUS OF ETHICAL AXIOMS

The preceding suggests agreeing with Einstein (1950, p. viii) that "Ethical axioms are found and tested not very differently from the axioms of science. *Die Wahrheit liegt in der Bewährung.* Truth is what stands the test of experience". In the cases of moral norms and ethical principles, the relevant empirical data concern human welfare. For this reason the biological and social indicators, such as life expectancy, infant mortality rate, number of school years, and median income, are more relevant to moral norms and ethical principles than the academic discussions on exceptional but minor moral problems, such as those posed by blasphemy, masturbation, abortion, test tube babies, or even suicide. All this pales by comparison with misery, war, and tyranny. (See Waddington 1960.)

Assuming that we have solved the problem of validation of moral rules, what can we say about the testing of ethical *theories*, i.e., systems of ideas about the nature, root, and function of moral norms? I hold that such theories are testable much in the same way as scientific and technological theories, i.e., by their agreement with the facts and their compatibility with other theories. Let me explain.

Any ethical theory should meet the following conditions. First, internal consistency, i.e., noncontradiction. Second, external consistency, i.e., compatibility with the bulk of scientific and technological knowledge about human nature and institutions. Third, ability to account for viable (livable) moral codes. Fourth, usefulness in suggesting some social reforms necessary for the exercise of the moral judgment. Fifth, usefulness in analyzing moral and ethical concepts and principles. Sixth, usefulness in resolving moral problems and in settling ethical disputes.

The condition of internal consistency disqualifies all the irrationalist ethical theories. The condition of external consistency disqualifies ethical emotivism (such as Hume's) and ethical intuitionism (such as Scheler's), since these views overlook the rational and empirical inputs of all moral deliberations. The same condition disqualifies utilitarianism, for ignoring the reality of moral feelings and for making use of subjective values and probabilities. The same condition disqualifies all the nonconsequentialist ethical theories, in particular Kant's. The fourth condition disqualifies ethical naturalism, for ignoring the social root and social efficiency of morals. The fifth disqualifies all the moral doctrines associated with philosophies alien to conceptual analysis. And the sixth disqualifies all the moral doctrines that are but arbitrary inventions of intellectuals removed from social reality—which, alas, has been the case with most moral philosophers.

6. OFFSHOOTS

The thesis that there are moral facts and truths has at least three consequences. First: If there are moral facts, then the moral principles are not dogmas but hypotheses, as Ingenieros (1917) proposed. And, being hypotheses, they must be confronted with the relevant facts, and revised if they do not match the corresponding facts. However, it does not follow that moral facts are just as inevitable as astronomic events. Whether or not they happen depends on us, moral agents.

Second: If there are moral facts, then moral judgments are not totally subjective and relative. Far from it, it is possible to evaluate such judgments in the light of experience, as well as to discuss them rationally. Moreover, it is possible to adjust action to morals and conversely. In particular, it is possible and desirable to reform society so as to minimize the frequency of destructive moral facts, such as armed violence, oppression, and exploitation.

Third: If the moral and ethical principles are empirically testable hypotheses as well as guides for individual and collective action, then it is possible and desirable to reconstruct moral codes as well as ethical theories. In particular, it is possible and desirable to reconstruct morality and its theory in such a manner that the following conditions be met.

1. *Realism*: Adjustment to the basic needs and legitimate aspirations of flesh-and-blood people placed in concrete social situations.

2. *Social utility*: Ability to inspire prosocial conduct and progressive social policies, as well as to discourage the antisocial ones.

3. *Flexibility*: Adaptability to new personal and social circumstances.

4. *Equity*: Efficiency in the task of decreasing social inequalities (though without imposing uniformity).

5. *Compatibility* with the best available knowledge of human nature and society.

In sum, there are moral facts and truths. Consequently, moral philosophers ought to pay less attention to metaethical disputes and become involved in the moral problems of real people, particularly the most vulnerable ones.

28

RIGHTS IMPLY DUTIES
(1999)

■

Moral and social progress has largely coincided with the expansion of rights, particularly from the time of the American Declaration of Independence of 1776. However, a rights-only political philosophy, such as contemporary libertarianism, is anything but progressive. In fact, it is regressive in being selfish and therefore morally corrosive and socially dissolving. In other words, an overemphasis on rights can backfire: if I only care about my own rights, then I am bound to trample over those of my fellow human beings, who will in turn have no qualms in infringing upon my own rights.

Does it follow that the struggle for an ever-expanding circle of rights, and the passing of legislation entrenching such rights, has gone too far? Not at all. In fact, most human beings do not enjoy some basic human rights, such as the rights to work, to free speech, and to association. Even in political democracies there are bulky pockets of people who cannot exercise some of their rights for lack of means: think of the jobless, the illiterate, and the politically marginal.

What is to be done to prevent the struggle for rights from degenerating into an antisocial fight for individual or group privileges? I submit the platitude that the concern for rights should be combined with a concern for the concomitant duties. Moreover, I submit that the pairing of rights to duties is necessary in a sustainable society: that rights and duties imply each other.

In the following we begin by elucidating the concepts of right and of duty, and go on to proposing some moral norms concerning both. We then proceed to prove that every right implies a duty. We end up by stating a further moral principle that occurs as the maximal principle of a morality designed to overcome the limitations of both individualism (and the egocentrism associated with it) and holism (and the concomitant sociocentrism).

1. RIGHTS AND DUTIES

Rights and duties may be partitioned into natural, moral, and legal. A *natural right* is one that an animal has conquered or been given. For example, a wolf acquires by force the right to control a certain territory, and it gives her cubs the right to play only within her sight. Moreover, the animal sees to it automatically, without any need for social control, that her offspring get the care they need: she thereby discharges her *natural duty*.

The concepts of natural right and natural duty become highly problematic in the case of humans, because there are no human beings in a state of nature. Indeed, humans are largely artifactual, i.e., products of their upbringing and their activity in a given society. Let homes or societies be changed, and different human beings will develop. This is why, unlike all the other animals, humans need a more or less explicit recognition of their moral and legal rights and duties. This is also why the concepts of right and of duty must be analyzed as high degree predicates: Animal a, when in state b, has the right (or duty) to perform (or refrain from performing) action c in society d when in state e. Shorter: rights and duties are not intrinsic but contextual.

We start by elucidating the concepts of legal right and legal duty by stipulating

Definition 1. If a is an action that a member b of a human group c is (physically) capable of performing, then

(i) b has the *legal right* to do a in c $=_{df}$ according to the legal codes in force in c, b is free to do or not to do a in c (i.e., b is neither compelled to do a nor prevented from doing a in c);

(ii) b has the *legal duty* to do a in c $=_{df}$ b has no legal right to refrain from doing a in c.

Note in the first place that a legal right is not the same as the effective exercise of it: ordinarily, rights do not come with the means to exercise them. (Think of the right to work.) The same holds for duties: having the legal duty to do x does not guarantee that x will be done. Secondly, according to the above definitions, rights and duties are vested in individuals, because they concern actions performable by individuals or their proxies (humans, animals, or robots). To put it negatively, there are no collective rights or duties, just as there are no social choice and no social action. The expression 'collective right (or duty)' must therefore be read as 'the right (or duty) of every member of a certain social group'. This is not to say that rights and duties are infrasocial (natural) or suprasocial (God-given). It just means that only individuals can be assigned rights and duties or deprived of them. The idea of collective rights and duties is a holistic myth that has been used to dilute personal responsibility and to discriminate in favor or against special groups. It has so been used by nationalists to deprive minorities from some of their rights.

The above definition tells us what legal rights and duties are, but not which they are in a given society. The answer to the which-question is not philosophical but legal: it is found in the codes of law and the jurisprudence of a land.

In addition to legal rights and duties there are moral ones. The former are specified by the law, whereas the latter are accepted or self-imposed independently. There is no moral obligation to obey the law of the land as such. Moreover, if a particular law is bad, then it is our moral duty to fight it. (See also essay 29.) Nor is every moral duty enshrined in a code of law. For example, there is no law ordering us to help our neighbor. The most praiseworthy actions are those performed beyond the call of (legal) duty. However, in every society the legal and moral codes have a nonempty intersection. For instance, the moral rights to live and love, to work and learn, and many others, are enshrined in most codes of law. A society where this were not the case would be hardly viable.

I submit that a moral right is the ability to meet a basic need, such as eating and consorting with other people, or a legitimate want, such as improving one's lot without violating other people's basic rights. Likewise, I shall propose that a requirable moral duty is to help someone else to exercise her legitimate moral rights. More precisely, we stipulate

NORM 1. If a is a human being in society b, and c is a thing or process in or out of a, then

(i) *a* has a *basic moral right* to *c* in *b* if, and only if, *c* contributes to the well-being of *a* without hindering anyone else in *b* from attaining or keeping items of the same kind as *c*;

(ii) *a* has a *secondary moral right* to *c* in *b* if, and only if, *c* contributes to the reasonable happiness of *a* without interfering with the exercise of the primary rights of anyone else in *b*.

According to this norm, everyone is entitled to enjoy well-being. Belief in such entitlement is instrumental in designing, proposing, and implementing social reforms aiming at the effective exercise of human rights.

The above norm suggests that the pursuit of (reasonable) happiness is legitimate, provided it does not hinder others from enjoying a state of well-being. But the norm does not state that everyone is entitled to reasonable happiness, because getting hold of the required means may involve entering into some competition, and in all competitions there are losers as well as winners. When the play is fair, the winners make use of special abilities which not everyone possesses or can acquire, as well as of a bit of luck, which does not strike everyone. (Luck = $_{df}$ Unexpected opportunity to gain something.) Of course, rational people will not persist along lines where they have no chance of winning honestly, but will try some alternative line instead. However, not everyone is rational, let alone lucky. Consequently, even the best of societies are bound to contain some unhappy people: some because they have failed to realize all their ambitions, others because of neuroendocrine dysfunctions, still others because they have been victims of an education that stresses personal success (or, on the contrary, selflessness) above all, and so on.

Our next norm concerns duties:

Norm 2. If *a* and *b* are human beings in society *c*, and *d* is an action that *a* can perform (by herself or with the help of others) without jeopardizing her own well-being, then:

(i) if *b* has a primary right in *c* to *d* or to an outcome of *d*, then *a* has the *primary moral duty* to do *d* for *a*, if and only if *a* alone in *c* can help *b* exercise her primary moral right to *d* or to an outcome of *d*;

(ii) if *b* has a secondary right in *c* to *d* or to an outcome of *d*, then *a* has the *secondary moral duty* to do *d* for *b*, if and only if *a* alone in *c* can help *b* exercise her secondary moral right to *d* or to an outcome of *d*.

Examples: Normally parents are the only people, or at least the best-placed ones, to take good care of their children; hence they have the moral duty to do so. Likewise, in most societies aged parents are dependent on their adult children for their well-being, whence the filial duty.

Since scarce resources can be had, if at all, in limited quantities or in exceptional circumstances, some needs and wants can be met only partially. In other words, the effective exercise of rights and performance of duties is limited to the availability of resources and their distribution in each society. Note also that, whereas some duties can be delegated to others, such as employees or robots, most rights are not transferable except with adequate compensation. For example, a life prisoner may wish to be given the chance of freedom in exchange for his volunteering in a risky medical experiment. The key word here is, of course, 'volunteering'.

In manipulating rights and freedoms, particularly in trading rights for obligations, we should abide by

NORM 3. All of the basic moral rights and duties are inalienable, except for trade-offs contracted between conscious and consenting adults under the supervision of a third party capable of having the contract observed.

Examples: The rights to live, work for compensation, choose one's friends, and have a say in matters of public interest, are morally inalienable. (Whence the immorality of military draft, involuntary unemployment, choosing other people's friends, and vote buying.) So are the duties to help our relatives, friends, and colleagues; to respect other people's rights; and to preserve the environment. On the other hand, property rights are typically alienable, whence they have no moral status.

Caution: the alienable-inalienable distinction is not a dichotomy, for there is a gray zone composed of such rights as selling parts of one's own body, e.g., blood and a kidney. Another is the right to pollute, which is tacitly recognized by any laws or rules that impose fines on polluters without forcing them not to dump wastes. However, these particular cases, though legally gray, are morally black, for the sale of parts of one's own body puts life at risk and in the hands of the better off, and environmental pollution is bad for everyone.

Sometimes certain moral rights or duties are enshrined in legal codes. At other times they conflict with the law. For example, a political prisoner who has not committed any violent act has the moral right to try and escape from prison, and we all have the duty to report on those who plan criminal actions, in particular aggressive wars. We generalize and make

NORM 4. Morality overrides the law of the land.

Rights and duties can be grouped into five main categories: environmental, biopsychological, cultural, economic, and political. Although these are different, they are not mutually independent. For example, the free exercise of civic rights, e.g., voting, and duties, e.g., protecting public goods, is necessary to protect or facilitate the access to economic, cultural, biopsychological, and environmental resources; and the enjoyment and protection of the environment are necessary for both sanitary and long-term economic reasons. Shorter: the moral rights and duties form a *system*.

The systemic character of moral rights and duties is not universally understood. On the one hand, liberals emphasize political and cultural rights, as well as certain economic rights, such as those of hiring and firing. On the other hand, socialists emphasize the biopsychological rights and the right to work, together with security and education. The former overlook the fact that it is hardly possible to play an active part in politics or to enjoy culture on an empty stomach; and some socialists overlook the fact that only political liberty and participation can conquer or protect other rights. To press for rights or duties of certain kinds while overlooking the rest, is a mark of naïveté or worse. And to forget about duties altogether, or to put the defense of flag, altar, or party before interpersonal mutual help and international cooperation, is to invite jingoism and fanaticism, hence bloodshed.

2. RIGHTS IMPLY DUTIES

Though neither of the norms proposed in the preceding section is deducible from another, they are related in various ways. An obvious relation is this: A duty cannot be performed unless one has both the freedom and the means to exercise the right to do so. Other relations between rights and duties are far from obvious, and they vary from one social group to the next. For example, the members of a pleasure-oriented group will tend to neglect their duties, whereas those of a stern religious or political group will tend to forego some of their rights. We stipulate the following relations among the four items in question:

NORM 5. (i) Primary rights take precedence over secondary rights. (ii) Primary duties take precedence over secondary duties. (iii) Primary duties take precedence over secondary rights. (iv) An individual faced with a conflict between a right and a duty is morally free to choose either, subject only to condition (iii).

In other words, first things should come first, and survival of self and others first of all. In particular, the meeting of basic needs ought to dominate the satisfaction of wants; and among the latter the legitimate ones should take precedence over all others. Also, duty comes before pleasure. However (see: [iv]), everybody has the right to sacrifice herself, even to the point of putting her well-being or even her own life on the line. At the same time, nobody has the right to demand such sacrifice, or even to remonstrate with anyone who fails to make it. In particular, nobody has the right to force or trick people into their death.

The above norms have been couched in the declarative, not in the imperative mode: they are propositions, not commandments. (However, they function as injunctions or exhortations, not as descriptions.) Being propositions, they can be conjoined together or with others to entail further propositions. (To put it negatively: we do not need any system of deontic logic.) A first consequence of Norms 1 and 2 is

THEOREM 1. Every right implies a duty.

Proof Take a two-person universe and call N_i a basic need, and R_{ii} the corresponding right, of person i, with $i = 1, 2$. Likewise, call D_{12} the duty of person 1 toward person 2 with regard to her need N_2, and D_{21} the duty of person 2 toward individual 1 with regard to her need N_1. Our first two norms in the preceding section can then be abbreviated to

Norm 1. (If N_1 then R_1) & (If N_2 then R_2)

Norm 2. [If R_1, then (If R_1 then D_{21})] & [If R_2 then (If R_2 then D_{12})]

By the principle of the hypothetical syllogism it follows that

[If N_1 then (If R_1 then D_{21})] & [If N_2 then (If R_2, then D_{12})].

And, since everyone has basic needs, it finally follows, by *modus ponens*, that

(If R_1, then D_{21}) & (If R_2, then D_{12})   Q.E.D.

This theorem, and the postulate that all persons have certain moral rights (Norm 1), jointly entail

COROLLARY 1. Everyone has some duties.

We can extract further consequences if we adjoin further premises. We shall add now the anthropological, sociological, and economic truism that no human being is fully self-reliant or self-sufficient:

LEMMA. Everybody needs the help of someone else to meet all of her basic needs and some of her wants.

This proposition entails the mutual help, reciprocity, or *quid pro quo* principle:

THEOREM 2. Helping implies being helped and conversely.

Proof As in the proof of the previous theorem, imagine a two-person universe. In this case our Lemma reads: (If N_1, then H_{21}) & (If N_2, then H_{12}), where 'H_{21}' stands for "person 2 helps person 1 meet need N_1", and similarly for H_{12}. Suppose now that 1 helps 2, but 2 does not reciprocate, i.e., H_{12} & not-H_{21}. This is equivalent to the denial of "If H_{12} then H_{21}". Transposing the Lemma we get: "If not-H_{21}, then not-N_1". But N_1 is the case. Hence, by *modus tollens*, H_{21}, which contradicts our supposition. Therefore: If H_{12}, then H_{21}.

To prove the second conjunct, we exchange the indices in every line of the preceding proof. Conjoining the two results we obtain the desired theorem, i.e.,

H_{12} if and only if H_{21}.   Q.E.D.

3. THE SUPREME AGATHONIST PRINCIPLE

Life in a viable society requires a balanced mix of self-help with mutual help. This is why the most universal (least ethnocentric) moral maxims combine rights with duties. Witness the Golden Rule. Any moral doctrine that, like hedonism, admits only rights or, like Kantianism, admits only duties may be said to be *unbalanced*, hence unviable. Only psychopaths, professional criminals, warmongers, robber barons, tyrants, and members of societies in the process of disintegration refuse to engage in mutual help—to their own long-term detriment and that of their society.

Normal social behavior is prosocial, i.e., a matter of give and take, of balancing rights with duties, and conversely. For example, every able-bodied person has the right to work and the duty to work well, and every citizen has the right to have a say in the manner he is being ruled, and the duty to participate in the political process, if only to see to it that his rights will be respected.

Although basic needs and legitimate wants generate rights, and these duties, the latter in turn restrict rights. In particular, every one of my liberties ends where someone else's begins. The law of reciprocity, or give and take, is neither more nor less than a trade-off between rights and duties. More precisely, there is a negative feedback between these.

The exclusive pursuit of one's own interests with disregard for the well-being of others is not only socially destructive. It is also chimerical, for nobody is fully self-reliant, and a person does not get help from others on a regular basis unless she is willing to do something for them. (Even slavery and serfdom involve trade-offs, in particular the exchange of labor for security.) As for the exclusive dedication to the welfare of others, without regard for one's own well-being, it is self-destructive and therefore inefficient. A sick saint is as useless to society as a sated pig.

All of the preceding can be summarized into a moral norm that constitutes the supreme principle of our morality:

NORM 6. Enjoy life and help live (enjoyable lives).

The first conjunct is of course the devise of hedonism or individualistic utilitarianism. It is tempered by the second conjunct, the lemma of philanthropism. (The latter must not be mistaken for the negative utilitarian maxim "Let live" or "Do no harm", which goes well with the egoistic maxim "Do not become concerned, do not become involved".) Taken together the two parts of Norm 6 synthesize egoism and altruism, self-interest and other-interest, egocentrism and sociocentrism, autonomy and heteronomy. It may therefore be called the *selftuist* principle. I submit that this principle, far from being just a pious thought, is the most effective guide for the design or redesign, building or maintenance of a good society, namely one in which everyone can meet his or her basic needs and legitimate wants.

IX

SOCIAL AND POLITICAL PHILOSOPHY

■

29

MORALITY IS THE BASIS OF LEGAL AND POLITICAL LEGITIMACY

(1992)

■

1. THE PROBLEM

What can be legitimate or illegitimate, i.e., what are the possible bearers of the predicate "legitimate"? All social human actions and all the norms (moral or legal) and institutions that regulate such actions can be said to be legitimate or illegitimate in some respect or other. In particular, governments and nongovernmental organizations, as well as their actions, can be legitimate or illegitimate relative to certain standards.

In legal and political practice one says that an action, a parliamentary bill, a governmental decree, or the activity of an organization is legitimate if, and only if, it conforms to the law of the land and, in particular, to its constitution. But why should the law of the land, or even its supreme code of law or constitution, be exempt from suspicions of illegitimacy? Just because it is propped up by custom and the law and order forces?

For that matter, why should anything social be immune to a possible charge of illegitimacy? After all, every society is in a process of flux, and all social change, whether progressive or regressive, involves changes in what is judged to be legitimate or illegitimate. Remember what was said of British social progress in the nineteenth century: that it consisted mainly in repealing unfair laws. If our "perception" of fairness is changeable, our judgment of legitimacy is bound to be subject to change as well. As for politics, the cases of Charles I, Louis XVI, and so many other erstwhile legitimate heads of state are a good reminder of the ephemeral nature of political legitimacy.

Does it follow that legitimacy is one more *filia temporis*, i.e., that there are no well-defined and fairly constant standards helping us judge in a rational way whether anything is objectively legitimate regardless of the situation? Shorter: Are we forced to adopt a historicist, relativist, and perhaps even irrationalist view of legitimacy just because judgments of legitimacy are bound to change in the course of history?

I submit that the historicist-relativist view on legitimacy is mistaken. It is mistaken for the same reason that it is an error to think that, because particular truths about the world are likely to change as a result of research, the very concepts and criteria of (factual) truth are bound to zigzag along with the history of science. Whoever adopts this view cannot claim it to be true, and he cannot undertake to find out whether today's scientific data and theories are truer than their predecessors. Likewise, the historicist-relativist view in the matter of legitimacy is bound to bow to the powers that be. Hence, it won't be willing or even able to condemn cannibalism, torture, slavery, conquest, or exploitation, provided that such practices are or were condoned by some entrenched custom or code of law.

I will attempt to defend the thesis that, ultimately, legitimacy is based on morality, and that even the latter can only be justified in terms of the axiological principle that only what helps meet basic human needs or legitimate aspirations can be judged to be good. In a nutshell, my main thesis is that good precedes just precedes legitimate.

2. FORMAL LEGITIMACY OF NORMS

Before tackling the question 'When is a norm legitimate?' it will be convenient to recall a few technical concepts. Firstly, we shall take it that every norm couched in the imperative mode can be propositionalized, i.e., translated into the declarative mode. Thus, "Don't kill" can be propositionalized as "It is wrong to kill". The equivalence we claim is epistemic, not pragmatic. Obviously, an imperative, when uttered by a person in authority, carries far more weight than a declarative: it is a command, not just an information.

The propositionalization of norms, proposed independently by Miró-Quesada (1951), Leonard (1959), and Bunge (1973c), allows one to use ordinary logic throughout, and hence to dispense with deontic logic, which, as is well known, is ridden with paradox. (See, e.g., Weinberger 1985.) In particular, the trick allows one to investigate the logical consistency of systems of norms in the same way one examines the consistency of theories.

Secondly, we shall suppose that there are no stray norms: that every norm is a constituent of some normative system, i.e., of a collection of norms that are held together by logical relations, such as that of implication. More precisely, we adopt the following definition (Bunge 1989a, p. 298): "A *normative system* is a set of norms, descriptive propositions, and definitions, that (a) share some referents (e.g., human beings or actions of certain types); (b) can be disjoined, conjoined, instantiated, and generalized to yield further norms in the system; (c) contains (potentially) all of the logical consequences of any one or more norms in the system, and (d) can be enriched, without contradiction, with further nonnormative formulas, in particular data and natural or social laws, i.e., it is an incomplete system." (For an alternative definition see Alchourrón and Bulygin 1971.)

Thirdly, we define a *metanorm* as a norm about norms, i.e., one that refers to one or more norms. The categorical imperative and the legal principle *No crime without law* are classical examples of metanorms. Like norms, metanorms come in clusters.

We are now ready to examine the question 'When is a norm legitimate?'. We begin by elucidating the notion of formal legitimacy, and leave the matter of substantive legitimacy for the next section. We stipulate that *a norm in a system is formally legitimate* if, and only if, it is consistent with the basic norms in the same system, and it satisfies the previously stipulated metanorms. Obviously, the second conjunct becomes redundant if the whole system complies with the given metanorms.

Normative systems fare similarly. We stipulate that *a normative system is formally legitimate* if, and only if, it is internally consistent and it satisfies the previously stipulated metanorms. Thus, a constitution is formally legitimate if it is coherent and abides by the underlying metanorms. Some of these are laid down in the preamble to the constitution or in an accompanying document, such as the American Declaration of Independence or in the *Déclaration des droits de l'homme et du citoyen*.

Obviously, formal legitimacy is insufficient, for a monstrous moral or legal code can be formally legitimate, as was the case with the Nazi criminal code. Where shall we turn for substantive legitimacy? I suggest that only morals can supply it, though not all of them, but only those that promote well-being.

3. SUBSTANTIVE LEGITIMACY OF NORMS

Before tackling the problem of substantive legitimacy we must introduce two concepts that, though crucial, are alien to most moral and legal philosophies: those of basic need and legitimate want. We stipulate that a *basic human need* is one that must be met for a human being to keep or regain physical and mental health in his or her society. Adequate food and shelter, feeling loved and needed, and enjoying company and security are obvious examples of basic needs. Although the means for meeting these needs vary from one society to the next, the needs themselves are universal: they characterize human nature.

We explicate next what we mean by a socially legitimate want, wish, or desire, in contrast to an unqualified desire. Let x be a want of a human being b in circumstance c in society d. We say that x is a *legitimate want* (or *desire* or *aspiration*) of b in circumstance c in society d if, and only if, x can be met in d (i) without hindering the satisfaction of any basic need of any other members of d, and (ii) without endangering the integrity of any valuable subsystem of d, much less that of d as a whole. Since different people may have different desires in one and the same society, and since different societies have different resources, it is clear that legitimate wants, unlike basic needs, are place- and time-bound rather than universal.

When can we say that a norm is legitimate regardless of its relations to other members of the same system or to a collection of metanorms? In other words, when is the content of a norm legitimate? I submit that a norm is *substantively legitimate* if, and only if, its implementation helps meet either a basic need or a legitimate want of all human beings. For example, the laws that impose on a government the obligation to protect the environment, to police the streets, to issue food stamps to the unemployed, and to provide free and universal elementary education and medical care are substantively legitimate, because they help meet some of the basic or universal human needs. On the other hand, any laws that enshrine privileges without compensation are illegitimate. Likewise, any bills or private initiatives promoting the sciences, the humanities, or the arts are substantively legitimate as long as they help some people realize their legitimate aspirations without jeopardizing anyone else's well-being. But if such aspirations were satisfied at the expense of other people's well-being, then those governmental or private initiatives would be controversial if not plainly illegitimate.

(As a matter of fact, one of the arguments employed in the Third

World to fight the support of basic research boils down to the proposition that science is a luxury that poor nations cannot afford. That this argument is fallacious for ignoring the benefits of science for education, technology, and industry is beside the point. The point is that socially aware people draw a distinction that escapes most moral philosophers, namely that between need and want.)

What holds for individual norms holds, mutatis mutandis, for normative systems as well. A system of norms will be said to be *substantively legitimate* if, and only if, every member of it is substantively legitimate. A few exceptions may be unavoidable, but once identified they can be corrected. As a matter of fact, in all advanced countries the codes of law are under constant scrutiny and are reformed from time to time in the light of experience and in tune with new aspirations and new possibilities.

4. FULL LEGITIMACY OF NORMS AND ACTIONS

Having elucidated the concepts of the formal and substantive legitimacy of norms, we are now prepared to define the notion of norm legitimacy *tout court*. We stipulate that a norm, or a normative system, is *legitimate* if, and only if, it is both formally and substantively legitimate. The import of this definition for ordinary, legal, or political practice is clear. A norm, law, code, decree, or regulation is *legitimate* if, and only if, (a) it is consistent with the entire legal corpus as well as with the underlying metanorms, and (b) it helps meet some basic human needs or legitimate aspirations. The reader is invited to supply his own examples and counterexamples.

Having reached this point, the reader may wonder whether I have not confused legitimacy with justice, in particular social justice as different from mere legal justice. Indeed, I plead guilty to having deliberately introduced a concept of legitimacy which is ostensively coextensive (and covertly cointensive as well) with a broad concept of justice that includes both social and international justice. The conflation I advocate makes it easier to defend the view that matters of legitimacy are ultimately matters of justice.

What is to be done about a norm, law, code, decree, or regulation that is illegitimate in some respect? Obviously, it must be contested or, if necessary, even disobeyed. In any event, illegitimacy calls for action of some kind, individual or collective. Therefore, it is time to move from norm to action. We shall limit our considerations to deliberate or purposeful actions, for these are the only ones that can be ruled by norms. We stipulate that *a deliberate action is legitimate* if, and only if, it fits some legitimate norm—with the proviso that this norm may be moral rather than legal, and unwritten rather than written. The proviso is important because the legal or moral code in force at a given place and time may be deficient, e.g., for favoring some special interests. In this case we may be forced to challenge a law or decree in the name of a metanorm or of a higher moral principle. For example, there is no moral obligation to observe a law or decree that is applied retroactively. And there is a moral obligation to challenge any laws establishing the death penalty, the political disenfranchisement of a minority, or unfair taxation.

We have just used in a tacit manner a metanorm, namely that, in case of conflict between a legal disposition and a moral principle, the latter shall prevail. This metanorm is in tune with the moral underpinning of substantive legitimacy. But this is not the only principle underlying our *dicta* concerning the death penalty, political discrimination, and unfair taxation. The tacit principle in question is the supreme norm of agathonism, viz., *Enjoy life and help live* (Bunge 1989a, see also essay 28).

Thus, in the last analysis we are to attribute legitimacy (= justice), or its dual, in the light of that supreme norm. That is, a norm or an action is legitimate (just) if, and only if, it jibes with the principle enjoining us to enjoy life and helping others enjoy their own lives. Interestingly enough, correct (true) attribution of legitimacy in individual cases calls for some empirical investigation, for we must ascertain whether in fact the norm or action in question enhances the quality of life of the people involved.

5. ILLEGITIMATE POWER AND LEGITIMATE REVOLT

There are four basic kinds of power: biological, economic, cultural, and political. (Examples: physical force, control of a natural resource, religious influence, and parliamentary majority, respectively.) For every one of these kinds of power we stipulate that it is legitimate if, and only if, it abides by legitimate norms. Equivalently: a power is illegitimate if it is exerted in accordance with illegitimate rules—which may be corruptions of legitimate norms. In particular, a legitimate government is one that abides by the legitimate legal and moral norms. And an illegitimate government is one that abides by illegitimate rules, whether legal or in violation of legitimate laws. For example, a government that oppresses its citizens, as military dictatorships are wont to do, or loots or oppresses those of other nations, as imperialist powers do by definition, is illegitimate.

The next question is: What legitimate means can people use to countervail an illegitimate government? Passive disobedience, including the refusal to pay taxes, is clearly one of them. But, unless well organized, passive disobedience may turn into apathy, which can only benefit the illegitimate power; or it may end up in anarchy, to the benefit of no one.

To be efficient, political protest must set people in motion with a definite purpose, such as obtaining the removal of some political leaders or even the replacement of the incumbent administration.

However, we must face the possibility that the politicians in power resort to violence in order to stay in power. We all know how easy it is for an illegitimate government to escalate violence. The process may start by denying permission to demonstrate in the streets; water cannons may be used to disperse the demonstration; if jets of water fail, beatings may be resorted to; and if beatings prove insufficient, guns or even tanks will come in handy. The case of international violence and the response to it is of course parallel.

Is it justified to resort to violence in order to respond to violence? It may be argued that it is: that, when all peaceful means to attain a legitimate goal fail, the use of violence is legitimate. (After all, Marx and Weber defined the State as the legal monopoly of violence.) But, as we all know, the use of violence poses two problems, one practical and the other moral. The practical problem is that a violent action, such as a popular armed insurrection or a military intervention, may be crushed, and if victorious it may cost an unpredictably high number of lives, or it may end up by installing an equally oppressive regime. The moral problem is, of course, that the taking of lives goes against a basic moral norm.

The defender of violent collective action will dismiss the moral objection and counter with a utilitarian argument. She will say that a well-organized revolution, merciless repression, or "surgical" military strike, as the case may be, is bound to cost fewer lives than the bloody dictatorship in the saddle. But the truth is that, except in the extreme case of a weak, corrupt, and discredited dictatorship, such as Fulgencio Batista's in Cuba, no one can predict the number of victims, combatants or not, caused by an armed insurrection or a military strike. Nor can anyone predict whether the new government won't be tempted to continue to use violence even after its predecessor has been crushed. In short, for both moral and practical reasons revolution is to be avoided— whenever possible. Revolution should only be resorted to when all attempts at reforming a thoroughly unjust and violent order have repeatedly failed—and even so it should minimize bloodshed.

How can power won by violence, perhaps at the cost of many lives, be legitimate? This is a loaded question, for it presupposes that the matter of legitimacy is of the all-or-none type. An initially illegitimate government may gradually earn legitimacy, just as an initially legitimate government may gradually lose it. For example, the young Soviet government earned legitimacy and popular support almost overnight by making peace with the Central Powers, implementing a land reform, initiating a vast educational campaign, and defending the country against foreign military intervention. But it lost its legitimacy as it dissolved the soviets (popular councils) and became a bloody dictatorship which suppressed all criticism, discouraged initiative, and stifled creativity. In short, legitimacy is a matter of degree, and this degree is changeable.

6. THE TECHNOLOGICAL INPUT

Suppose that, by some means or other, peaceful or violent, we got a legitimate government in place. Would this suffice? Clearly not, because the administration, though honest and driven by the best of intentions, may be incompetent. In other words, we expect more than legitimacy from good government: we also expect technical competence. In particular, we expect sound management of the treasury and of the education and healthcare systems, and much more.

Suppose next that the unlikely has come to pass: that we finally got a government which is competent as well as legitimate. Should this satisfy us? I expect not, for the two conditions may be satisfied by parliamentary democracy or even by enlightened despotism, provided the latter be mild. Assuming that the reader shares the writer's dislike of all kinds of despotism, he or she may wonder what is wrong with purely representative democracy. Let us address briefly this important question.

Let me begin by declaring that I believe political democracy to be necessary as a means to defend and expand our rights, and as a school of civic duty. But I submit that political democracy is insufficient to assure that the basic needs and the legitimate aspirations of everyone be met. In fact, it may mask economic and cultural privileges of all kinds.

Political democracy is a good starting point for a process leading to integral democracy, a social order where everyone has access to the economic and cultural resources necessary to satisfy his or her basic needs and legitimate desires. An integral or social democracy is a participative or grass roots one, one where there is no concentration of economic, cultural, or political power. Such a social order would be legitimate according to our definition of "legitimacy" in sections 2 and 3.

But might such an order work? Not unless it were to combine popular participation with technical competence. Indeed, it takes technical expertise and organizational skills, as well as widespread respect for such competence, to prevent participative democracy from degenerating into unending meetings or, worse, mob rule. The intelligent design and the effective implementation of plans calls for sociotechnology as much as it requires enthusiastic and orderly participation of all the stakeholders involved. The combination of integral democracy with technological expertise may be called *technoholodemocracy* (see essay 30).

7. CONCLUSION

The concept of legitimacy occurring in legal theory is too narrow to question the legitimacy of laws, governments, and actions that are blatantly unjust. Therefore we need a broader concept of legitimacy. We obtain it by resorting to the concepts of basic human need and legitimate aspiration, which are in turn related to the biopsychosocial features of human beings and to the structure of the societies they live in. By exposing these roots we come up with a concept of legitimacy that is coextensive with the broad concept of justice. And when asking what kind of social order could possibly be fully legitimate or just, we propose this answer: technoholodemocracy, or a combination of integral (social) democracy with technology.

30

TECHNOHOLODEMOCRACY
AN ALTERNATIVE TO
CAPITALISM AND SOCIALISM

(1994)

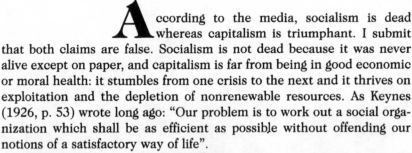

According to the media, socialism is dead whereas capitalism is triumphant. I submit that both claims are false. Socialism is not dead because it was never alive except on paper, and capitalism is far from being in good economic or moral health: it stumbles from one crisis to the next and it thrives on exploitation and the depletion of nonrenewable resources. As Keynes (1926, p. 53) wrote long ago: "Our problem is to work out a social organization which shall be as efficient as possible without offending our notions of a satisfactory way of life".

What used to be called "real socialism" was not authentically socialist, because it involved dictatorship instead of self-government, and nationalization instead of cooperative ownership. As for capitalism, it is subject to unavoidable market disequilibria and destructive business cycles, and it prospers through exploitation, in particular the looting of the Third World. Furthermore, both capitalism and so-called real socialism have been degrading the environment and they must grow and expand to flourish—and economic growth impoverishes the environment, while expansion leads to war. For these reasons, neither of these social orders is functional, sustainable, or morally justifiable.

Because neither corporate capitalism nor "real socialism" works well, and because neither is morally justifiable, we must try to design and construct a better social order. This alternative society should have a moral rationale: it should make it possible for everyone to enjoy life, and morally imperative to help others live an enjoyable life. I suggest that this new social order should constitute an extension of political democracy (popular representation *cum* participation) to encompass biological (gender and ethnic) democracy, economic democracy (cooperative ownership and self-management), and cultural democracy (cultural autonomy and universal and lifelong access to culture and education). Given the complexity of modern society, this goal calls for technical expertise and national—nay, worldwide—control and coordination.

In short, I submit that the new social order we need in order to implement welfare, social justice, democracy, peace, and a sustainable economy, is integral democracy informed, though not dominated, by science-based sociotechnology. I propose to call it *technoholodemocracy*. (Recall essay 29.)

1. CAPITALISM: PALEO AND NEO

Classical or raw capitalism proved economically too unstable and politically too vulnerable. It has therefore been tempered by a large number of government controls and relief measures adopted over the past hundred years. However, even reformed capitalism and the accompanying welfare state have notorious defects, such as ecological unsustainability and pronounced inequalities in the access to wealth, political power, and cultural resources. Moreover, welfare capitalism seems to have reached its limit: it cannot be improved by instituting further social programs, for these would only increase welfare dependency, bureaucracy, tax burdens, and the fiscal deficit. Still, welfare capitalism seems to be the best social order invented so far. Hence, it should make a good launching pad for a socially just and environmentally sustainable social order.

Because capitalism is technically and morally flawed, we need to question its very basis and, in particular, its philosophical underpinnings. The philosophy behind the ideology that extols capitalism is individualism, both ontological and moral. According to it, society is an aggregate of autonomous individuals, everyone of whom is, or ought to be, free to pursue her own interests. This view is contrary to fact: As a matter of fact, society is a supersystem composed of social networks, and as such it has features that their individual components lack. And moral individualism, i.e., egoism, is morally wrong, because it overlooks our duties to others and thus weakens the social networks that support and control individual agency. (Recall essays 21 and 28.)

The rejection of individualism does not amount to the espousal of holism (or collectivism). Holism is false because it underplays individual action, and it is morally wrong for regarding the individual as a pawn of so-called higher instances, such as the race, the nation, or the party. No wonder that holism was the philosophy behind both the traditional Right (in particular fascism) and communism. (The New Right is extremely individualistic, to the point that it calls itself libertarian.)

The alternative to both individualism and holism is systemism. This is the view that society is a system of systems, every one of which is composed by interacting individuals, and is characterized by systemic or emergent properties. Thus, systemism combines the valid points of both individualism and collectivism. In particular, it encourages combining competition with cooperation, as well as balancing rights with duties. Its maximal moral principle is "Enjoy life and help live (an enjoyable life)".

2. SOCIALISM: BOGUS AND AUTHENTIC

What passed for socialism in the nearly defunct Marxist-Leninist camp was never such. Marxism-Leninism mistakes nationalization for socialization, and party dictatorship for the rule of the people: it is statist and undemocratic. It has aptly been characterized as bureaucratic capitalism. Engels himself would not have approved of it, because in his *Anti-Dühring* (1878) he called the state ownership of business 'spurious socialism'. Besides, neither he nor Marx ever expressed any enthusiasm for central planning.

I submit that the two pillars of authentic socialism are cooperative ownership and self-government. More precisely, I define *genuine socialism* as the classless social order characterized by (a) the cooperative ownership of the means of production, exchange, and finance; and (b) the self-management of all business firms—that is, democracy in the workplace. Shorter: Socialism = Cooperativism.

This kind of socialism is more in line with Mill's than with Marx's or Lenin's. In fact, in Mill's *Principles of Political Economy* (1891, p. 775) we read: "The form of association, however, which if mankind continue to improve, must be expected in the end to predominate, is not that which can exist between a capitalist as chief, and workpeople without a voice in the management, but the association of the laborers themselves on terms of equality, collectively owning the capital with which they carry on their operations, and working under managers elected and removable by themselves." (More on economic democracy and market socialism in Miller 1977 and Dahl 1985.)

In principle, cooperativist socialism should be more satisfying and therefore more sustainable than other social orders, and this for the following reasons. First, for putting an end to class divisions and hence to class conflicts. Second, for relying on reciprocity, a far stronger social bond than either subordination or impersonal market relations. (See Kropotkin 1914.) Third, for substituting personal initiative, responsibility, and open debate for blind obedience to management. And fourth, for involving the decentralization required not only to make grassroots participation possible, but also to endow firms with the elasticity required by a rapidly changing environment, as well as to cut the bureaucratic waste and the excessive salaries and perks paid to (though not always earned by) top-level executives in corporations. The spectacular success of the Mondragón network of about a hundred highly diversified cooperatives in the Basque country shows that cooperativism can be made to work well even in a capitalist environment (Whyte and Whyte 1988).

Cooperativism is also morally correct, because it is fair to every cooperant and because it gives her a say in the way her own life is to be run, particularly at her workplace. In fact, whereas under alternative social orders only a few are propertied (or control everything even when they do not own anything), under cooperative socialism everyone owns and controls a slice of the pie—or, to put it in paradoxical terms, everyone becomes a capitalist and has a say in management. She is not only a shareholder, but is also represented in the boardroom. This is not only morally desirable; it is also psychologically effective, for no one looks better after property than its owner.

Is cooperative socialism desirable and viable? I submit that it is neither, and this not in being excessive, but in being insufficient. In fact, cooperative socialism (a) is sectoral rather than systemic, for it only concerns the ownership and management of the means of production: it says nothing about politics, culture, or the environment; (b) it ignores the inevitability, nay desirability, of (bounded) competition alongside cooperation; in particular, it plays down personal interests, aspirations, and initiatives, all of which are bound to originate rivalries and conflicts; (c) it does not tackle the problem of coordinating the cooperatives on the local, national, and global scales; in particular, it does not address the problem of the unequal receipts of different cooperatives, e.g., those that mine diamonds and those that make classical music; and (d) it keeps silent about the management and public control of public goods such as security, public health, education, science, the humanities, and the arts—that is, about the state. In short, cooperative property and self-management are necessary but insufficient, except on a small scale.

Cooperative socialism, in sum, is neither desirable nor viable because it is narrowly economicist, hence shortsighted. We need a social order based on a more comprehensive view of society—one inspired by the systemic view of society. The systemic alternative to the known social orders will be sketched anon.

3. A THIRD WAY: INTEGRAL TECHNOHOLODEMOCRACY

I submit that both political and economic democracy are necessary to build a just and viable society. But I also suggest that they are insufficient, because the former only concerns the polity, whereas the latter only concerns the economy. I claim that, in view of the fact that society is composed of three interlocked artificial systems embedded in nature, namely the economy, the culture, and the polity, we should strive for integral democracy combined with technical expertise and regard for the environment. In other words, the idea is to expand democracy and join it with management science, normative macroeconomics, the law, and other branches of sociotechnology. Let me outline this proposal.

Let us recall that *technoholodemocracy* (or integral technodemocracy) means the social order that promotes equal access to wealth, culture, and political power. This is the goal. The means is, to paraphrase Lincoln, the enlightened rule of the people, by the people, and for the people, in all and only social matters—economic, cultural, political, and environmental—with the advice of experts. Shorter: Holotechnodemocracy is equality through cooperative property, self-management, political democracy, and technical expertise. The latter enters in the design and evaluation of the policies and plans required at all levels for the efficient functioning of the cooperatives, their federations, and the state. This schema calls for a few clarifications.

First, integral technodemocracy does not involve literal egalitarianism, let alone forced equalization à la Pol Pot: it only involves what may be called *qualified* equality, a combination of egalitarianism with meritocracy. This results from combining three principles: (a) the socialist maxim "To each according to his needs, from each according to his abilities"; (b) Locke's principle of the rightful ownership of the fruits of one's labors; and (c) Rawls's principle, according to which the sole inequalities justified in the distribution of goods and services are those that are likely to benefit everybody, namely the reward of merit and the correction of misdeed (Rawls 1971).

Second, holotechnodemocracy involves combining cooperation with competition. In particular, it joins cooperative ownership with the market. We need the former to strengthen solidarity and avert exploitation and privilege. And we need competition to encourage emulation, initiative, creativity, excellence, selectivity, and advancement, as well as to keep costs down and minimize waste. However, competition ought to serve the public interest as well as personal interests. This kind of competition may be called *coordinated* or *managed* competition, in contrast with the individualist or cut-throat competition typical of paleocapitalism. And coordinated competition requires some planning, regulations, bargaining, and strategic alliances. In short, we need the social or controlled market rather than the free market. (For a mathematical model of competitive cooperation, or regulated competition, see Bunge 1976a.)

Third, the efficient management of large-scale systems calls for the central coordination of their component units: it requires setting up federations and states, and eventually a world government. All of these macrosystems should deliberate democratically with the help of expert advice, and they should act in the interests of their components. In particular, the state should be the neutral umpire and technical advisor and assistant in the service of all. It should facilitate coordination rather than impose it bureaucratically or militarily from above, it should mediate in cases of conflict, and it should act sometimes as a catalyzer and at others as a buffer. Of course, nothing of this is new: what would be new, and needs to be done, is the actual performance of such activities.

Fourth, integral technodemocracy would require a far smaller and weaker state than any other social order since the birth of industrial civilization. Indeed, the good society does not need big government, because (a) a full-employment economy has no need for relief programs; (b) equity, reciprocity, solidarity, intelligent bargaining, cooperative property, and grass roots political participation render strong law-enforcing agencies all but redundant; (c) well-educated and morally upright citizens can participate competently and honestly in the administration of the common good; and (d) a peaceful world society needs no professional armed forces.

Fifth, liberty, which is very restricted in a class society, and contract, which is largely a fiction in it, should flourish in an integral democracy. Indeed, freedom and, in particular, the liberty to enter into symmetrical contracts, can only flourish among equals: freedom and fair contracts are largely illusory wherever power is concentrated in the state, corporations, parties, or churches.

4. CONCLUSION

Is integral technodemocracy one more utopia? Perhaps, but who said that all utopias are necessarily bad? We need utopias to plan for a better future, particularly in our time, as the survival of humankind is threatened by overpopulation, depletion of resources, environmental degradation, new plagues, war (nuclear as well as conventional), and fanaticism. Since capitalism and bogus socialism, each supported by an obsolete philosophy, have brought us to the brink of global disaster, we must find a third way.

The next problem is to design feasible and minimally destructive political procedures for effecting the transition from the old to the desirable social order. The transition should be gradual, because revolutions cause too much pain and division, and because it takes time to reform institutions and learn new habits. And the policies and plans guiding the transition should be systemic rather than purely economic, if only because no economy can move in a political, cultural, and environmental vacuum.

In the capitalist democracies, the new social order might grow from a fragmentation of the megacorporations, which have become dysfunctional anyway. The resulting fragments could be reorganized as cooperatives, perhaps with the financial and technical assistance of progressive governments. The main obstacles to a peaceful transition are intellectual inertia, political apathy, and the military power behind economic power.

In the ex-communist nations, the transition should be even easier, because they have no capital to speak of: they only have resources (human and natural) and capital goods, and at present their main goal is survival rather than short-term profit maximization. The main obstacles here are the lack of a democratic tradition and the advice of Western economic consultants to adopt a "shock therapy", which is destructive of the safety-net and of the solidarity and other good habits and institutions that may remain.

However, these are political speculations. What matters from a philosophical viewpoint is that holotechnodemocracy is supported by a systemic social ontology—more realistic than either individualism or holism—as well as by a selftuistic morality centered around the principle" Enjoy life and help live an enjoyable life", which combines rights with duties and pleasure with work.

Pseudosocialism is dead: Long live authentic socialism enriched with political, cultural, and biological democracy, as well as with sociotechnology and concern for the environment.

BIBLIOGRAPHY

Agassi, J., and R. S. Cohen, eds. 1982. *Scientific Philosophy Today: Essays in Honor of Mario Bunge*. Dordrecht: D. Reidel.

Albert, D. Z. 1994. "Bohm's Alternative to Quantum Mechanics." *Scientific American* 270, no. 5: 58–67.

Albert, H. 1988. "Hermeneutics and Economics: A Criticism of Hermeneutical Thinking in the Social Sciences." *Kyklos* 41: 573–602.

Alchourrón, C. E., and E. Bulygin. 1971. *Normative Systems*. Wien: Springer-Verlag.

Allais, M., and O. Hagen, eds. 1979. *Expected Utility Hypotheses and the Allais Paradox*. Dordrecht: D. Reidel.

Archer, M. 1987. "Resisting the Revival of Relativism." *International Sociology* 2: 235–50.

Armstrong, D. M. 1968. *A Materialist Theory of the Mind*. London: Routledge & Kegan Paul.

Ashby, W. R. 1956. *An Introduction to Cybernetics*. New York: Wiley.

Aspect, A., J. Dalibard, and G. Roger. 1982. "Experimental Test of Bell's Inequalities Using Time-varying Analyzers." *Physical Review Letters* 49: 1804–1807.

Aspect, A., P. Grangier, and G. Roger. 1981. "Experimental Test of Realistic Local Theories via Bell's Theorem." *Physical Review Letters* 47: 460–63.

413

Ayala, F. J. 1970. "Teleological Explanation in Evolutionary Biology." *Philosophy of Science* 37: 1–15.

Bandura, A. 1974. "Behavior Theory and the Models of Man." *American Psychologist* 29: 859–69.

Barker, S. F. 1957. *Induction and Hypothesis*. Ithaca, N.Y.: Cornell University Press.

Barnes, B. 1977. *Interests and the Growth of Knowledge*. London: Routledge & Kegan Paul.

———. 1983. "On the Conventional Character of Knowledge and Cognition." In *Science Observed*, edited by K. D. Knorr-Cetina and M. Mulkay, pp. 19–51. London: Sage.

Becker, G. S. 1976. *The Economic Approach to Human Behavior*. Chicago: University of Chicago Press.

Bell, J. S. 1964. "On the Einstein-Podolsky-Rosen Paradox." *Physics* 1: 195–220.

———. 1966. "The Problem of Hidden Variables in Quantum Mechanics." *Review of Modern Physics* 38: 447–52.

Berger, P. L., and T. Luckmann. 1966. *The Social Construction of Reality*. New York: Doubleday.

Bernard, C. [1865] 1952. *Introduction à l'étude de la médecine expérimentale* (Introduction to the Study of Experimental Medicine). Reprint, Paris: Flammarion.

Bindra, D. 1976. *A Theory of Intelligent Behavior*. New York: Wiley Interscience.

———. 1984. "Cognitivism: Its Origin and Future in Psychology." *Annals of Theoretical Psychology* 1: 1–29.

Black, M. 1962. *Models and Metaphors*. Ithaca, N.Y.: Cornell University Press.

Blalock, H. M., Jr., and A. B. Blalock, eds. 1968. *Methodology in Social Research*. New York: McGraw-Hill.

Bloor, D. 1976. *Knowledge and Social Imagery*. London: Routledge & Kegan Paul.

Bôcher, M. 1905. "The Fundamental Conceptions and Methods of Mathematics." *Bulletin of the American Mathematical Society* 11: 115.

Bohm, D. 1952. "A Suggested Interpretation of the Quantum Theory in Terms of 'Hidden Variables.'" *Physical Review* 85: 166–80.

———. 1957. *Causality and Chance in Modern Physics*. London: Routledge & Kegan Paul.

Bohm, D., and B. Hiley. 1975. "On the Intuitive Understanding of Non-Locality as Implied by Quantum Theory." *Foundations of Physics* 5: 93–109.

Bohr, N. 1935. "Can Quantum-Mechanical Description of Physical Reality Be Considered Complete?" *Physical Review* 48: 696–702.

———. 1949. "Discussion with Einstein on Epistemological Problems in Atomic Physics." In *Albert Einstein: Philosopher-Scientist*, edited by P. A. Schilpp, pp. 199–241. Evanston, Ill.: Library of Living Philosophers.

———. 1958. *Atomic Physics and Human Knowledge*. New York: Wiley.

Borst, C. V., ed. 1970. *The Mind-Body Identity Theory*. London: Macmillan.

Boudon, R. 1967. *L'analyse mathématique des faits sociaux* (The Mathematical Analysis of Social Facts). Paris: Plon.

————. 1981. *The Logic of Social Action*. London: Routledge & Kegan Paul.

————. 1990a. *L'art de se persuader des idées fausses, fragiles ou douteuses* (The Art of Convincing Oneself of False, Fragile, or Dubious Idas). Paris: Fayard.

————. 1990b. "On Relativism." In *Studies on Mario Bunge's Treatise*, edited by P. Weingartner and G. J. W. Dorn, pp. 229–43. Amsterdam: Rodopi.

Braudel, F. 1976. *The Mediterranean*. 2 vols. New York: Harper & Row.

Breit, W. 1984. "Galbraith and Friedman: Two Versions of Economic Reality." *Journal of Post Keynesian Economics* 7: 18–28.

Brink, D. O. 1989. *Moral Realism and the Foundations of Ethics*. Cambridge: Cambridge University Press.

Bunge, M. 1951. "What Is Chance?" *Science and Society* 15: 209–31.

————. 1955. "Strife about Complementarity." *British Journal for the Philosophy of Science* 6: 1–12, 141–54.

————. 1959a. *Causality*. Cambridge, Mass.: Harvard University Press. Revised edition, New York: Dover, 1979.

————. 1959b. "¿Como sabemos que existe la atmósfera?" (How Do We Know That the Atmosphere Exists?) *Revista de la Universidad de Buenos Aires* 4: 246–60.

————. 1959c. *Metascientific Queries*. Springfield, Ill.: Charles C. Thomas.

————. 1961. "Laws of Physical Laws." *American Journal of Physics* 29: 432–48.

————. 1962. *Intuition and Science*. Englewood Cliffs, N.J.: Prentice-Hall.

————. 1963. *The Myth of Simplicity*. Englewood Cliffs, N.J.: Prentice-Hall.

————. 1967a. *Scientific Research*. Vol. 1, *The Search for System*. Berlin, Heidelberg, and New York: Springer-Verlag. Reissued and revised 1998 as *Philosophy of Science*. Vol. 1, *From Problem to Theory*. New Brunswick, N.J.: Transaction Publishers.

————. 1967b. *Scientific Research*. Vol. 2, *The Search for Truth*. Berlin, Heidelberg, and New York: Springer-Verlag. Reissued and revised 1998 as *Philosophy of Science*. Vol. 2, *From Explanation to Justification*. New Brunswick, N.J.: Transaction Publishers.

————. 1967c. *Foundations of Physics*. Berlin, Heidelberg, and New York: Springer-Verlag.

————. 1967d. "A Ghost-Free Axiomatization of Quantum Mechanics." In *Quantum Theory and Reality*, edited by M. Bunge, pp. 105–17. New York: Springer-Verlag.

————. 1969. "What Are Physical Theories About?" *American Philosophical Quarterly Monograph* 3: 61–99.

————. 1973a. *Method, Model and Matter*. Dordrecht: D. Reidel.

————. 1973b. *Philosophy of Physics*. Dordrecht: D. Reidel.

————. 1973c. "Normative Wissenschaft ohne Normen—aber mit Werten" (Normative Science without Norms—But with Values). *Conceptus* 7: 57–64.

————. 1973d. Review of *Probabilistic Theory of Causality*, by P. Suppes. *British Journal for the Philosophy of Science* 24: 409–10.

————, ed. 1973. *Exact Philosophy: Problems, Tools, and Goals*. Dordrecht: D. Reidel.

————. 1974a. *Treatise on Basic Philosophy*. Vol. 1, *Sense and Reference*. Dordrecht: D. Reidel.

Bunge, M. 1974b. *Treatise on Basic Philosophy*. Vol. 2, *Interpretation and Truth*. Dordrecht: D. Reidel.

———. 1974c. "The Concept of Social Structure." In *Developments in the Methodology of Social Science*, edited by W. Leinfellner and E. Köhler, pp. 175–215. Dordrecht: D. Reidel.

———. 1974d. "The Relations of Logic and Semantics to Ontology." *Journal of Philosophical Logic* 3: 195–210.

———. 1976a. "A Model for Processes Combining Competition with Cooperation." *Applied Mathematical Modeling* 1: 21–23.

———. 1976b. "Possibility and Probability." In *Foundations of Probability Theory, Statistical Inference, and Statistical Theories of Science*, vol. 3, edited by W. L. Harper and C. A. Hooker, pp. 17–33. Dordrecht: D. Reidel.

———. 1976c. "The Philosophical Richness of Technology." *PSA* 2: 153–72.

———. 1977a. *Treatise on Basic Philosophy*. Vol. 3, *The Furniture of the World*. Dordrecht: D. Reidel

———. 1977b. "Levels and Reduction." *American Journal of Physiology: Regulatory, Integrative and Comparative Physiology* 2: 75–82.

———. 1979a. *Treatise on Basic Philosophy*. Vol. 4, *A World of Systems*. Dordrecht: D. Reidel.

———. 1979b. "The Einstein-Bohr Debate over Quantum Mechanics: Who Was Right about What?" *Lecture Notes in Physics* 100: 204–19.

———. 1980a. *The Mind-Body Problem*. Oxford: Pergamon Press.

———. 1980b. *Ciencia y desarrollo* (Science and Development). Buenos Aires: Siglo Veinte.

———. 1981a. *Scientific Materialism*. Dordrecht: D. Reidel.

———. 1981b. "Four Concepts of Probability." *Applied Mathematical Modeling* 5: 306–12.

———. 1982a. "Is Chemistry a Branch of Physics?" *Zeitschrift für allgemeine Wissenschaftstheorie* 13: 209–23.

———. 1982b. *Economía y filosofía* (Economics and Philosophy). Madrid: Tecnos.

———. 1983a. *Treatise on Basic Philosophy*. Vol. 5, *Exploring the World*. Dordrecht: D. Reidel.

———. 1983b. *Treatise on Basic Philosophy*. Vol. 6, *Understanding the World*. Dordrecht: D. Reidel.

———. 1985a. *Treatise on Basic Philosophy*. Vol. 7, *Philosophy of Science and Technology, Part 1; Formal and Physical Sciences*. Dordrecht: Reidel.

———. 1985b. *Treatise on Basic Philosophy*. Vol. 7, *Philosophy of Science and Technology, Part 2; Life Science, Social Science and Technology*. Dordrecht: Reidel.

———. 1986. "Considérations d'un philosophe sur l'économie du néoconservatisme (néoliberalisme)" (A Philosopher's Considerations on the Economics of Neoconservatism [Neoliberalism]). In *Néoconservatisme et restructuration de l'Etat* (Neoconservatism and the Restructuring of the State), edited by L. Jalbert and L. Lepage, pp. 49–70. Sillery: Presses de l'Université du Québec.

———. 1987. "Seven Desiderata for Rationality." In *Rationality: The Critical View*, edited by J. Agassi and I. Jarvie, pp. 5–15. Dordrecht: Martinus Nijhoff.

———. 1989a. *Treatise on Basic Philosophy*. Vol. 8, *Ethics: The Good and the Right*. Dordrecht: D. Reidel.

———. 1989b. "Reduktion and Integration, Systeme and Niveaus, Monismus and Dualismus" (Reduction and Integration, Systems and Levels, Monism and Dualism). In *Gehirn and Bewusstsein* (Brain and Consciousness), edited by E. Pöppel, pp. 87–104. Weinheim: VCH.

———. 1990. "What Kind of Discipline Is Psychology: Autonomous or Dependent, Humanistic or Scientific, Biological or Sociological?" *New Ideas in Psychology* 8: 121–37. (Comments and author's replies: pp. 139–88, 375–79.)

———. 1991a. "A Critical Examination of the New Sociology of Science, Part 1." *Philosophy of the Social Sciences* 21: 524–60.

———. 1991b. "A Skeptic's Beliefs and Disbeliefs." *New Ideas in Psychology* 9: 131–49.

———. 1991c. "What Is Science? Does It Matter to Distinguish It from Pseudoscience? A Reply to My Commentators." *New Ideas in Psychology* 9: 245–83.

———. 1992. "A Critical Examination of the New Sociology of Science, Part 2." *Philosophy of the Social Sciences* 22: 46–76.

———. 1996. *Finding Philosophy in Social Science*. New Haven: Yale University Press.

———. 1998. *Social Science under Debate*. Toronto: University of Toronto Press.

———. 1999. *The Sociology-Philosophy Connection*. New Brunswick, N.J.: Transaction Publishers.

Bunge, M., and R. Ardila. 1987. *Philosophy of Psychology*. New York: Springer-Verlag.

Bunge, M., and M. García-Sucre. 1976. "Differentiation, Participation, and Cohesion." *Quality and Quantity* 10: 171–78.

Bunge, M., and A. J. Kálnay. 1983a. "Solution to Two Paradoxes in the Quantum Theory of Unstable Systems." *Nuovo Cimento* 77B: 1–9.

———. 1983b. "Real Successive Measurements on Unstable Quantum Systems Take Nonvanishing Time Intervals and Do Not Prevent Them from Decaying." *Nuovo Cimento* 77B: 10–18.

Calvin, M. 1969. *Chemical Evolution*. New York: Oxford University Press.

Campbell, D. T. 1974. " 'Downward Causation' in Hierarchically Organized Biological Systems." In *Studies in the Philosophy of Biology*, edited by F. J. Ayala and T. Dobzhansky, pp. 179–86. Los Angeles: University of California Press.

Carnap, R. 1938. "Logical Foundations of the Unity of Science." In *International Encyclopedia of Unified Science*, vol. 1, no. 1, pp. 42–62. Chicago: University of Chicago Press.

———. 1950. *Logical Foundations of Probability*. Chicago: University of Chicago Press.

Carr, E. H. 1967. *What Is History?* New York: Vintage Books.

Cicourel, A. V. 1974. *Cognitive Sociology: Language and Meaning in Social Interaction*. New York: Free Press.

Cini, M. 1983. "Quantum Theory of Measurement without Wave Packet Collapse." *Nuovo Cimento* 73B: 27–56.

Cini, M., and J.-M. Lévy-Leblond, eds. 1990. *Quantum Theory without Reduction*. Bristol: Adam Hilger.

Clauser, J. F., and A. Shimony. 1978. "Bell's Theorem: Experimental Tests and Implications." *Reports on Progress in Physics* 41: 1881–1927.

Claverie, P., and S. Diner. 1977. "Stochastic Electrodynamics and Quantum Theory." *International Journal of Quantum Chemistry* 12, Supplement 1: 41–82.

Coleman, J. S. 1990. *Foundations of Social Theory*. Cambridge, Mass.: Belknap Press.

Collins, H. M. 1983. "An Empirical Relativist Programme in the Sociology of Scientific Knowledge." In *Science Observed*, edited by K. D. Knorr-Cetina and M. Mulkay, pp. 85–113. London: Sage.

Cooper, L. 1973. "A Possible Organization of Animal Memory and Learning." In *Collective Properties of Physical Systems*, edited by B. Lundqvist and S. Lundqvist, pp. 252–64. New York: Academic Press.

Cowan, J. D., and G. B. Ermentrout. 1979. "A Mathematical Theory of Visual Hallucination Patterns." *Biological Cybernetics* 34: 137–50.

Czerwinski, Z. 1958. "Statistical Inference, Induction and Deduction." *Studia Logica* 7: 243.

D'Abro, A. 1939. *The Decline of Mechanism (in Modern Physics)*. New York: D. Van Nostrand.

Dahl, R. 1985. *A Preface to Economic Democracy*. Berkeley: University of California Press.

Danto, A. C. 1965. *Analytical Philosophy of History*. Cambridge: Cambridge University Press.

Davidson, D. 1967. "Causal Relations." *Journal of Philosophy* 64: 691–703.

de Finetti, B. 1972. *Probability, Induction and Statistics*. New York: John Wiley.

de la Peña-Auerbach, L. 1969. "New Formulation of Stochastic Theory and Quantum Mechanics." *Journal of Mathematical Physics* 10: 1620–30.

d'Espagnat, B. 1979. "The Quantum Theory and Reality." *Scientific American* 241, no. 5: 158–81.

Dirac, P. A. M. 1958. *The Principles of Quantum Mechanics*. Oxford: Clarendon Press.

Dobzhansky, T. 1974. "Chance and Creativity in Evolution." In *Studies in the Philosophy of Biology*, edited by F. J. Ayala and T. Dobzhansky, pp. 307–38. Los Angeles: University of California Press.

Dobzhansky, T., et al. 1977. *Evolution*. San Francisco: Freeman.

Duhem, P. 1914. *La théorie physique* (Physical Theory). Paris: Rivière.

Dummett, M. 1977. *Elements of Intuitionism*. Oxford: Clarendon Press.

du Pasquier, G. 1926. *Le calcul des probabilitités, son évolution mathématique et philosophique* (The Probability Calculus and Its Mathematical and Philosophical Evolution). Paris: Hermann.

Eberhard, P. H. 1982. "Constraints of Determinism and of Bell's Inequalities Are Not Equivalent." *Physical Review Letters* 49: 1474–77.

Eccles, J. C. 1951. "Hypotheses Relating to the Mind-Body Problem." *Nature* 168: 53–64.

———. 1980. *The Human Psyche*. New York: Springer International.

Edelen, D. G. B. 1962. *The Structure of Field Space*. Berkeley: University of California Press.

Einstein, A. 1934. *Mein Weltbild* (My Worldview). Amsterdam: Querido Verlag.

———. 1948. "Quantum Theory and Reality." *Dialectica* 2: 320–24.

———. 1949. "Autobiographical Notes." In *Albert Einstein: Philosopher-Scientist*, edited by P. A. Schilpp. Evanston, Ill.: Library of Living Philosophers.

———. 1950. "The Laws of Science and the Laws of Ethics." Preface to P. Frank, *Relativity: A Richer Truth*. Boston: Beacon Press.

———. 1951. "Remarks on the Theory of Knowledge of Bertrand Russell." In *The Philosophy of Bertrand Russell*, edited by P. A. Schilpp, pp. 277–91. New York: Tudor.

Einstein, A., B. Podolsky, and N. Rosen. 1935. "Can Quantum-Mechanical Description of Reality Be Considered Complete?" *Physical Review* 47: 777–80.

Engels, F. [1878] 1954. *Anti-Dühring*. Moscow: Foreign Languages Publishers.

Feigl, H. 1967. *The 'Mental' and the 'Physical'*. Minneapolis: University of Minnesota Press.

Ferrater Mora, J. M. 1990. "On Mario Bunge's Semantical Realism." In *Studies on Mario Bunge's Treatise*, edited by P. Weingartner and G. J. W. Dorn, pp. 29–37. Amsterdam: Rodopi.

Feyerabend, P. K., and G. Maxwell, eds. 1966. *Mind, Matter, and Method*. Minneapolis: University of Minnesota Press.

Feynman, R. P. 1989. *What Do You Care What Other People Think?* New York: W. W. Norton.

Field, H. H. 1980. *Science Without Numbers*. Princeton, N.J.: Princeton University Press.

Fine, A. 1982. "Hidden Variables, Joint Probabilities and the Bell Inequalities." *Physical Review Letters* 48: 291–95.

Fine, T. L. 1973. *Theories of Probability: An Examination of Foundations*. New York: Academic Press.

Fleck, L. [1935] 1979. *Genesis and Development of a Scientific Fact*. Chicago: University of Chicago Press.

Fodor, J. 1975. *The Language of Thought*. New York: Crowell.

———. 1981. "The Mind-Body Problem." *Scientific American* 244, no. 1: 114–23.

Fréchet, M. 1946. "Les définitions courantes de la probabilité" (Current Definitions of Probability). Reprinted in *Les mathématiques et le concret* (Mathematics and the Concrete), pp. 157–204. Paris: Presses Universitaires de France.

Friedman, M. 1953. "The Methodology of Positive Economics." In *Essays in Positive Economics,* pp. 3–43. Chicago: University of Chicago Press.

———. 1976. *Price Theory*. New York: Aldine.

Gardner, M. 1992. "Probability Paradoxes." *Skeptical Inquirer* 16: 129–32.

Garfinkel, H. 1967. *Studies in Ethnomethodology*. Englewood Cliffs, N.J.: Prentice-Hall.

Geertz, C. 1973. *The Interpretation of Cultures*. New York: Basic Books.

Goodman, N. 1955. *Fact, Fiction and Forecast*. Cambridge, Mass.: Harvard University Press.

———. 1958. "The Test of Simplicity." *Science* 128: 1064–69.

Granovetter, M. 1985. "Economic Action and Social Structure: The Problem of Embeddedness." *American Journal of Sociology* 91: 481–510.

Griffith, J. S. 1967. *A View of the Brain*. Oxford: Clarendon Press.

Grünbaum, A. 1963. *Philosophical Problems of Space and Time*. New York: Alfred Knopf.

Harman, G. 1977. *The Nature of Morality*. New York: Oxford University Press.

Harré, R., and E. H. Madden. 1975. *Causal Powers: A Theory of Natural Necessity*. Oxford: Blackwell.

Harris, M. 1968. *The Rise of Anthropological Theory*. New York: Crowell.

———. 1979. *Cultural Materialism*. New York: Random House.

Harrison, M. 1965. *Introduction to Switching and Automata Theory*. New York: McGraw-Hill.

Hart, H. L. A., and A. M. Honoré. 1959. *Causation in the Law*. Oxford: Clarendon Press.

Hayek, F. A. 1955. *The Counter-Revolution of Science*. Glencoe, Ill.: Free Press.

Hebb, D. O. 1949. *The Organization of Behavior*. New York: Wiley.

———. 1980. *Essay on Mind*. Hillsdale, N.J.: Erlbaum.

———. 1982. *The Conceptual Nervous System*, edited by H. A. Buchtel. Oxford: Pergamon Press.

Heisenberg, W. 1969. *Der Teil und das Ganze* (The Part and the Whole). München: Piper.

Hesse, M. 1966. *Models and Analogies in Science*. Notre Dame, Ind.: University of Notre Dame Press.

Holton, G. 1973. *Thematic Origins of Scientific Thought*. Cambridge, Mass.: Harvard University Press.

Homans, G. C. 1974. *Social Behavior: Its Elementary Forms*. New York: Harcourt Brace Jovanovich.

Hook, S., ed. 1960. *Dimensions of Mind*. New York: New York University Press.

Husserl, E. [1931] 1950. *Cartesianische Meditationen* (Cartesian Meditations). In *Husserliana: Gesammelte Werke* (Husserliana: Collected Works), vol. 1. The Hague: Martinus Nijhoff.

Ingenieros, J. 1917. *Hacia una moral sin dogmas* (Towards a Morality without Dogma). Buenos Aires: L. J. Rosso y Cía.

Jasso, G. 1980. "A New Theory of Distributive Justice." *American Sociological Review* 45: 3–32.

Jeffreys, H. 1957. *Scientific Inference*. Cambridge: Cambridge University Press.

Kahnemann, D., J. L. Knetsch, and R. H. Thaler. 1986. "Fairness as a Constraint on Profit Seeking: Entitlements in the Market." *American Economic Review* 76: 728–41.

Katz, J. J. 1981. *Language and Other Abstract Objects*. Totowa, N.J.: Rowman & Littlefield.

Keynes, J. M. 1921. *A Treatise on Probability*. London: Macmillan.

———. 1926. *The End of Laissez-faire*. London: Hogarth Press.

Kilmer, W. L., W. S. McCulloch, and J. Blum. 1968. "Some Mechanisms for a Theory of the Reticular Formation." In *System Theory and Biology*, edited by M. D. Mesarovic. New York: Springer-Verlag.

Klir, G. J., ed. 1972. *Trends in General Systems Theory*. New York: Wiley Interscience.

Kneale, W. 1949. *Probability and Induction*. Oxford: Oxford University Press.

Knorr-Cetina, K. D., and M. Mulkay, eds. 1983. *Science Observed*. London: Sage.

Koch, S. 1978. "Psychology and the Future." *American Psychologist* 33: 631–47.

Koestler, A., and J. R. Smythies, eds. 1969. *Beyond Reductionism*. Boston: Beacon Press.

Kolmogoroff, A. 1933. *Grundbegriffe der Wahrscheinlichkeitsrechnung* (Principles of the Probability Calculus). Berlin: Springer-Verlag.

Kotarbinski, T. 1965. *Praxiology*. Oxford: Pergamon.

Kropotkin, P. 1914. *Mutual Aid*. Boston: Porter Sargent Publishers.

Lalande, A. 1938. *Vocabulaire technique et critique de la philosophie* (Technical and Critical Vocabulary of Philosophy). 4th ed. 3 vols. Paris. Félix Alcan.

Latour, B. 1983. "Give Me a Laboratory and I Will Raise the World." In *Science Observed*, edited by K. D. Knorr-Cetina and M. Mulkay, pp. 141–70. London: Sage.

———. 1988. "A Relativistic Account of Einstein's Relativity." *Social Studies of Science* 18: 3–44.

Latour, B., and S. Woolgar. 1979. *Laboratory Life: The Social Construction of Scientific Facts*. London: Sage.

Leonard, H. S. 1959. "Interrogatives, Imperatives, Truth, Falsity, and Lies." *Philosophy of Science* 26: 172–86.

Lerner, D., ed. 1965. *Cause and Effect*. New York: Free Press.

Lévy, M. 1979. "Relations entre chimie et physique" (Relationships between Chemistry and Physics). *Epistemologia* 2: 337–70.

Lévy-Leblond, J. M. 1977. "Towards a Proper Quantum Theory." In *Quantum Mechanics: A Half Century Later*, edited by J. L. Lopes and M. Paty, pp. 171–206. Dordrecht: D. Reidel.

Livingston, P. 1988. *Literary Knowledge*. Ithaca, N.Y.: Cornell University Press.

Lovinbond, S. 1983. *Realism and Imagination in Ethics*. Minneapolis: University of Minnesota Press.

Luce, R. D., R. R. Bush, and E. Galanter, eds. 1963–65. *Handbook of Mathematical Psychology*. 3 vols. New York: Wiley.

Luhmann, N. 1990. *Die Wissenschaft der Gesellschaft* (The Science of Society). Frankfurt: Suhrkamp.

MacKay, D. M. 1978. "Selves and Brains." *Neuroscience* 3: 599–606.

Mackie, J. L. 1975. *The Cement of the Universe: A Study of Causation*. Oxford: Clarendon Press.

Mac Lane, S. 1986. *Mathematics Form and Function*. New York: Springer-Verlag.

Mahner, M., and M. Bunge. 1997. *Foundations of Biophilosophy*. Berlin, Heidelberg, and New York: Springer-Verlag.

Mayr, E. 1965. "Cause and Effect in Biology." In *Cause and Effect*, edited by D. Lerner, pp. 33–50.

Merton, R. K. 1957. *Social Theory and Social Structure*. Rev. ed. Glencoe, Ill.: Free Press.

Mill, J. S. 1871. *Principles of Political Economy*. 7th ed. In *Collected Works*, vol. 3. Toronto: University of Toronto Press.

Miller, D. 1977. "Socialism and the Market." *Political Theory* 5: 473–89.

Miller, G. A. 1967. *The Psychology of Communication*. New York: Basic Books.

Milner, P. 1970. *Physiological Psychology*. New York: Holt, Rinehart and Winston.

Miró-Quesada, F. [1951] 1986. "La lógica del deber ser y su eliminabilidad" (The Logic of "Ought" and Its Eliminability). In *Ensayos de filosofía del derecho* (Essays on the Philosophy of Law), pp. 114–24. Lima: Universidad de Lima.

Monod, J. 1970. *Le hasard et la nécessité* (Chance and Necessity). Paris: Editions du Seuil.

Monroe, C., et al. 1996. "A 'Schrödinger Cat' Superposition State of an Atom." *Science* 272: 1131–36.

Nino, C. S. 1985. *Etica y Derechos humanos* (Ethics and Human Rights). Buenos Aires: Paidos.

Nordin, I. 1979. "Determinism and Locality in Quantum Mechanics." *Synthese* 42: 71–90.

O'Connor, J., ed. 1969. *Modern Materialism: Readings on Mind-Body Identity*. New York: Harcourt, Brace and World.

Olson, M. 1971. *The Logic of Collective Action*. Cambridge, Mass.: Harvard University Press.

Oparin, A. I. 1938. *The Origin of Life on Earth*. New York: Macmillan.

Ostwald, W. 1902. *Vorlesungen über Naturphilosophie* (Lectures on the Philosophy of Nature). Leipzig: Veit & Co.

Padulo, L., and M. Arbib. 1974. *System Theory*. Philadelphia: W. B. Saunders.

Parsons, T. 1951. *The Social System*. New York: Free Press.

Pellionisz, A., and R. Llinás. 1979. "Brain Modeling by Tensor Network Theory and Computer Simulation." *Neuroscience* 4: 323–48.

Pérez, R., L. Glass, and R. Shlaer. 1975. "Development of Specificity in the Cat Visual Cortex." *Journal of Mathematical Biology* 1: 275–88.

Piaget, J. 1974. *Understanding Causality*. New York: Norton.

Place, U. T. 1956. "Is Consciousness a Brain Process?" *British Journal of Psychology* 67: 44–50.

Platts, M. 1979. *Ways of Meaning: An Introduction to a Philosophy of Language*. London: Routledge & Kegan Paul.

Poincaré, H. 1903. *La science et l'hypothèse* (Science and Hypothesis). Paris: Flammarion.

Polanyi, J. C., and J. L. Schreiber. 1973. "The Dynamics of Bimolecular Collisions." In *Physical Chemistry: An Advanced Treatise*, edited by H. Eyring, J. Henderson, and W. Jost, vol. 6A, pp. 383–487. New York: Academic Press.

Pólya, G. 1954. *Mathematics and Plausible Reasoning*. 2 vols. Princeton, N.J.: Princeton University Press.

Popper, K. R. 1945. *The Open Society and Its Enemies.* 2 vols. London: Routledge.

———. 1953. "A Note on Berkeley as Precursor of Mach." *British Journal for the Philosophy of Science* 4: 26–36.

———. 1957a. "Probability Magic or Knowledge Out of Ignorance." *Dialectica* 11: 354.

———. 1957b. "The Propensity Interpretation of the Calculus of Probability and the Quantum Theory." In *Observation and Interpretation,* edited by S. Körner, pp. 65–70. London: Butterworths Science Publishers.

———. 1959. *The Logic of Scientific Discovery.* London: Hutchinson.

———. 1960. "Probabilistic Independence and Corroboration by Empirical Tests." *British Journal for the Philosophy of Science* 10: 315.

———. 1962. *Conjectures and Refutations.* London: Routledge & Kegan Paul.

———. 1967. "Quantum Mechanics without 'The Observer'." In *Quantum Theory and Reality,* edited by M. Bunge, pp. 7–44. Berlin, Heidelberg, and New York: Springer-Verlag.

———. 1968. "Epistemology without a Knowing Subject." In B. van Rootselaar and J. F. Staal (eds.), *Logic, Methodology and Philosophy of Science III,* pp. 333–73. Amsterdam: North-Holland.

———. 1972. *Objective Knowledge.* Oxford: Clarendon Press.

———. 1974. "Autobiography." In *The Philosophy of Karl Popper,* edited by P. A. Schilpp, vol. 1. La Salle, Ill.: Open Court.

Popper, K. R., and J. C. Eccles. 1977. *The Self and Its Brain.* New York: Springer International.

Pribram, K. 1971. *Language of the Brain.* Englewood Cliffs, N.J.: Prentice-Hall.

Puterman, Z. 1977. *The Concept of Causal Connection.* Uppsala: Filosofiska Studier.

Putnam, H. 1975. *Mind, Language and Reality.* Cambridge: Cambridge University Press.

Pylyshyn, Z. 1978. "Computational Models and Empirical Constraints." *Behavioral and Brain Sciences* 1: 93–99.

Quine, W. V. [1936] 1949. "Truth by Convention." In *Readings in Philosophical Analysis,* edited by H. Feigl and W. Sellars, pp. 250–73. New York: Appleton-Century-Crofts.

Rapoport, A. 1972. "The Uses of Mathematical Isomorphism in General Systems Theory." In *Trends in General Systems Theory,* edited by G. J. Klir. New York: Wiley Interscience.

Rawls, J. 1971. *A Theory of Justice.* Cambridge, Mass.: Belknap Press.

Reichenbach, H. 1949. *The Theory of Probability.* Los Angeles: University of California Press.

Restivo, S. 1992. *Mathematics in Society and History.* Dordrecht: Kluwer.

Ricoeur, P. 1975. *La métaphore vive* (Metaphor Is Alive). Paris: Editions du Seuil.

Rodríguez-Consuegra, F. 1992. "Un inédito de Gödel contra el convencionalismo: historia, análisis, contexto y traducción" (An Unpublished Paper by Gödel Against Conventionalism: History, Analysis, Context, and Translation). *Arbor* 142 (558-559-560): 323–48.

Rohrlich, F. 1983. "Facing Quantum Mechanical Reality." *Science* 221: 1251–55.

Rosenberg, A. 1976. *Microeconomic Laws*. Pittsburgh: University of Pittsburgh Press.

Rosenblueth, A. 1970. *Mind and Body*. Cambridge, Mass.: MIT Press.

Routley, R. 1980. *Exploring Meinong's Jungle and Beyond*. Interim edition. Canberra: Australian National University.

Russell, B. [1914] 1952. *Our Knowledge of the External World*. London: Allen & Unwin.

———. 1948. *Human Knowledge*. London: Allen & Unwin.

Samuelson, P. 1965. "Some Notions on Causality and Teleology in Economics." In *Cause and Effect*, edited by D. Lerner, pp. 99–144. New York: Free Press.

Savage, L. J. 1954. *The Foundations of Statistics*. New York: Wiley.

Schaffner, K. 1969. "The Watson-Crick Model and Reductionism." *British Journal for the Philosophy of Science* 20: 325–48.

Schrödinger, E. 1935a. "Discussion of Probability Relations between Separated Systems." *Proceedings of the Cambridge Philosophical Society* 31: 555–63.

———. 1935b. "Die gegenwärtige Situation in der Quantentheorie" (The Current Status of Quantum Theory). *Naturwissenschaften* 23: 807–12, 823–28, 844–49. Translated by J. D. Trimmer. *Proceedings of the American Philosophical Society* 124: 323–38 (1980).

Schütz, A. [1932] 1967. *The Phenomenology of the Social World*. Evanston, Ill.: Northwestern University Press.

Scriven, M. 1971. "The Logic of Cause." *Theory and Decision* 2: 49–66.

Sen, A. K. 1977. "Rational Fools: A Critique of the Behavioral Foundations of Economic Theory." *Philosophy & Public Affairs* 6: 317–44.

Shankland, R. S. 1963. "Conversations with Einstein." *American Journal of Physics* 31: 47–57.

———. 1973. "Conversations with Einstein II." *American Journal of Physics* 47: 895.

Shultz, T. R. 1978. *The Principles of Causal Inference and Their Development*. Montreal: McGill University, Department of Psychology.

Siegel, H. 1987. *Relativism Refuted: A Criticism of Contemporary Epistemological Relativism*. Dordrecht: D. Reidel.

Skinner, B. F. 1938. *The Behavior of Organisms: An Experimental Analysis*. New York: Appleton-Century-Crofts.

Smart, J. J. C. 1959. "Sensations and Brain Processes." *Philosophical Reviews* 68: 141–56.

Sober, E. 1993. *Philosophy of Biology*. Boulder, Colo.: Westview Press.

Stone, L. 1972. *The Causes of the English Revolution 1529–1942*. London: Routledge and Kegan Paul.

Stove, D. C. 1991. *The Plato Cult and Other Philosophical Follies*. Oxford: Blackwell.

Suppes, P. 1969. "Stimulus-Response Theory of Finite Automata." *Journal of Mathematical Psychology* 6: 327–55.

———. 1970. *A Probabilistic Theory of Causality*. Amsterdam: North-Holland.

———. 1974. *Probabilistic Metaphysics*. Uppsala: Filosofiska Studier.

Svechnikov, G. A. 1971. *Causality and the Relation of States in Physics.* Moscow: Progress Publishers.

Taylor, R. 1967. "Causation." In *The Encyclopaedia of Philosophy*, edited by P. Edwards, vol. 2, pp. 56–66. New York: Macmillan and Free Press.

Thom, R. 1972. *Stabilité structurelle et morphogenèse* (Structural Stability and Morphogenesis). Reading, Mass.: Benjamin.

Tuomela, R. 1977. *Human Action and Its Explanation.* Dordrecht: Reidel.

Turing, A. 1950. "Can a Machine Think?" *Mind* NS 59: 433–60.

Vaihinger, H. 1920. *Die Philosophie des Als-Ob* (The Philosophy of 'As If'). Leipzig: Meiner.

van Fraassen, B. 1982. "The Charybdis of Realism: Epistemological Implications of Bell's Inequality." *Synthese* 52: 25–38.

Vendler, Z. 1967. "Causal Relations." *Journal of Philosophy* 64: 704–13.

Venn, J. 1888. *The Logic of Chance.* 3d ed. London: Macmillan.

Ville, J. 1939. *Etude critique de la notion de collectif* (A Critical Study of the Notion of the Collective). Paris: Gauthier-Villars.

Vollmer, G. 1975. *Evolutionäre Erkenntnistheorie* (Evolutionary Epistemology). Stuttgart: Hirzel.

von der Malsburg, C. 1973. "Self-Organization of Orientation-Sensitive Cells in the Striate Cortex." *Kybernetik* 14: 85–100.

von Mises, R. 1972. *Wahrscheinlichkeit, Statistik und Wahrheit* (Probability, Statistics and Truth). 4th ed. Wien: Springer-Verlag.

von Neumann, J. [1932] 1955. *Mathematische Grundlagen der Quantenmechanik.* Berlin: Springer-Verlag. Translated by R. T. Beyer, *Mathematical Foundations of Quantum Mechanics*. Princeton, N.J., Princeton University Press.

von Neumann, J., and O. Morgenstern. 1953. *Theory of Games and Economic Behavior.* 3d ed. Princeton, N.J.: Princeton University Press.

von Smoluchowski, M. 1918. "Über den Begriff des Zufalls und den Ursprung der Wahrscheinlichkeitsgesetze in der Physik" (On the Notion of Chance and the Origin of the Probabilistic Laws in Physics). *Naturwissenschaften* 6: 253.

von Wright, G. H. 1957. *The Logical Problem of Induction.* 2d ed. London: Blackwell.

———. 1971. *Explanation and Understanding.* Ithaca, N.Y.: Cornell University Press.

Waddington, C. H. 1960. *The Ethical Animal.* London: Allen and Unwin.

Wald, A. 1950. *Statistical Decision Functions.* New York: Wiley.

Wallace, W. 1972. *Causality and Scientific Explanation I: Medieval and Early Classical Science.* Ann Arbor: University of Michigan Press.

———. 1974. *Causality and Scientific Explanation II: Classical and Contemporary Science.* Ann Arbor: University of Michigan Press.

Watson, J. B. 1925. *Behaviorism.* New York: People's Institute.

Weinberger, O. 1985. "Freedom, Range for Action, and the Ontology of Norms." *Synthese* 65: 307–24.

Weingartner, P., and G. J. W. Dorn, eds. 1990. *Studies on Mario Bunge's Treatise.* Amsterdam: Rodopi.

Wheeler, J. A. 1978. "Not Consciousness but the Distinction between the Probe and the Probed as Central to the Elemental Act of Observation." In *The Role of Consciousness in the Physical World,* edited by R. G. Jahn, pp. 87–111. Boulder, Colo.: Westview Press.

Whewell, W. 1858. *History of the Inductive Sciences*. New York: Appleton.

Whyte, W. F., and K. K. Whyte. 1988. *Making Mondragón*. Ithaca, N.Y.: ILR Press, Cornell University.

Wiggins, D. 1976. "Truth, Invention and the Meaning of Life." *Proceedings of the British Academy* 62: 331–78.

Wilson, H. R. 1975. "A Synaptic Model for Spatial Frequency Adaptation." *Journal of Theoretical Biology* 50: 327–52.

Winch, P. 1958. *The Idea of a Social Science*. London: Routledge & Kegan Paul.

Wisdom, J. O. 1952. *Foundations of Inference in Natural Science*. London: Methuen.

Wold, H. 1969. "Mergers of Economics and Philosophy of Science: A Cruise in Deep Seas and Shallow Waters." *Synthese* 20: 427–82.

Woods, J. 1974. *The Logic of Fiction*. The Hague: Mouton.

Zahler, R. S., and H. Sussmann. 1977. "Claims and Accomplishments of Applied Catastrophe Theory." *Nature* 269: 759–63.

NAME INDEX

SUBJECT INDEX

433

CPSIA information can be obtained
at www.ICGtesting.com
Printed in the USA
BVHW082312170721
612102BV00007B/29/J

27 - 5'7
135